Lecture Notes in Computer Science   12542

More information about this subseries at http://www.springer.com/series/7407

Jean Claude Bajard · Alev Topuzoğlu (Eds.)

Arithmetic of Finite Fields

8th International Workshop, WAIFI 2020
Rennes, France, July 6–8, 2020
Revised Selected and Invited Papers

 Springer

Editors
Jean Claude Bajard 🆔
Sorbonne Paris
Paris, France

Alev Topuzoğlu 🆔
Sabancı University
Istanbul, Turkey

ISSN 0302-9743   ISSN 1611-3349   (electronic)
Lecture Notes in Computer Science
ISBN 978-3-030-68868-4   ISBN 978-3-030-68869-1   (eBook)
https://doi.org/10.1007/978-3-030-68869-1

LNCS Sublibrary: SL1 – Theoretical Computer Science and General Issues

This Springer imprint is published by the registered company Springer Nature Switzerland AG
The registered company address is: Gewerbestrasse 11, 6330 Cham, Switzerland

Preface

The 8th International Workshop on the Arithmetic of Finite Fields (WAIFI 2020) was quite exceptional. It was originally planned to be held at the University of Rennes 1, France. However, like most meetings in 2020, it ended up as a virtual workshop, a shift due to the COVID-19 pandemic.

Without doubt, we all missed the face-to-face interaction that we value so much. On the other hand, the unusual format of the meeting made it accessible to a wider research community. Indeed, WAIFI 2020 attracted over 200 registered participants from all around the world.

The program consisted of five plenary and twelve contributed talks. The plenary speakers were André Chailloux (Inria, Paris, France), Elisa Gorla (University of Neuchâtel, Switzerland), Gary McGuire (University College Dublin, Ireland), Emmanuela Orsini (KU Leuven, Belgium) and Eric Schost (University of Waterloo, Canada). We invited the plenary speakers to contribute survey papers to the proceedings volume. We are very glad that Elisa Gorla, Gary McGuire and Emmanuela Orsini were able to allocate the time to prepare the manuscripts that are included in this volume. An extended abstract of the talk of André Chailloux is also included here. Video recordings of the talks of André Chailloux, Elisa Gorla, Emmanuela Orsini and Eric Schost can be found at http://waifi.org/program.html.

The number of submissions to WAIFI 2020 was rather low, which was not surprising, considering the uncertainties surrounding the COVID-19 pandemic. Out of the 22 fine papers, which received at least three single-blind reviews by PC members or external reviewers chosen by the members, 12 were selected after a discussion online. We are grateful to the Program Committee (PC) members and external reviewers for ensuring a rigorous reviewing process despite all the difficulties caused by the pandemic and the confinement measures. We are also grateful to the authors for agreeing to make video recordings presenting highlights of their papers, which are available on http://waifi.org/program.html.

We worked very closely and harmoniously with the general chairs Sylvain Duquesne and Arnaud Tisserand. Their engagement and hard work in leading the overall organization are much appreciated. Special thanks go to José Luis Imaña, the publicity chair, who also maintained the website with great care. Indeed, the website was viewed over 1600 times in the two weeks starting on 29 June 2020, when the programme was announced and the pre-recorded video presentations of selected papers were posted. We are also thankful to the Steering Committee for their continual support and acknowledge the brilliant work done by the Organizing Committee.

University of Rennes 1 provided the essential infrastructure. We are particularly thankful to INRIA, Lab-STICC CNRS and Centre Henri Lebesgue. We acknowledge the support of Pôle d'excellence cyber and GDR Sécurité Informatique in publicizing the workshop. The program ran very smoothly, for which we are indebted to Sylvain

Duquesne for his handling of the software platform SVI esolutions and for the tireless support he offered to each and everyone who had to use the platform.

As with the previous workshops, Springer agreed to publish the proceedings of WAIFI 2020 as an LNCS volume. We thank Alfred Hoffman and Anna Kramer at Springer for all their help. The EasyChair conference management system was helpful, once again, during submission and selection phases.

With almost no prior experience in organizing online workshops of this size, it was challenging at times to put together this event. Over 100 emails per week, exchanged between the (general, PC, publicity) chairs, organizing committee, authors, speakers, PC members and session chairs, especially during the weeks leading to the meeting, may indicate the indispensable support and understanding we received from all. We express our gratitude to them and all the participants of WAIFI 2020.

November 2020   Alev Topuzoğlu
                                                       Jean Claude Bajard

Organization

General Chairs

Sylvain Duquesne   IRMAR, Rennes 1, France
Arnaud Tisserand   CNRS, Lab-STICC, Lorient, France

Program Committee Chairs

Jean Claude Bajard   Sorbonne Université, France
Alev Topuzoğlu   Sabancı University, Turkey

Publicity Chair

José Luis Imaña   Complutense University of Madrid, Spain

Steering Committee

Lilya Budaghyan   University of Bergen, Norway
Claude Carlet   University of Paris 8, France
Anwar Hasan   University of Waterloo, Canada
José Luis Imaña   Complutense University of Madrid, Spain
Çetin Kaya Koç   University of California Santa Barbara, USA
Sihem Mesnager   University of Paris 8, France
Ferruh Özbudak   Middle East Technical University, Turkey
Svetla Petkova-Nikova   KU Leuven, Belgium
Francisco Rodríguez-Henríquez   CINVESTAV-IPN, Mexico
Erkay Savaş   Sabancı University, Turkey

Program Committee

Nurdagül Anbar   Sabancı University, Turkey
Diego Aranha   Aarhus University, Denmark
Lejla Batina   Radboud University, The Netherlands
Peter Beelen   Technical University of Denmark, Denmark
Karim Bigou   Université de Bretagne Occidentale, Brest, France
Joppe Bos   NXP Semiconductors, Leuven, Belgium
Claude Carlet   University of Paris 8, France
Robert Coulter   University of Delaware, USA
Massimo Giulietti   University of Perugia, Italy
Guang Gong   University of Waterloo, Canada
Robert Granger   University of Surrey, UK

Aurore Guillevic	Inria Nancy - Grand Est, France
Daniel Katz	California State University, Northridge, USA
Yinan Kong	Macquarie University, Australia
Gohar Kyureghyan	University of Rostock, Germany
Sihem Mesnager	University of Paris 8, France
Lucia Moura	University of Ottawa, Canada
Daniel Panario	Carleton University, Canada
Thomas Plantard	University of Wollongong, Australia
Alexander Pott	Otto von Guericke University Magdeburg, Germany
John Sheekey	University College of Dublin, Ireland
Antonia Wachter-Zeh	Technical University of Munich, Germany
Arne Winterhof	Johann Radon Institute for Computational and Applied Mathematics, Austria

Organizing Committee

Elisa Lorenzo Garcìa	IRMAR, Rennes 1, France
Felix Ulmer	IRMAR, Rennes 1, France
Pierre-Alain Fouque	IRISA, Rennes 1, France
Adeline Roux-Langlois	IRISA, Rennes 1, France
Karim Bigou	Université de Bretagne Occidentale, Brest, France

Additional Reviewers

Juliano Bandeira Lima	Georg Maringer
Daniele Bartoli	Bruno Martin
Matteo Bonini	Maria Montanucci
Cunsheng Ding	Alessandro Neri
Masaya Fujisawa	Ferruh Özbudak
Amparo Fúster-Sabater	Marco Timpanella
Anna-Maurin Graner	Peter Schwabe
Somphong Jitman	Arnaud Sipasseuth
Lukas Kölsch	Pietro Speziali
Giorgos Kapetanakis	Qiang Wang
Frieder Ladisch	Zilong Wang
Stefano Lia	Giovanni Zini
Petr Lisonek	Vincent Zucca

Abstracts of Invited Talks/Papers

Solving Multivariate Polynomial Systems and an Invariant from Commutative Algebra

Alessio Caminata[1] (ID) and Elisa Gorla[2]

[1] Dipartimento di Matematica, Università degli Studi di Genova,
via Dodecaneso 35, 16146, Genova, Italy
caminata@dima.unige.it
[2] Institut de Mathématiques,
Université de Neuchâtel, Rue Emile-Argand 11, 2000,
Neuchâtel, Switzerland
elisa.gorla@unine.ch

Abstract. The complexity of computing the solutions of a system of multivariate polynomial equations by means of Gröbner bases computations is upper bounded by a function of the solving degree. In this paper, we discuss how to rigorously estimate the solving degree of a system, focusing on systems arising within public-key cryptography. In particular, we show that it is upper bounded by, and often equal to, the Castelnuovo-Mumford regularity of the ideal generated by the homogenization of the equations of the system, or by the equations themselves in case they are homogeneous. We discuss the underlying commutative algebra and clarify under which assumptions the commonly used results hold. In particular, we discuss the assumption of being in generic coordinates (often required for bounds obtained following this type of approach) and prove that systems that contain the field equations or their fake Weil descent are in generic coordinates. We also compare the notion of solving degree with that of degree of regularity, which is commonly used in the literature. We complement the paper with some examples of bounds obtained following the strategy that we describe.

Linearized Polynomials and Their Adjoints, and Some Connections to Linear Sets and Semifields

Gary McGuire⬤ and John Sheekey⬤

UCD School of Mathematics and Statistics,
University College Dublin, Dublin, Ireland
gary.mcguire@ucd.ie
john.sheekey@ucd.ie

Abstract. For a q-linearized polynomial function L on a finite field, we give a new short proof of a known result, that $L(x)/x$ and $L^*(x)/x$ have the same image, where $L^*(x)$ denotes the adjoint of L. We give some consequences for semifields, recovering results first proved by Lavrauw and Sheekey. We also give a characterization of planar functions.

Efficient, Actively Secure MPC with a Dishonest Majority: a Survey

Emmanuela Orsi ⓘ

imec-COSIC, KU Leuven,
Leuven, Belgium
emmanuela.orsini@kuleuven.be

Abstract. The last ten years have seen a tremendous growth in the interest and practicality of secure multiparty computation (MPC) and its possible applications. Secure MPC is indeed a very hot research topic and recent advances in the field have already been translated into commercial products world-wide. A major pillar in this advance has been in the case of active security with a dishonest majority, mainly due to the SPDZ-line of work protocols. This survey gives an overview of these protocols, with a focus of the original SPDZ paper (Damgård et al. CRYPTO 2012) and its subsequent optimizations.

Introduction to Quantum Computing

André Chailloux

Inria Paris, France

Abstract. The goal of this invited talk was to present an introduction to Quantum Computing for computer scientists which are *not specialists* in the field. Here we present a brief summary of the contents of this talk, available at http://www-labsticc. univ-ubs.fr/waifi2020/videos/waifi2020-video-plenary-chailloux.mp4.

After a small introduction to the field, the talk is divided into 4 parts: basic notions of quantum computing, quantum error correction, quantum algorithms and perspectives.

Basic Notions of Quantum Computing. I first present textbook knowledge on the foundations of Quantum Computing. Here, bits are replaced by qubits which can be represented by vectors in a complex Hilbert space, and computational gates are replaced by unitary matrices that act on these qubits. The talk goes through these notions not only by describing the mathematical rules behind quantum bits and operations but also trying to give an intuition behind fundamental notions of quantum computing: what does it mean to be in a superposition of states? What does it mean that measuring a state alters it?

Quantum Error Correction. I then briefly mention one important theorem: the Threshold Theorem. Qubits are indeed very fragile and become noisy very fast. There are ways to perform quantum error correction but this requires adding more qubits, which themselves create more errors. The Threshold Theorem states that it is possible to correct these errors faster than they occur when adding new qubits, so stable quantum computations are in theory possible, even though they require much more resources than those we have today.

Quantum Algorithms. Then, I present some of the most iconic quantum algorithms: Shor's algortihm and Grover's algorithm. Shor's quantum algorithm shows that with a fully working quantum computer, one can solve the factoring and the discrete logarithm problems in polynomial time. As a consequence, this would break most of today's public key cryptography and we need to design new public key cryptosystems if we want to avoid this weakness.

Perspectives for Quantum Computing. Finally, I present existing technologies for quantum computing and those we can expect in the near and less near future. Quantum Key Distribution for performing unconditional key exchange is a mature technology that is already commercially available and can be used for highly sensitive data. On the other hand, quantum computers are still at a very early stage. Very recently however, several private companies managed to construct small quantum computers that have up to around 60 qubits. While we can't perform any useful computation with these, we arrived at a point where these small quantum computers cannot be simulated with usual computers so there is indeed some strong computational power here that needs to be further improved in order to see real speedups promised by quantum computing.

Contents

Invited Papers

Solving Multivariate Polynomial Systems and an Invariant from
Commutative Algebra   3
   Alessio Caminata and Elisa Gorla

Linearized Polynomials and Their Adjoints, and Some Connections
to Linear Sets and Semifields   37
   Gary McGuire and John Sheekey

Efficient, Actively Secure MPC with a Dishonest Majority: A Survey   42
   Emmanuela Orsini

Finite Field Arithmetic

A HDL Generator for Flexible and Efficient Finite-Field Multipliers
on FPGAs   75
   Joël Cathébras and Roselyne Chotin

Trisymmetric Multiplication Formulae in Finite Fields   92
   Hugues Randriambololona and Édouard Rousseau

Coding Theory

A Construction of Self-dual Skew Cyclic and Negacyclic Codes of Length
n over $\mathbb{F}_{p^n}$ 115
   Aicha Batoul, Delphine Boucher, and Ranya Djihad Boulanouar

Decoding up to 4 Errors in Hyperbolic-Like Abelian Codes
by the Sakata Algorithm   134
   José Joaquín Bernal and Juan Jacobo Simón

Dihedral Codes with Prescribed Minimum Distance   147
   Martino Borello and Abdelillah Jamous

Sequences

Recursion Polynomials of Unfolded Sequences   163
   Ana I. Gomez, Domingo Gomez-Perez, and Andrew Tirkel

Finding Linearly Generated Subsequences   174
   Claude Gravel, Daniel Panario, and Bastien Rigault

Special Functions over Finite Fields

Generalization of a Class of APN Binomials to Gold-Like Functions.   195
  D. Davidova and N. Kaleyski

On Subspaces of Kloosterman Zeros and Permutations of the Form
$L_1(x^{-1}) + L_2(x)$.   207
  Faruk Göloğlu, Lukas Kölsch, Gohar Kyureghyan, and Léo Perrin

Explicit Factorization of Some Period Polynomials   222
  Gerardo Vega

Improved Lower Bounds for Permutation Arrays Using Permutation
Rational Functions   234
  Sergey Bereg, Brian Malouf, Linda Morales, Thomas Stanley,
  and I. Hal Sudborough

Bases

Existence and Cardinality of k-Normal Elements in Finite Fields.   255
  Simran Tinani and Joachim Rosenthal

Author Index   273

Invited Papers

Solving Multivariate Polynomial Systems and an Invariant from Commutative Algebra

Alessio Caminata[1] and Elisa Gorla[2]

[1] Dipartimento di Matematica, Università degli Studi di Genova,
via Dodecaneso 35, 16146 Genova, Italy
caminata@dima.unige.it
[2] Institut de Mathématiques, Université de Neuchâtel,
Rue Emile-Argand 11, 2000 Neuchâtel, Switzerland
elisa.gorla@unine.ch

Abstract. The complexity of computing the solutions of a system of multivariate polynomial equations by means of Gröbner bases computations is upper bounded by a function of the solving degree. In this paper, we discuss how to rigorously estimate the solving degree of a system, focusing on systems arising within public-key cryptography. In particular, we show that it is upper bounded by, and often equal to, the Castelnuovo-Mumford regularity of the ideal generated by the homogenization of the equations of the system, or by the equations themselves in case they are homogeneous. We discuss the underlying commutative algebra and clarify under which assumptions the commonly used results hold. In particular, we discuss the assumption of being in generic coordinates (often required for bounds obtained following this type of approach) and prove that systems that contain the field equations or their fake Weil descent are in generic coordinates. We also compare the notion of solving degree with that of degree of regularity, which is commonly used in the literature. We complement the paper with some examples of bounds obtained following the strategy that we describe.

Keywords: Gröbner basis · Solving degree · Degree of regularity · Castelnuovo-Mumford regularity · Generic coordinates · Multivariate cryptography · Post-quantum cryptography

Introduction

Polynomial system solving plays an important role in many areas of mathematics. In this paper, we discuss how to solve a system of multivariate polynomial equations by means of Gröbner bases techniques and estimate the complexity of polynomial system solving. Our motivation comes from public-key cryptography, where the computational problem of solving polynomial systems of equations plays a major role.

© Springer Nature Switzerland AG 2021
J. C. Bajard and A. Topuzoğlu (Eds.): WAIFI 2020, LNCS 12542, pp. 3–36, 2021.
https://doi.org/10.1007/978-3-030-68869-1_1

In multivariate cryptography, the security relies on the computational hardness of finding the solutions of a system of polynomial equations over a finite field. One can use similar strategies in order to produce public-key encryption schemes and digital signature algorithms, whose security relies on this problem. For signature schemes, e.g., the public key takes the form of a polynomial map

$$\mathcal{P} : \mathbb{F}_q^n \longrightarrow \mathbb{F}_q^r$$
$$(a_1, \ldots, a_n) \longmapsto (f_1(a_1, \ldots, a_n), \ldots, f_r(a_1, \ldots, a_n))$$

where $f_1, \ldots, f_r \in \mathbb{F}_q[x_1, \ldots, x_n]$ are multivariate polynomials with coefficients in a finite field $\mathbb{F}_q$. The secret key allows Alice to easily invert the system $\mathcal{P}$. In order to sign the hash b of a message, Alice computes $a \in \mathcal{P}^{-1}(b)$ and sends it to Bob. Bob can readily verify the validity of the signature by checking whether $\mathcal{P}(a) = b$. An illegitimate user Eve who wants to produce a valid signature without knowing Alice's secret key is faced with the problem of solving the polynomial system of r equations in n variables

$$\begin{cases} f_1(x_1, \ldots, x_n) = b_1 \\ \quad \vdots \\ f_r(x_1, \ldots, x_n) = b_r \end{cases}$$

Even without knowing Alice's secret key, Eve may be able to exploit the structure of $\mathcal{P}$ in order to solve the system. Such an approach is largely used and the adopted strategies vary significantly from one cryptographic scheme to another. Moreover a direct attack is always possible, i.e., Eve may try to solve the system by computing a Gröbner basis of it. Therefore, being able to estimate the computational complexity of solving a multivariate polynomial system gives an upper bound of the security of the corresponding cryptographic scheme, and is therefore highly relevant. In this context, the complexity of solving a polynomial system is typically large enough to make the computation unfeasible, since being able to compute a solution would enable the attacker to forge a digital signature or to decrypt an encrypted message. We emphasize that the security of multivariate cryptographic schemes is a theme of high current interest. For example, the National Institute of Standards (NIST) is in the process of selecting post-quantum cryptographic schemes for standardization. Three digital signature algorithms were selected as finalists in Round 3 by NIST in July 2020 [NIST], one of which is a multivariate scheme.

Multivariate polynomial systems also appear in connection with the Discrete Logarithm Problem (DLP) on an elliptic or hyperelliptic curve. An index calculus algorithm for solving the DLP on an abelian variety was proposed in [Gau09]. The relation-collection phase of the algorithm relies on Gröbner bases computations to solve a large number of polynomial systems. These systems usually do not have any solutions, but, whenever they have one, they produce a decomposition of a point of the abelian variety over the chosen factor base. In contrast with polynomial systems arising within multivariate cryptography, it is feasible to solve the polynomial systems arising within index calculus algorithms.

Nevertheless, it is important to be able to accurately estimate the complexity of solving them. In fact, the complexity of solving these systems has a direct impact on the complexity of the corresponding index calculus algorithm to solve the DLP.

Estimating the complexity of solving multivariate polynomial systems is relevant within public-key cryptography. In this context, we usually wish to compute the solutions over a finite field of a system of multivariate polynomial equations. Typically, the systems have one, or few, or no solutions, not only over the chosen finite field, but also over its algebraic closure. Moreover, the equations are usually not homogeneous. The degrees of the equations are often small for systems coming from multivariate cryptography, but they can be large for systems arising within index calculus algorithms. Similarly, the number of equations and of variables can vary. Therefore, in this paper we concentrate on finite fields and on non homogeneous systems, which have a finite number of solutions over the algebraic closure. We however do not make assumptions on the number of variables, the number of equations and their degrees.

This paper is devoted to an in-depth discussion of how to estimate the complexity of computing a Gröbner basis for a system of multivariate polynomial equations. As said before, our focus is on finite fields and on systems that have a finite number of solutions over the algebraic closure. At the same time, we try to keep the discussion more general, whenever possible. We often concentrate on systems which are not homogeneous, not only because this is the relevant case for cryptographic applications, but also because it is the most difficult case to treat.

After recalling in Sect. 1 the commutative algebras preliminaries that will be needed throughout the paper, in Sect. 2 we discuss in detail the relation between computing Gröbner bases and solving polynomial systems. This connection is often taken for granted within the cryptographic community, as are the necessary technical assumptions. In Sect. 2 we discuss in detail what these technical assumptions are and what can be done when they are not satisfied. We also show in Theorem 3 that, under the usual assumptions, solving a polynomial system of equations is polynomial-time-equivalent to computing a Gröbner basis of it. We conclude with Subsect. 2.1, where we discuss the feasibility of adding the field equations to a system.

Section 3 is the core of the paper. After establishing the setup that we will be adopting, we prove some results on Gröbner bases and homogenization/dehomogenization. They allow us to compare, in Theorem 7, the solving degree of a system, the solving degree of its homogenization, and the solving degree of the homogenization of the ideal generated by its equations. Combining these results with a classical theorem by Bayer and Stillman [BS87], we obtain Theorem 9 and Theorem 10, where we show that the Castelnuovo-Mumford regularity upper bounds the solving degree of a system, and recover Macaulay's Bound in Corollary 2. These results hold under the assumption that the homogenized system of equations is in generic coordinates, an assumption that is often overlooked in the cryptographic literature and that we discuss in Sect. 1.

In Theorem 11 we prove that any system that contains the field equations or their fake Weil descent is in generic coordinates.

In Sect. 4 we discuss the relation between solving degree and degree of regularity. The latter concept is commonly used in the cryptographic literature and often used as a proxy for the solving degree. In Sect. 4 we discuss the limitations of this approach. In particular, Example 11 and Example 12 are examples of systems coming from index calculus for which, respectively, the degree of regularity is strictly smaller than the solving degree and the degree of regularity is not defined.

Finally, Sect. 5 is meant as an example of how the results from Sect. 3, in combination with known commutative algebra results, easily provide estimates for the solving degree. In particular, Theorem 13 and Theorem 14 give bounds for the solving degree of polynomial systems coming from the MinRank Problem.

1 Preliminaries

In this section we introduce the basic notations and terminology from commutative algebra that we need in the rest of the paper. All the definitions and the proofs of the results that we quote here are extensively covered in the books [KR00, KR05, KR16, CLO07].

1.1 Polynomial Rings and Term Orders

We work in a polynomial ring $R = k[x_1, \ldots, x_n]$ in n variables over a field k. An element $f \in R$ is a polynomial, and may be written as a finite sum $f = \sum_\nu a_\nu x^\nu$, where $\nu \in \mathbb{N}^n$, $a_\nu \in k$, and $x^\nu = x_1^{\nu_1} \cdots x_n^{\nu_n}$. A polynomial of the form $a_\nu x^\nu$ is called a monomial of degree $|\nu| = \nu_1 + \cdots + \nu_n$. In particular, every polynomial f is a sum of monomials. The degree of f, denoted by $\deg(f)$, is the maximum of the degrees of the monomials appearing in f. If all these monomials have the same degree, say d, then f is *homogeneous* of degree d. A monomial $a_\nu x^\nu$ with $a_\nu = 1$ is *monic*. A monic monomial is also called a *term*.

Notation. Given a system of polynomials $\mathcal{F} = \{f_1, \ldots, f_r\} \subseteq R$ we denote by $(\mathcal{F}) = (f_1, \ldots, f_r)$ the ideal that they generate, that is $(f_1, \ldots, f_r) = \{\sum_{i=1}^r p_i f_i : p_i \in R\}$.

The list $\mathcal{F} = \{f_1, \ldots, f_r\}$ is called a system of generators of the ideal $I = (\mathcal{F})$. $\mathcal{F}$ is a *minimal system of generators* for I if the ideal generated by any non-empty proper subset of $\mathcal{F}$ is strictly contained in I. If the polynomials $f_1, \ldots, f_r$ are homogeneous, then we say that the system $\mathcal{F}$ and the ideal I are *homogeneous*.

Remark 1. Let I be an ideal of R minimally generated by homogeneous polynomials $f_1, \ldots, f_r$. Then every homogeneous minimal system of generators of I consists of r polynomials of the same degrees as $f_1, \ldots, f_r$.

For any degree $d \in \mathbb{Z}_+$, denote by R_d the d-th homogeneous component of R. R_d is generated as a k-vector space by the monomials of R of degree d. If $I \subseteq R$ is homogeneous, we let $I_d = I \cap R_d$ be the k-vector space of homogenous polynomials of degree d in I.

We denote by $\mathbb{T}$ the set of terms of R. A *term order* on R is a total order τ on the set $\mathbb{T}$, which satisfies the following additional properties:

1. $m \leq_\tau n$ implies $p \cdot m \leq_\tau p \cdot n$ for all $p, m, n \in \mathbb{T}$;
2. $1 \leq_\tau m$ for all $m \in \mathbb{T}$.

If in addition $m <_\tau n$ whenever $\deg(m) < \deg(n)$, we say that the term order τ is *degree-compatible*.

Example 1 (Lexicographic order). Let x^α and x^β be two terms in R. We say that $x^\alpha >_{LEX} x^\beta$ if the leftmost non-zero entry in the vector $\alpha - \beta \in \mathbb{Z}^n$ is positive. This term order is called *lexicographic* and it is not degree-compatible. We denote it by LEX.

Example 2 (Degree reverse lexicographic order). Let x^α and x^β be two terms in R. We say that $x^\alpha >_{DRL} x^\beta$ if $|\alpha| > |\beta|$, or $|\alpha| = |\beta|$ and the rightmost non-zero entry in $\alpha - \beta \in \mathbb{Z}^n$ is negative. This term order is called *degree reverse lexicographic* (DRL for short) and it is degree-compatible.

Let $f = \sum_{i \in \mathcal{I}} a_i m_i \in R \setminus \{0\}$ be a polynomial, where $a_i \in k \setminus \{0\}$, and $m_i \in \mathbb{T}$ are distinct terms. We fix a term order τ on R. The *initial term* or *leading term* of f with respect to τ is the largest term appearing in f, that is $\mathrm{in}_\tau(f) = m_j$, where $m_j > m_i$ for all $i \in \mathcal{I} \setminus \{j\}$. The *support* of f is $\mathrm{supp}(f) = \{m_i : i \in \mathcal{I}\}$. Given an ideal I of R, the *initial ideal* of I is

$$\mathrm{in}_\tau(I) = (\mathrm{in}_\tau(f) : f \in I \setminus \{0\}).$$

Definition 1. *Let I be an ideal of R. A set of polynomials $\mathcal{G} \subseteq I$ is a Gröbner basis of I with respect to τ if $\mathrm{in}_\tau(I) = (\mathrm{in}_\tau(g) : g \in \mathcal{G})$. A Gröbner basis is reduced if $m \notin (\mathrm{in}_\tau(h) : h \in \mathcal{G} \setminus \{g\})$ for all $g \in \mathcal{G}$ and $m \in \mathrm{supp}(g)$.*

Sometimes we will need to consider a field extension. At the level of the ideal, this corresponds to looking at the ideal generated by the equations in a polynomial ring over the desired field extension.

Definition 2. *Let $I = (f_1, \ldots, f_r) \subseteq R = k[x_1, \ldots, x_n]$, let $K \supseteq k$ be a field extension. We denote by $IK[x_1, \ldots, x_n]$ the extension of I to $K[x_1, \ldots, x_n]$, i.e., the ideal of $K[x_1, \ldots, x_n]$ generated by $f_1, \ldots, f_r$. In symbols, $IK[x_1, \ldots, x_n] = (f_1, \ldots, f_r) \subseteq K[x_1, \ldots, x_n]$.*

1.2   Zero Loci of Ideals

We are mostly interested in ideals, whose zero locus is finite.

Definition 3. *The* affine zero locus *of an ideal $I = (f_1, \ldots, f_r) \subseteq R$ over the algebraic closure $\bar{k}$ of k is*

$$\mathcal{Z}(I) = \{P \in \bar{k}^n : f(P) = 0 \ \text{for all } f \in I\} = \{P \in \bar{k}^n : f_1(P) = \ldots = f_r(P) = 0\}.$$

We also denote it by $\mathcal{Z}(f_1, \ldots, f_r)$.

Definition 4. *The* projective zero locus *of a homogeneous ideal $I = (f_1, \ldots, f_r) \subseteq R$ over the algebraic closure $\bar{k}$ of k is*

$$\mathcal{Z}_+(I) = \{P \in \mathbb{P}(\bar{k})^n : f(P) = 0 \ \text{for all } f \in I\}$$
$$= \{P \in \mathbb{P}(\bar{k})^n : f_1(P) = \ldots = f_r(P) = 0\}.$$

We also denote it by $\mathcal{Z}_+(f_1, \ldots, f_r)$.

Remark 2. The following are equivalent for a homogeneous ideal $I \subseteq R$:

$$|\mathcal{Z}(I)| < \infty \Leftrightarrow \mathcal{Z}(I) = \{(0, \ldots, 0)\} \Leftrightarrow \mathcal{Z}_+(I) = \emptyset.$$

These conditions are equivalent to the fact that the *Krull dimension* of R/I is zero. This is in turn equivalent to R/I being a finite dimensional k-vector space.

In Definition 3 and Definition 4 it is important to look at the zero locus of I or $\mathcal{F}$ over the algebraic closure of the base field. For cryptographic applications, often the base field k is a finite field. In this case the condition that the zero locus is finite over k is trivially satisfied by any ideal or system of equations.

1.3   Infinite Fields and the Zariski Topology

Let k be a field. The *Zariski topology* on the affine space k^n is the set of complements of solution sets of systems of polynomial equations over R, that is $\{k^n \setminus \mathcal{Z}(f_1, \ldots, f_r) \mid f_1, \ldots, f_r \in R\}$. If k is an algebraically closed field, or at least an infinite field, then every non-empty open set in the Zariski topology is dense, i.e., its closure is equal to the entire space. A non-empty open subset of k^n is often called a *generic set* and a property which holds on a non-empty open set is *generic*. Intuitively, a generic set is almost the whole space and a generic property holds almost everywhere in k^n.

If k is a finite field, on the other side, the Zariski topology is the discrete topology on k^n. In other words, any subset of k^n is both open and closed, and the algebraic-geometric intuition of genericity fails. In particular, one can no longer say that a non-empty open subset of k^n is almost the whole space, as the closure of any subset of k^n is the subset itself. Therefore, as genericity loses its meaning over a finite field, we always will need to assume that the ground field is infinite when dealing with generic sets or properties.

1.4   Generic Changes of Coordinates

Fix a term order τ on $R = k[x_1, \ldots, x_n]$. We denote by $\mathrm{GL}(n, k)$ the general linear group of $n \times n$ invertible matrices with entries in k. This group acts on R via linear changes of coordinates. Namely, a matrix $g = (g_{i,j}) \in \mathrm{GL}(n, k)$ acts on the variable x_j as $g(x_j) = \sum_{i=1}^{n} g_{i,j} x_i$. We refer to g also as a *linear change of coordinates*. We observe that $\mathrm{GL}(n, k) \subseteq k^{n^2}$ is an open subset with respect to the Zariski topology.

It is easy to find examples of $g \in \mathrm{GL}(n, k)$ such that $\mathrm{in}_\tau(gI) \neq \mathrm{in}_\tau(I)$, that is, initial ideals are not independent of coordinate changes. However, a famous theorem by Galligo states that, applying a generic change of coordinates to an ideal I, the initial ideal stays the same.

Theorem 1. [Gal74] *Assume that k is infinite. Let I be a homogeneous ideal of R, then there exist a non-empty Zariski-open set $U \subseteq \mathrm{GL}(n, k)$ and a monomial ideal J such that $\mathrm{in}_\tau(gI) = J$ for all $g \in U$.*

This motivates the following definition.

Definition 5. *Let k be an infinite field. An ideal $I \subseteq R$ is* in generic coordinates *if $1 \in U$, i.e., if*

$$\mathrm{in}_\tau(gI) = \mathrm{in}_\tau(I)$$

for all $g \in U$.

Let k be any field and let $K \supseteq k$ with K infinite. I is in generic coordinates *over K if $IK[x_1, \ldots, x_n] \subseteq K[x_1, \ldots, x_n]$ is in generic coordinates.*

Notice that, over an infinite field k, gI is by definition in generic coordinates for any ideal I and $g \in U$, that is, for any ideal I and for a generic g. Informally, any homogeneous ideal can be put in generic coordinates by applying a random change of coordinates to it. If k is finite, it suffices to apply to I a random change of coordinates over a field extension of sufficiently large cardinality.

1.5   Homogeneous Ideals Associated to a System

Let $R = k[x_1, \ldots, x_n]$ and let $S = R[t]$. Given a polynomial $f \in R$, we denote by $f^h \in S$ the homogenization of f with respect to the new variable t. For $\mathcal{F} = \{f_1, \ldots, f_r\} \subseteq R$, we let $\mathcal{F}^h \subseteq S$ denote the system obtained from $\mathcal{F}$ by homogenizing each f_i with respect to t, that is $\mathcal{F}^h = \{f_1^h, \ldots, f_r^h\}$.

For an ideal $I \subseteq R$, the *homogenization* of I with respect to t, or simply the homogenization of I, is the ideal

$$I^h = (f^h : f \in I) \subseteq S.$$

If $I = (\mathcal{F}) \subseteq R$, then I^h is a homogeneous ideal of S which contains $(\mathcal{F}^h)$. It is easy to produce examples where the containment is strict.

Remark 3. Let $\mathcal{G}$ be a Gröbner basis of I with respect to a degree-compatible term order on R. It can be shown that $\mathcal{G}^h = \{g^h : g \in \mathcal{G}\}$ is a Gröbner basis of I^h with respect to a suitable term order on S, see e.g. [KR05, Section 4.3]. In particular $I^h = (g^h : g \in \mathcal{G})$, hence the degrees of a minimal system of generators of I^h are usually different from those of a minimal system of generators of I. Instead, the degrees of a minimal system of generators of $(\mathcal{F}^h)$ coincide with the degrees of $f_1, \ldots, f_r$.

The *dehomogenization map* ϕ is the standard projection on the quotient $\phi : S \to R \cong S/(t-1)$. For any system of equations $\mathcal{F} \subseteq R$ generating an ideal $I = (\mathcal{F})$ we have $\phi(I^h) = (\phi(\mathcal{F}^h)) = I$. Notice that one also has $\phi((\mathcal{F}^h)) = (\phi(\mathcal{F}^h)) = I$.

For a polynomial $f \in R$, we denote by f^{top} its homogeneous part of highest degree. For a system of equations $\mathcal{F} = \{f_1, \ldots, f_r\}$ we denote by

$$\mathcal{F}^{\mathrm{top}} = \{f_1^{\mathrm{top}}, \ldots, f_r^{\mathrm{top}}\}.$$

Both the ideal $(\mathcal{F}^h)$ and the ideal $(\mathcal{F}^{\mathrm{top}})$ depend on $\mathcal{F}$, and not only on the ideal $I = (\mathcal{F})$.

2 The Importance of Being *LEX*

The main goal of this section is clarifying the relation between solving a system of polynomial equations $\mathcal{F}$ and computing a Gröbner basis of the ideal I generated by the system. In the cryptographic literature it is often stated that, thanks to the Shape Lemma, the problem of finding the solutions of $\mathcal{F}$ can be reduced to that of computing a lexicographic Gröbner basis of I. This statement is however not rigorous, since the Shape Lemma only holds under certain assumptions, which are not always verified for cryptographic systems.

We start by stating the assumptions under which the Shape Lemma holds and showing that, when they are satisfied, the problem of solving the system $\mathcal{F}$ is polynomial-time-equivalent to that of computing a lexicographic Gröbner basis of I. Then we discuss what can be done in the case when the assumptions of the Shape Lemma are not satisfied. We come to the conclusion that, in all situations, one can easily compute the solutions of $\mathcal{F}$ from a lexicographic Gröbner basis of I. We stress that we are not stating that directly computing the reduced lexicographic Gröbner basis is the most efficient way to solve a system (see also Sect. 3). We conclude the section with a brief discussion of when it is feasible to add the field equations to a system $\mathcal{F}$ and how that affects the computation of a Gröbner basis of it.

Throughout the section we focus on systems of equations which have a finite number of solutions over the algebraic closure of the field of definition, since systems that arise in public key cryptography are usually of this kind. Moreover, we always assume that our systems have at least one solution. In fact, if the system has no solutions, the corresponding ideal is equal to the polynomial ring, that is the reduced Gröbner basis with respect to any term order is equal to

{1}. In this case, therefore, computing the reduced lexicographic Gröbner basis allows us to decide that the system has no solutions, without any additional work.

We start by recalling the Shape Lemma.

Theorem 2 (Shape Lemma – [KR00], Theorem 3.7.25). *Let k be a field and let $f_1, \ldots, f_r \in R$ be such that the corresponding ideal $I = (f_1, \ldots, f_r)$ is radical, in normal x_n-position, and $|\mathcal{Z}(I)| = d < \infty$. The reduced lexicographic Gröbner basis of I is of the form*

$$\{g_n(x_n), x_{n-1} - g_{n-1}(x_n), \ldots, x_1 - g_1(x_n)\},$$

where $g_1, \ldots, g_n$ are univariate polynomials in x_n and $\deg(g_1), \ldots, \deg(g_{n-1}) < \deg(g_n) = d$.

The Shape Lemma assumes that the ideal I is radical and in normal x_n-position. An ideal I is *radical* if $f^\ell \in I$ for some $\ell > 0$ implies $f \in I$. This assumption is not always verified for ideals generated by systems arising in cryptography. Later in the section, we will show how one can use a more general version of the Shape Lemma in order to overcome this problem.

Being in *normal x_n-position* means that any two distinct zeros $(a_1, \ldots, a_n)$, $(b_1, \ldots, b_n) \in \mathcal{Z}(I)$ satisfy $a_n \neq b_n$. Notice that every ideal I with finite affine zero locus can be brought into normal x_n-position by a suitable linear change of coordinates, passing to a field extension if needed (see [KR00, Proposition 3.7.22]). A field extension may indeed be needed, as the next example shows.

Example 3. Let $\mathcal{F} = \{x_1^2 + x_1, x_1 x_2, x_2^2 + x_2\} \subseteq R = \mathbb{F}_2[x_1, x_2]$. Then $I = (x_1^2 + x_1, x_1 x_2, x_2^2 + x_2)$ is a radical ideal and $\mathcal{Z}(I) = \{(0,0), (0,1), (1,0)\}$. We claim that I cannot be brought in normal x_2-position by a linear change of coordinates over $\mathbb{F}_2$. In fact, a linear change of coordinates over $\mathbb{F}_2$ sends x_2 to either x_1, x_2, $x_1 + x_2$, $x_1 + 1$, $x_2 + 1$, or $x_1 + x_2 + 1$. However, all these linear forms take the same value on at least two of the elements of $\mathcal{Z}(I)$.

Finally, the Shape Lemma assumes that $|\mathcal{Z}(I)| < \infty$. If k is a finite field, then one can add the field equations to I and obtain an ideal J which is radical and such that $\mathcal{Z}(J) = \mathcal{Z}(I) \cap k^n$, in particular $|\mathcal{Z}(J)| < \infty$. This is however not always advantageous or even feasible, as we discuss in Sect. 2.1.

Whenever the assumptions of the Shape Lemma are satisfied, computing the solutions of a system of equations has the same complexity as computing the reduced lexicographic Gröbner basis of the ideal generated by the system.

Theorem 3. *Let $\mathcal{F} = \{f_1, \ldots, f_r\} \subseteq R$ be a polynomial system such that the corresponding ideal $I = (f_1, \ldots, f_r)$ is radical and in normal x_n-position. Assume that $|\mathcal{Z}(I)| = d < \infty$ and $\mathcal{Z}(I) \subseteq \mathbb{F}_q^n$. Consider the LEX order. The set of solutions of $\mathcal{F}$ can be computed from the reduced Gröbner basis of I probabilistically in time polynomial in $\log q, n$ and d. Conversely, the reduced Gröbner basis of I can be computed from the set of solutions of $\mathcal{F}$ deterministically in time polynomial in $\log q, n$ and d.*

Proof. By the Shape Lemma, the reduced lexicographic Gröbner basis of I has the form:

$$\{g_n(x_n), x_{n-1} - g_{n-1}(x_n), \ldots, x_1 - g_1(x_n)\}, \tag{1}$$

where $g_i(x_n)$ are polynomials in the variable x_n only, and $\deg(g_j) < \deg(g_n) = d$ for $1 \le j < n$.

If we know the reduced lexicographic Gröbner basis of I, then we can factor the polynomial $g_n(x_n)$ to find its roots. Each root α of $g_n(x_n)$ corresponds to a solution $(g_1(\alpha), \ldots, g_{n-1}(\alpha), \alpha)$ of $f_1 = \ldots = f_r = 0$. Notice that the only operation required, apart from the arithmetic over $\mathbb{F}_q$, is factoring univariate polynomials, which can be done in probabilistic polynomial time over a finite field.

Vice versa, assume that we know $\mathcal{Z}(I) = \{P_1, \ldots, P_d\} \subseteq \mathbb{F}_q^n$ of $\mathcal{F}$. Write $P_i = (a_{i,1}, \ldots, a_{i,n})$ for $i = 1, \ldots, d$. We wish to compute the reduced lexicographic Gröbner basis of I, knowing that it is of the form (1). Since the roots of g_n are exactly $a_{1,n}, \ldots, a_{d,n}$ we can compute $g_n(x_n) = \prod_{i=1}^d (x_n - a_{i,n})$. Now fix $j \in \{1, \ldots, n-1\}$. Since $g_j(a_{i,n}) = a_{i,j}$ for $i = 1, \ldots, d$ and $\deg(g_j) < d$, we can compute $g_j(x_n)$ by using Lagrange interpolation:

$$g_j(x_n) = \sum_{i=1}^d \left(\prod_{\substack{1 \le \lambda \le d \\ \lambda \ne i}} \frac{x_n - a_{\lambda,n}}{a_{i,n} - a_{\lambda,n}} \right) a_{i,j}.$$

$\square$

We now discuss the situation in which the assumptions of the Shape Lemma do not hold. In particular, we consider the case when I is not radical. Some authors state that, since $I + (x_1^q - x_1, \ldots, x_n^q - x_n) \subseteq \mathbb{F}_q[x_1, \ldots, x_n]$ is always radical, up to adding the field equations one may assume without loss of generality that I is radical. However, adding the field equations to the system is not always computationally feasible, even in the case of systems coming from cryptography. Therefore, being able to deal with the situation when the ideal I is not radical is relevant for cryptographic applications. We discuss this issue in more detail in Sect. 2.1.

Before continuing our discussion, we give an example of system coming from multivariate cryptography for which the corresponding ideal is not radical, adding the field equations to the system is not feasible, and one ends up with a reduced lexicographic Gröbner basis which does not have the shape predicted by the Shape Lemma. Indeed, this was the case for most of the instances of the ABC cryptosystem [TDTD13, TXPD15] that we computed. Since the field sizes proposed in [TXPD15] for achieving 80-bits security are 2^8, 2^{16}, and 2^{32}, adding the field equations to the system is not feasible. In our next example we disregard the linear transformations used in the ABC cryptosystem to disguise the private key, since they do not affect the property of the system to generate a radical ideal.

Example 4. We consider $R = \mathbb{F}_2[x_1, x_2, x_3, x_4]$ with the *LEX* term order and a toy instance of an ABC cryptosystem with

$$A = \begin{pmatrix} x_1 & x_2 \\ x_3 & x_4 \end{pmatrix}, \ B = \begin{pmatrix} x_1 + x_2 + x_3 & x_1 + x_2 \\ x_1 + x_3 + x_4 & x_3 \end{pmatrix}, \ C = \begin{pmatrix} x_1 + x_2 + x_3 + x_4 & x_1 + x_4 \\ x_1 + x_4 & x_1 \end{pmatrix}.$$

We let $p_1, \ldots, p_8$ be the entries of the matrices AB and AC. We take a random plaintext $b = (0, 1, 1, 0) \in \mathbb{F}_2^4$ and we evaluate the polynomials $p_1, \ldots, p_8$ at b to obtain the ciphertext $a = (1, 1, 0, 1, 0, 0, 0, 0) \in \mathbb{F}_2^8$. We then consider the system $\mathcal{F} = \{p_i - a_i : i = 1, \ldots, 8\}$ and the corresponding ideal $I = (\mathcal{F}) \subseteq R$. The ideal I is not radical as $(x_3 + 1)^2 \in I$, but $x_3 + 1 \notin I$. A computation with MAGMA shows that the reduced lexicographic Gröbner basis of I is $\{x_1, x_2+x_3, x_3^2+1, x_4\}$.

We now discuss how one can efficiently compute the solutions of a polynomial system from its lexicographic Gröbner basis, without assuming that the ideal generated by the equations is radical. We stress that we always assume that the system has finitely many solutions over the algebraic closure. The next result will be central to our discussion, as we will use it as a substitute of the Shape Lemma.

Theorem 4 (Elimination Theorem – [CLO07], Chap. 3.1, Theorem 2). *Let $I \subseteq R$ be an ideal and let $\mathcal{G}$ be a lexicographic Gröbner basis of I. Then for every $1 \leq \ell \leq n - 1$ the set $\mathcal{G} \cap k[x_{\ell+1}, \ldots, x_n]$ is a Gröbner basis of $I \cap k[x_{\ell+1}, \ldots, x_n]$ with respect to the LEX order on $k[x_{\ell+1}, \ldots, x_n]$.*

In the next result we use Theorem 4 to prove that one can easily compute the solutions of $\mathcal{F}$ from the reduced lexicographic Gröbner basis of I.

Theorem 5. *Let I be a proper ideal of $R = k[x_1, \ldots, x_n]$ with finite affine zero locus. The reduced lexicographic Gröbner basis of I has the form*

$$p_{n,1}(x_n),$$
$$p_{n-1,1}(x_{n-1}, x_n), \ldots, p_{n-1,t_{n-1}}(x_{n-1}, x_n),$$
$$p_{n-2,1}(x_{n-2}, x_{n-1}, x_n), \ldots, p_{n-2,t_{n-2}}(x_{n-2}, x_{n-1}, x_n),$$
$$\ldots$$
$$p_{1,1}(x_1, \ldots, x_n), \ldots, p_{1,t_1}(x_1, \ldots, x_n),$$

where $p_{i,t_j} \in k[x_i, \ldots, x_n]$ for every index $i \in \{1, \ldots, n\}, j \in \{1, \ldots, t_i\}$ and $t_1, \ldots, t_{n-1} \geq 1$. Moreover, for any $1 \leq \ell \leq n$, let $a = (a_{\ell+1}, \ldots, a_n) \in k^{n-\ell}$ be a solution of the equations

$$p_{n,1}(x_n),$$
$$p_{n-1,1}(x_{n-1}, x_n), \ldots, p_{n-1,t_{n-1}}(x_{n-1}, x_n),$$
$$\ldots$$
$$p_{\ell+1,1}(x_{\ell+1}, \ldots, x_n), \ldots, p_{\ell+1,t_{\ell+1}}(x_{\ell+1}, \ldots, x_n),$$

and let

$$p_\ell(x_\ell) = \gcd\{p_{\ell,1}(x_\ell, a_{\ell+1}, \ldots, a_n), \ldots, p_{\ell,t_\ell}(x_\ell, a_{\ell+1}, \ldots, a_n)\}.$$

Then $p_\ell(x_\ell) \notin k$.

Proof. Let $\mathcal{G}$ be the reduced lexicographic Gröbner basis of I. The set $\mathcal{G} \cap k[x_\ell, \ldots, x_n]$ is of the form

$$\mathcal{G} \cap k[x_\ell, \ldots, x_n] = \{p_{i,j}(x_i, \ldots, x_n) \mid \ell \leq i \leq n, 1 \leq j \leq t_i\}$$

for some $t_1, \ldots, t_n \geq 0$. Moreover, for any $1 \leq \ell \leq n$ such that $p_\ell(x_\ell) \neq 0$, one has $t_\ell \geq 1$. Hence it suffices to show that $p_\ell(x_\ell) \notin k$ for $1 \leq \ell \leq n$.

We prove the claim by descending induction on $\ell \leq n$. Let $\ell = n$, then $\mathcal{G} \cap k[x_n]$ is the reduced lexicographic Gröbner basis of $I \cap k[x_n]$ by Theorem 4. Let $p_{n,1}(x_n)$ be a monic generator of $I \cap k[x_n]$, then $\mathcal{G} \cap k[x_n] = \{p_{n,1}(x_n)\}$ and $t_n = 1$. Since the affine zero locus of I is finite, $p_{n,1}(x_n) \neq 0$. Moreover, $p_n(x_n) = p_{n,1}(x_n) \notin k \setminus \{0\}$, since $\emptyset \neq \mathcal{Z}(I) \subseteq \mathcal{Z}(p_n)$.

We suppose now that the claim holds up to $\ell+1$ and we prove that $p_\ell(x_\ell) \notin k$. By Theorem 4, $\mathcal{G} \cap k[x_\ell, \ldots, x_n]$ is the reduced lexicographic Gröbner basis of $I \cap k[x_\ell, \ldots, x_n]$, in particular

$$I \cap k[x_\ell, \ldots, x_n] = (p_{i,j} \mid \ell \leq i \leq n, 1 \leq j \leq t_i).$$

Let $a \in \mathcal{Z}(I \cap k[x_{\ell+1}, \ldots, x_n]) \cap k^{n-\ell}$ and define

$$I(\ell, a) = (p_{\ell,1}(x_\ell, a_{\ell+1}, \ldots, a_n), \ldots, p_{\ell,t_\ell}(x_\ell, a_{\ell+1}, \ldots, a_n)) = (p_\ell(x_\ell)).$$

By [CLO07, Chapter 3.2, Theorem 3] and since $\mathcal{Z}(I)$ is a finite set, one has that

$$\mathcal{Z}(I \cap k[x_\ell, \ldots, x_n]) = \pi_{n-\ell+1}(\mathcal{Z}(I)),$$

where $\pi_i : k^n \to k^i$ is the projection on the last i coordinates. In particular, $\mathcal{Z}(I \cap k[x_\ell, \ldots, x_n])$ is finite. If $p_\ell(x_\ell)$ is the zero polynomial, then $\mathcal{Z}(I(\ell, a)) = \bar{k}$ and

$$\{(a_\ell, a_{\ell+1}, \ldots, a_n) \mid a_\ell \in \bar{k}\} \subseteq \mathcal{Z}(I \cap k[x_\ell, \ldots, x_n]),$$

contradicting the finiteness of $\mathcal{Z}(I \cap k[x_\ell, \ldots, x_n])$. If instead $p_\ell(x_\ell) \in k \setminus \{0\}$, then $\mathcal{Z}(I(\ell, a)) = \emptyset$. However, $a = (a_{\ell+1}, \ldots, a_n) \in \mathcal{Z}(I \cap k[x_{\ell+1}, \ldots, x_n]) = \pi_{n-\ell}(\mathcal{Z}(I))$, where equality holds by [CLO07, Chapter 3.2, Theorem 3]. So there exist $a_1, \ldots, a_\ell \in \bar{k}$ such that $(a_1, \ldots, a_n) \in \mathcal{Z}(I)$. Therefore, $\pi_{n-\ell+1}(a_1, \ldots, a_n) = (a_\ell, \ldots, a_n) \in \mathcal{Z}(I \cap k[x_\ell, \ldots, x_n])$, that is $a_\ell \in \mathcal{Z}(I(\ell, a)) = \emptyset$, a contradiction. $\square$

We use the previous result to build an algorithm which computes the affine zero locus of an ideal I from its reduced lexicographic Gröbner basis. We adopt the notation of Theorem 5.

Corollary 1. *Let $I \subseteq R = k[x_1, \ldots, x_n]$ be an ideal with finite affine zero locus $\mathcal{Z}(I)$. Then $\mathcal{Z}(I)$ can be computed as follows:*

1. *Compute the reduced lexicographic Gröbner basis $\mathcal{G}$ of I to obtain the monic polynomial $p_n \in k[x_n]$ such that $(p_n) = I \cap k[x_n]$.*
2. *If $p_n = 1$, then $\mathcal{Z}(I) = \emptyset$. Else, factor p_n.*

3. *For every root α of p_n compute*

$$p_{n-1}(x_{n-1}) = \gcd\{p_{n-1,1}(x_{n-1}, \alpha), \ldots, p_{n-1,t_{n-1}}(x_{n-1}, \alpha)\}.$$

4. *Factor p_{n-1}.*
5. *For every root β of p_{n-1} compute*

$$p_{n-2}(x_{n-2}) = \gcd\{p_{n-2,1}(x_{n-2}, \beta, \alpha), \ldots, p_{n-2,t_{n-2}}(x_{n-2}, \beta, \alpha)\}.$$

6. *Proceed similarly, until all the elements of $\mathcal{Z}(I)$ are found.*

Notice that the computation is even more efficient under the assumption that the system $\mathcal{F}$, or equivalently the ideal I, has only one zero over the algebraic closure. This is often the case for polynomial systems coming from multivariate cryptosystems, where we usually require that for each ciphertext b there is a unique plaintext a such that $f_i(a) = b$ for every $i = 1, \ldots r$.

In such a situation, one does not need to factor any univariate polynomial, since each one of them has exactly one solution, which, for a monic polynomial of degree d, can be computed by multiplying the coefficient of x^{d-1} by $(-1)^{d-1}d^{-1}$.

Remark 4. Assume that k is either a finite field or has characteristic zero. If I admits only one solution $(a_1, \ldots, a_n) \in \bar{k}^n$, then in fact $(a_1, \ldots, a_n) \in k^n$. This is true even if the solution has multiplicity higher than one. In fact, $g_n(x_n) = (x_n - a_n)^d \in k[x_n]$, hence $da_n \in k$. If k has characteristic zero, then $a_n \in k$. Else, let p be the characteristic of k and write $d = p^\ell e$ where $p \nmid e$. Then $g_n(x_n) = \left(x_n^{p^\ell} - a_n^{p^\ell}\right)^e \in k[x_n]$, so $ea_n^{p^\ell} \in k$. This implies $a_n^{p^\ell} \in k$, hence $a_n \in k$, since k is a finite field. One proceeds similarly to prove that $a_i \in k$ for all i.

Remark 5. By [CLO07, Chapter 3.2, Theorem 3] and since $\mathcal{Z}(I)$ is a finite set, one has that

$$\mathcal{Z}(I \cap k[x_\ell, \ldots, x_n]) = \pi_{n-\ell+1}(\mathcal{Z}(I))$$

for $1 \leq \ell \leq n$, where $\pi_i : k^n \to k^i$ is the projection on the last i coordinates. This implies that each path from the roots to the leaves in the tree-shaped computation of Corollary 1 produces a solution. In particular, Corollary 1 does not perform useless computations.

2.1   Adding the Field Equations to a System

Let $\mathcal{Q} = \{x_1^q - x_1, \ldots, x_n^q - x_n\}$ be the system consisting of the field equations relative to $\mathbb{F}_q$. Clearly, for any system of equations $\mathcal{F} = \{f_1, \ldots, f_r\} \subseteq R = \mathbb{F}_q[x_1, \ldots, x_n]$ one has

$$\mathcal{Z}(\mathcal{F} \cup \mathcal{Q}) = \mathcal{Z}(\mathcal{F}) \cap \mathbb{F}_q^n.$$

The systems $\mathcal{F}$ and $\mathcal{F} \cup \mathcal{Q}$, however, often have different algebraic properties. It is easy to show that the ideal generated by $\mathcal{F} \cup \mathcal{Q}$ is always radical, while the ideal generated by $\mathcal{F}$ may not be. The structure of the reduced Gröbner bases of the ideals generated by the two systems and the degrees of the elements appearing in them are often different as well. As a consequence, adding the field equations to a system often affects the complexity of computing a Gröbner basis.

Therefore, passing from $\mathcal{F}$ to $\mathcal{F} \cup \mathcal{Q}$ may or may not provide an advantage. It typically provides an advantage for fields of small size, since the equations of $\mathcal{Q}$ have low degree and adding them to $\mathcal{F}$ makes the ideal radical, a necessary hypothesis for the Shape Lemma (Theorem 2) to apply. Over fields of large size, however, adding the field equations may make the computation of a Gröbner basis practically infeasible. This is due to the fact that we are adding to the system equations of large degree, which are involved in the computation of a Gröbner basis, therefore increasing the degree of the computation. In the next example, we show that the solving degree may increase when passing from $\mathcal{F}$ to $\mathcal{F} \cup \mathcal{Q}$ (see Definition 6 for the definition of solving degree).

Example 5. Let $\mathcal{F} = \{x_3^2 - x_2, x_2^3 - x_1\} \subseteq \mathbb{F}_5[x_1, x_2, x_3]$ and let $I = (\mathcal{F})$. The affine zero locus of I over $\overline{\mathbb{F}}_5$ is infinite. If we add the field equations $\mathcal{Q} = \{x_1^5 - x_1, x_2^5 - x_2, x_3^5 - x_3\}$ of $\mathbb{F}_5$ to $\mathcal{F}$, we obtain the ideal $J = (\mathcal{F} \cup \mathcal{Q})$, which has $\mathcal{Z}(J) = \{(0,0,0),(1,1,1),(4,4,2),(4,4,3),(1,1,4)\}$. The elements of $\mathcal{F}$ are a Gröbner basis of I with respect to the LEX order, while the reduced Gröbner basis of J with respect to the same order also contains $x_3^5 - x_3$. In particular, the Gröbner basis of J contains a polynomial of higher degree and one can easily verify that

$$\mathrm{solv.\,deg}(\mathcal{F} \cup \mathcal{Q}) = 5 > 3 = \mathrm{solv.\,deg}(\mathcal{F}).$$

Even if we restrict our attention to polynomial systems arising in public-key cryptography, one may not always assume that the field equations can be added to the system. An example coming from multivariate cryptography was given in Example 4. Another example are systems coming from the relation-collection phase of index calculus on elliptic or hyperelliptic curves, since the field size is very large (e.g., the field size required for 80-bit security is at least $q \sim 2^{160}$ for an elliptic curve and $q \sim 2^{80}$ for a hyperelliptic curve of genus two). In such a situation, adding equations of degree q to the system would make it unmanageable.

3   Solving Degree of Polynomial Systems

In Sect. 2 we discussed how one can compute the solutions of a polynomial system, starting from a lexicographic Gröbner basis of the ideal that it generates. In this section, we address the problem of estimating the complexity of computing a lexicographic Gröbner basis. In practice, one observes that computing a Gröbner basis with respect to LEX is usually slower than with respect to any other term order. On the other hand, computing a Gröbner basis with respect to DRL is often faster than with respect to any other term order. Therefore, computing a degree reverse lexicographic Gröbner basis and converting it to a lexicographic Gröbner basis using FGLM or a similar algorithm is usually more efficient than computing a lexicographic Gröbner basis directly. For this reason, in this section we discuss the complexity of computing a Gröbner basis of an

ideal I in a polynomial ring $R = k[x_1, \ldots, x_n]$ over a field k with respect to the
DRL order. We refer the reader to [FGLM93] for a description of the FGLM
algorithm and an estimate of its complexity.

3.1   Macaulay Matrices and Solving Degree

We have two main classes of algorithms for computing Gröbner bases: *Buch-
berger's Algorithm* and *linear algebra based algorithms*, which transform the
problem of computing a Gröbner basis into one or more instances of Gaussian elimination. Examples of linear algebra based algorithms are: F_4 [Fau99],
F_5 [Fau02], the *XL Algorithm* [CKPS00], and *MutantXL* [DBMMW08]. Buch-
berger's Algorithm is older, and its complexity has been extensively studied. Linear algebra based algorithms are often faster in practice and have contributed
to breaking many cryptographic challenges. However, their complexity is less
understood, especially when the input consists of polynomials which are not
homogeneous.

   In this section, we discuss the complexity of linear algebra based algorithms,
which is dominated by Gaussian elimination on the *Macaulay matrices*. First
we describe them for homogeneous systems, following [BFS15, p. 54]. Let $\mathcal{F} =
\{f_1, \ldots, f_r\} \subseteq R$ be a system of homogeneous polynomials and fix a term order.
The *homogeneous Macaulay matrix* M_d of $\mathcal{F}$ has columns indexed by the terms
of R_d sorted, from left to right, according to the chosen order. The rows of M_d
are indexed by the polynomials $m_{i,j}f_j$, where $m_{i,j} \in R$ is a term such that
$\deg(m_{i,j}f_j) = d$. Then the entry (i, j) of M_d is the coefficient of the monomial
of column j in the polynomial corresponding to the i-th row.

   Now let $f_1, \ldots, f_r$ be any polynomials (not necessarily homogeneous). For
any degree $d \in \mathbb{Z}_+$ the *Macaulay matrix* $M_{\leq d}$ of $\mathcal{F}$ has columns indexed by the
terms of R of degree $\leq d$, sorted in decreasing order from left to right. The rows
of $M_{\leq d}$ are indexed by the polynomials $m_{i,j}f_j$, where $m_{i,j}$ is a term in R such
that $\deg(m_{i,j}f_j) \leq d$. The entries of $M_{\leq d}$ are defined as in the homogeneous case.
Notice that, if $f_1, \ldots, f_r$ are homogeneous, the Macaulay matrix $M_{\leq d}$ is just a
block matrix, whose blocks are the homogeneous Macaulay matrices $M_d, \ldots, M_0$
associated to the same equations. This is the reason for using homogeneous
Macaulay matrices in the case that $f_1, \ldots, f_r$ are homogeneous.

   The size of the Macaulay matrices $M_{\leq d}$ and M_d, hence the computational
complexity of computing their reduced row echelon forms, depends on the degree
d. Therefore, following [DS13], we introduce the next definition.

Definition 6. *Let $\mathcal{F} = \{f_1, \ldots, f_r\} \subseteq R$ and let τ be a term order on R. The
solving degree of $\mathcal{F}$ is the least degree d such that Gaussian elimination on the
Macaulay matrix $M_{\leq d}$ produces a Gröbner basis of $\mathcal{F}$ with respect to τ. We
denote it by $\mathrm{solv.\,deg}_\tau(\mathcal{F})$. When the term order is clear from the context, we
omit the subscript τ.*

   *If $\mathcal{F}$ is homogeneous, we consider the homogeneous Macaulay matrix M_d and
let the solving degree of $\mathcal{F}$ be the least degree d such that Gaussian elimination
on $M_0, \ldots, M_d$ produces a Gröbner basis of $\mathcal{F}$ with respect to τ.*

Some algorithms perform Gaussian elimination on the Macaulay matrix for increasing values of d. An algorithm of this kind has a termination criterion, which allows to decide whether a Gröbner basis has been found and the algorithm can be stopped. For example, F_5 uses the so-called signatures for this purpose. Other algorithms perform Gaussian elimination on just one Macaulay matrix, for a large enough value of d. For such an algorithm, a sharp bound on the solving degree provides a good estimate for the value of d to be chosen. In both cases, the solving degree produces a bound on the complexity of computing the desired Gröbner basis. In particular, one may choose to artificially stop a Gröbner basis computation in the degree corresponding to the solving degree. For this reason, we use the solving degree to measure the complexity of Gröbner bases computations and we do not discuss termination criteria.

Remark 6. If $\mathcal{F}$ is not homogeneous, then Gaussian elimination on $M_{\leq d}$ may produce a row that corresponds to a polynomial f such that $\deg(f) < d$ and $\mathrm{in}(f)$ was not the leading term of any row of $M_{\leq d}$ before performing Gaussian elimination. If this is the case, then some variants of the algorithms add to $M_{\leq d}$ the rows corresponding to the polynomials mf, where m is a monomial and $\deg(mf) \leq d$. Then they proceed to compute the reduced row echelon form of this larger matrix. If no Gröbner basis is produced in degree $\leq d$, then they proceed by adding to this matrix the appropriate multiples of its rows in the next degree and continue as before. This potentially has the effect of enlarging the span of the rows of $M_{\leq d}$, for all d. Introducing this variation may therefore reduce the computational cost of computing a Gröbner basis with respect to a given term order, since we might be able to obtain a Gröbner basis in a smaller degree than the solving degree, as defined in Definition 6. Throughout the paper, we consider the situation when *no extra rows are inserted*. Notice that the solving degree is an upper bound on the degree in which the algorithms adopting this variation terminate.

Definition 7. *Let $I \subseteq R$ be an ideal and let τ be a term order on R. We denote by* $\max. \mathrm{GB}. \deg_\tau(I)$ *the maximum degree of a polynomial appearing in the reduced τ Gröbner basis of I. If $I = (\mathcal{F})$, we sometimes write* $\max. \mathrm{GB}. \deg_\tau(\mathcal{F})$ *in place of* $\max. \mathrm{GB}. \deg_\tau(I)$.

It is clear that

$$\max. \mathrm{GB}. \deg_\tau(\mathcal{F}) \leq \mathrm{solv}. \deg_\tau(\mathcal{F}),$$

for any system of polynomials $\mathcal{F}$ and any term order τ. Equality does not hold in general, as we show in Example 8.

Remark 7. Assume that $\mathcal{F} = \{f_1, \ldots, f_r\}$ is homogeneous. Gaussian elimination on M_d exclusively produces rows that correspond to polynomials of degree d. Therefore

$$\mathrm{solv}. \deg_\tau(\mathcal{F}) = \max. \mathrm{GB}. \deg_\tau(\mathcal{F})$$

for any τ.

Notice moreover that the solving degree of a system $\mathcal{F}$ may be strictly smaller than the largest degree of an equation of $\mathcal{F}$. This may happen, e.g., when $\mathcal{F}$ contains redundant equations.

Example 6. Let $\mathcal{F} = \{x^2 + x, xy, y^2 + y, x^2y + x^2 + x\} \subseteq \mathbb{F}_2[x, y]$. The reduced *DRL* Gröbner basis of $I = (\mathcal{F})$ is $\{x^2 + x, xy, y^2 + y\}$ and solv. $\deg_{DRL} \mathcal{F} = 2$.

3.2   Homogenization of Ideals and Extensions of Term Order

We consider a polynomial ring $R = k[x_1, \ldots, x_n]$ and its extension $S = R[t]$ with respect to a new variable t. We compare term orders on R and S.

Definition 8. *Let σ be a term order on R, let τ be a term order on $S = R[t]$, and let $\phi : S \to R$ be the dehomogenization map. We say that τ ϕ-extends σ, or that τ is a ϕ-extension of σ, if $\phi(\mathrm{in}_\tau(f)) = \mathrm{in}_\sigma(\phi(f))$ for every homogeneous $f \in S$.*

The next theorem relates Gröbner basis and dehomogenization.

Theorem 6. *Let σ be a term order on R, and let τ be a ϕ-extension of σ on S. Let I be an ideal in R, let J be a homogeneous ideal in S such that $\phi(J) = I$. The following hold:*

1. *$\mathrm{in}_\sigma(I) = \phi(\mathrm{in}_\tau(J))$;*
2. *if $\{g_1, \ldots, g_s\}$ is a homogeneous τ Gröbner basis of J, then $\{\phi(g_1), \ldots, \phi(g_s)\}$ is a σ Gröbner basis of I.*

Proof. We prove *(1)*. Notice that $\mathrm{in}_\tau(J) = (\mathrm{in}_\tau(f) : f \in J, f \text{ homogeneous})$, because J is a homogeneous ideal. Then we have

$$\phi(\mathrm{in}_\tau(J)) = (\phi(\mathrm{in}_\tau(f)) : f \in J, f \text{ homogeneous})$$
$$= (\mathrm{in}_\sigma(\phi(f)) : f \in J, f \text{ homogeneous}).$$

To conclude the proof of *(1)*, it suffices to show that

$$\{\phi(f) : f \in J, f \text{ homogeneous}\} = I.$$

The inclusion from left to right follows from the assumption that $\phi(J) = I$. To prove the other inclusion, we fix a system of generators $f_1, \ldots, f_r$ of I and consider $f = \sum_{i=1}^r p_i f_i \in I$, with $p_i \in R$. Let $h_i \in J$ be homogeneous such that $\phi(h_i) = f_i$ for all i and define $\tilde{p} = \sum_{i=1}^r t^{\alpha_i} p_i^h h_i$. The polynomial $\tilde{p}$ belongs to J and it is homogeneous for a suitable choice of the α_i's. Since $\phi(\tilde{p}) = \sum_{i=1}^r \phi(t^{\alpha_i} p_i^h h_i) = \sum_{i=1}^r p_i f_i = f$, the inclusion follows.

To prove *(2)*, observe that

$$\phi(\mathrm{in}_\tau(J)) = (\phi(\mathrm{in}_\tau(g_i)) : i = 1, \ldots, s) = (\mathrm{in}_\sigma(\phi(g_i)) \; i = 1, \ldots, s),$$

since ϕ is a homomorphism and τ ϕ-extends σ. This shows that $\{\phi(g_1), \ldots, \phi(g_s)\}$ is a Gröbner basis of $\phi(\mathrm{in}_\tau(J))$ with respect to σ, which is equal to $\mathrm{in}_\sigma(I)$ by *(1)*. $\qquad\square$

There is a natural way to ϕ-extend a term order σ on R to a term order $\bar{\sigma}$ on S.

Definition 9. *Let m, n be terms in R, let σ be a term order on R. Define a term order $\bar{\sigma}$ on S via: $t^\alpha m >_{\bar{\sigma}} t^\beta n$ if and only if $(m >_\sigma n)$ or $(m = n$ and $\alpha > \beta)$.*

Lemma 1. $\bar{\sigma}$ *is a term order on S which ϕ-extends σ.*

Proof. First we prove that $\bar{\sigma}$ is a term order. The fact that $1 <_\sigma m$ for every term $m \in R$ implies $1 <_{\bar{\sigma}} m$. We have also $1 = t^0 <_{\bar{\sigma}} t$.

Now, let $t^\alpha m >_{\bar{\sigma}} t^\beta n$, with m, n terms in R, and $\alpha, \beta \in \mathbb{N}$. We show that $>_{\bar{\sigma}}$ respects multiplication by terms. We have two possibilities: *1) $m >_\sigma n$* or *2) $m = n$ and $\alpha > \beta$*. If *1)* holds, then we have $x_i m >_\sigma x_i n$ for every $i = 1, \dots, n$ since σ is a term order, which implies $x_i t^\alpha m >_{\bar{\sigma}} x_i t^\beta n$. Clearly $t^{\alpha+1} m >_{\bar{\sigma}} t^{\beta+1} n$. If *2)* holds, then $x_i m = x_i n$ for every $i = 1, \dots, n$, therefore $x_i t^\alpha m >_{\bar{\sigma}} x_i t^\beta n$ since $\alpha > \beta$. Moreover we have $t^{\alpha+1} m >_{\bar{\sigma}} t^{\beta+1} n$, because $m = n$ and $\alpha + 1 > \beta + 1$.

Now we prove that $\bar{\sigma}$ ϕ-extends σ, that is $\phi(\text{in}_{\bar{\sigma}}(f)) = \text{in}_\sigma(\phi(f))$ for every $f \in S$ homogeneous. Let $f = \sum_{i=1}^d a_i t^{\alpha_i} m_i$ be a homogeneous polynomial, with $m_i \in R$ distinct terms, $\alpha_i \in \mathbb{N}$, and $a_i \in k \setminus \{0\}$. Then $\phi(f) = \sum_{i=0}^d a_i m_i$ and $\deg m_i = \deg f - \alpha_i$. If there is any cancellation in the sum defining $\phi(f)$, then the monomials cancelling have the same degree, then they have already been cancelled in f. Hence, there is no cancellation in $\phi(f)$. Without loss of generality, let $m_1 = \text{in}_\sigma(\phi(f))$, that is $m_1 >_\sigma m_i$ for every $i = 2, \dots, d$. Then $t^{\alpha_1} m_1 = \text{in}_{\bar{\sigma}}(f)$, and $\phi(\text{in}_{\bar{\sigma}}(f)) = m_1 = \text{in}_\sigma(\phi(f))$. $\qquad\square$

Example 7. The equality $\phi(\text{in}_{\bar{\sigma}}(f)) = \text{in}_\sigma(\phi(f))$ does not necessarily hold for f not homogeneous. For example consider $f = tx - x + ty \in S = k[x, y, t]$, and let $\sigma = LEX$. Then $\text{in}_{\bar{\sigma}}(f) = tx$, $\phi(f) = y$, and $\text{in}_\sigma(\phi(f)) = y \neq x = \phi(\text{in}_{\bar{\sigma}}(f))$.

The next Lemma gives an important example of ϕ-extension of a term order.

Lemma 2. *Fix the DRL order on R and extend it to the DRL order on S by letting t be the smallest variable. Then the DRL order on S ϕ-extends the DRL order on R.*

Proof. Let $f = \sum_{i=1}^d a_i t^{\alpha_i} m_i$ be a homogeneous polynomial, with distinct terms $m_i \in R$, $\alpha_i \in \mathbb{N}$, and $a_i \in k \setminus \{0\}$. Then $\phi(f) = \sum_{i=0}^d a_i m_i$ and $\deg m_i = \deg f - \alpha_i$. As in the proof of Lemma 1 there is no cancellation in $\phi(f)$.

Without loss of generality, let $\text{in}_{DRL}(\phi(f)) = m_1$, that is $m_1 >_{DRL} m_i$ for all $i = 2, \dots, d$. For each $i \in \{2, \dots, d\}$ we have two possibilities: either $\deg m_1 > \deg m_i$ or $\deg m_1 = \deg m_i$. If $\deg m_1 > \deg m_i$ then we have $\alpha_1 < \alpha_i$, since $\deg m_j + \alpha_j = \deg f$ for every j. This implies $t^{\alpha_1} m_1 >_{DRL} t^{\alpha_i} m_i$. If $\deg m_1 = \deg m_i$ then we have $\alpha_1 = \alpha_i$, and $t^{\alpha_1} m_1 >_{DRL} t^{\alpha_i} m_i$ follows from $m_1 >_{DRL} m_i$. Therefore we have $\text{in}_{DRL}(f) = t^{\alpha_1} m_1$, and $\phi(\text{in}_{DRL}(f)) = m_1 = \text{in}_{DRL}(\phi(f))$. $\qquad\square$

Remark 8. Fix the *DRL* order on R. The *DRL* order on S is different from the order $\overline{DRL}$ obtained by applying Definition 9. For example, let $R = k[x, y]$ with $x > y$, $S = R[t]$, and consider the monomials $t^3 x$ and ty^2. We have $t^3 x <_{\overline{DRL}} ty^2$ because $x <_{DRL} y^2$ in R. In particular, $\overline{DRL}$ is not degree-compatible, while DRL is. Notice however that the two orders coincide on pairs of terms of the same degree.

3.3 Solving Degree and Homogenization

Let $R = k[x_1, \ldots, x_n]$ with the *DRL* order and let $S = R[t]$ with the *DRL* order with t as smallest variable. Let $\mathcal{F} = \{f_1, \ldots, f_r\} \subseteq R$, let $I = (\mathcal{F}) \subseteq R$, let $I^h \subseteq S$ be the homogenization of I with respect to t, and let $(\mathcal{F}^h) \subseteq S$ be the ideal generated by $\mathcal{F}^h = \{f_1^h, \ldots, f_r^h\}$. The goal of this section is comparing the solving degrees of $\mathcal{F}$, $\mathcal{F}^h$, and I^h with respect to the chosen term orders. We start with a preliminary result on Gröbner bases and homogenization.

Proposition 1. *Let $R = k[x_1, \ldots, x_n]$ and let $S = R[t]$. Fix the DRL term order on R and extend it to the DRL term order on S by letting t be the smallest variable. Let I be an ideal of R with Gröbner basis $\{g_1, \ldots, g_s\}$. Then $\{g_1^h, \ldots, g_s^h\}$ is a Gröbner basis of I^h.*

Proof. First we show that $g_1^h, \ldots, g_s^h$ generate I^h. Clearly we have $g_1^h, \ldots, g_s^h \in I^h$. For the other inclusion, consider $f \in I$ of degree d with standard representation $f = \sum_{i=1}^s f_i g_i$ for some $f_i \in R$, that is $\mathrm{in}(f) \geq \mathrm{in}(f_i g_i)$ for all $i = 1, \ldots, s$.

Since $\mathrm{in}(f) \geq \mathrm{in}(f_i g_i)$ and DRL is degree-compatible, we have $d \geq \deg f_i + \deg g_i$. Therefore we can write

$$f^h = \sum_{i=1}^s t^{d - \deg f_i - \deg g_i} f_i^h g_i^h, \tag{2}$$

which shows that $f^h \in (g_1^h, \ldots, g_s^h)$.

To prove that $\{g_1^h, \ldots, g_s^h\}$ is a Gröbner basis, it is enough to show that (2) is a standard representation for f^h, i.e., $\mathrm{in}(f^h) \geq \mathrm{in}(t^{d - \deg f_i - \deg g_i} f_i^h g_i^h)$ for all $i = 1, \ldots, s$. We observe that $\mathrm{in}(f^h) = \mathrm{in}(f)$ does not contain the variable t and we distinguish two cases.

1. If $d - \deg f_i - \deg g_i > 0$, then a power of t appears in $t^{d - \deg f_i - \deg g_i} f_i^h g_i^h$, and in its initial term as well. It follows that $\mathrm{in}(f^h) \geq \mathrm{in}(t^{d - \deg f_i - \deg g_i} f_i^h g_i^h)$ since t is the smallest variable in the *DRL* term order of S.
2. If $d - \deg f_i - \deg g_i = 0$, then no power of t appears in $\mathrm{in}(f_i^h g_i^h)$. Therefore we have $\mathrm{in}(f_i^h g_i^h) = \mathrm{in}(f_i g_i) \leq \mathrm{in}(f) = \mathrm{in}(f^h)$. □

The next result relates the solving degrees of $\mathcal{F}$ and $\mathcal{F}^h$. It also clarifies why the largest degree of an element in a reduced Gröbner basis of $\mathcal{F}$ may be smaller than its solving degree.

Theorem 7. *Let* $\mathcal{F} = \{f_1, \ldots, f_r\} \subseteq R = k[x_1, \ldots, x_n]$ *and consider the system* $\mathcal{F}^h = \{f_1^h, \ldots, f_r^h\} \subseteq S = R[t]$ *obtained from* $\mathcal{F}$ *by homogenizing* $f_1, \ldots, f_r$ *with respect to* t. *Let* $I^h \subseteq S$ *be the homogenization of* $I = (\mathcal{F}) \subseteq R$ *with respect to* t. *Consider the term order* DRL *on* R *and* S, *with* t *as smallest variable. Then*

$$\max. \mathrm{GB}. \deg(\mathcal{F}^h) = \mathrm{solv}. \deg(\mathcal{F}^h) = \mathrm{solv}. \deg(\mathcal{F})$$

$$\geq \max. \mathrm{GB}. \deg(\mathcal{F}) = \max. \mathrm{GB}. \deg(I^h) = \mathrm{solv}. \deg(I^h).$$

Proof. We claim that the Macaulay matrix $M_{\leq d}$ of $\mathcal{F}$ with respect to DRL is equal to the homogeneous Macaulay matrix M_d of $\mathcal{F}^h$ with respect to DRL, for every $d \geq 1$. In fact, the monomials of S of degree d are exactly the homogenizations of the monomials of R of degree $\leq d$. Similarly, if $m_{i,j} f_j^h$ is the index of a row of M_d, i.e., $\deg(m_{i,j} f_j^h) = d$, then $\phi(m_{i,j} f_j^h) = \phi(m_{i,j}) f_j$ has degree $\leq d$, hence it is the index of a row of $M_{\leq d}$. Conversely, every index $m_{i,j} f_j^h$ of a row of M_d, can be obtained from an index of a row of $M_{\leq d}$ by homogenizing and multiplying by an appropriate power of t. In a nutshell, the statement on the columns follows from the fact that $I_{\leq d} = \phi\left((\mathcal{F}^h)_d\right)$. One also needs to check that the order on the columns of M_d and $M_{\leq d}$ is the same. We consider $M_{\leq d}$. Since DRL is degree-compatible, the columns are ordered in non-increasing degree order from left to right. The columns of the same degree $j \in \{1, \ldots, d\}$ are then ordered according to DRL. Similarly, since t is the smallest variable in the DRL order on S, the columns of M_d are ordered in increasing order (from left to right) of powers of t, which is equivalent to decreasing order of the degree of the variables $x_1, \ldots, x_n$. Then, the columns with the same power of t are ordered according to DRL on the variables $x_1, \ldots, x_n$. This proves that the matrices $M_{\leq d}$ and M_d coincide.

Let $I = (\mathcal{F})$ and $J = (\mathcal{F}^h)$. Since the matrices $M_{\leq d}$ and M_d coincide and since the dehomogenization of a Gröbner basis of $\mathcal{F}^h$ produces a Gröbner basis of $\mathcal{F}$ by Theorem 6, one has

$$\mathrm{solv}. \deg_{DRL}(\mathcal{F}) \leq \mathrm{solv}. \deg_{DRL}(\mathcal{F}^h).$$

To check that they are equal, for each minimal generator m of $\mathrm{in}(I)$, we consider the least degree d for which a polynomial f with $\mathrm{in}(f) = m$ appears among the rows of the reduced row echelon form of $M_{\leq d}$. Since $M_d = M_{\leq d}$, the polynomial $t^{d-\deg(f)} f^h$ appears among the rows of the reduced row echelon form of M_d. We claim that no polynomial g with $\mathrm{in}(g) \mid t^{d-\deg(f)} m = \mathrm{in}(t^{d-\deg(f)} f^h)$ appears as a row of the reduced row echelon form of M_e for some $e < d$. In fact, if this were the case then, by Theorem 6, the dehomogenization of $\mathrm{in}(g)$ would be equal to m and appear as a row of M_e. This contradicts the assumption that d is the least degree for which a polynomial with leading term m appears among the rows of the reduced row echelon form of $M_{\leq d}$. This shows that the least degree d in which the leading terms of the rows of the reduced row echelon form of the matrix $M_{\leq d}$ generate the initial ideal of I is the same as the the least degree e in which the leading terms of the rows of the reduced row echelon form of the matrix M_e generate $\mathrm{in}(J)_e$. Therefore

$$\mathrm{solv}. \deg_{DRL}(\mathcal{F}) = \mathrm{solv}. \deg_{DRL}(\mathcal{F}^h).$$

The equality $\max.\mathrm{GB}.\deg(\mathcal{F}) = \max.\mathrm{GB}.\deg(I^h)$ follows from the following two facts:

- By Lemma 2 and Theorem 6 the dehomogenization of a DRL Gröbner basis of I^h produces a DRL Gröbner basis of I.
- The homogenization of a DRL Gröbner basis of I produces a DRL Gröbner basis of I^h by Proposition 1.

In particular, no leading term of an element of the reduced Gröbner basis of I^h is divisible by t, so dehomogenization does not decrease the degrees of the elements of the Gröbner basis.

Finally, the two equalities

$$\max.\mathrm{GB}.\deg(\mathcal{F}^h) = \mathrm{solv}.\deg(\mathcal{F}^h) \text{ and } \max.\mathrm{GB}.\deg(I^h) = \mathrm{solv}.\deg(I^h)$$

follow from Remark 7.   □

Remark 9. Theorem 7 clarifies why, when the system $\mathcal{F}$ is not homogeneous, the largest degree of an element in a reduced Gröbner basis may be strictly smaller than the solving degree. This is due to the difference between the ideals $(\mathcal{F}^h)$ and I^h, and more specifically between $\max.\mathrm{GB}.\deg(\mathcal{F}^h)$ and $\max.\mathrm{GB}.\deg(I^h)$.

The following is an example where $\mathrm{solv}.\deg(\mathcal{F}) > \max.\mathrm{GB}.\deg(\mathcal{F})$. See also Example 11 for a cryptographic example.

Example 8. Let $R = k[x,y]$ and let $S = R[t] = k[x,y,t]$, both with the DRL order. We consider the system $\mathcal{F} = \{f_1, f_2\} \subseteq R$ with $f_1 = x^2 - 1$, $f_2 = xy + x$, and let $I = (\mathcal{F})$. Then $\mathcal{F}^h = \{f_1^h, f_2^h\} = \{x^2 - t^2, xy + xt\}$, and $I^h = (x^2 - t^2, y + t)$. Writing the Macaulay matrices of $\mathcal{F}$, $\mathcal{F}^h$, and $\{x^2 - t^2, y + t\}$ and doing Gaussian elimination, one sees that $\mathrm{solv}.\deg(\mathcal{F}) = \mathrm{solv}.\deg(\mathcal{F}^h) = 3$. By computing Gröbner bases, one can check that $\max.\mathrm{GB}.\deg(\mathcal{F}^h) = 3$ and $\max.\mathrm{GB}.\deg(\mathcal{F}) = \max.\mathrm{GB}.\deg(I^h) = 2$.

3.4 Solving Degree and Castelnuovo-Mumford Regularity

In what follows, we compare the solving degree of a homogeneous ideal with a classical invariant from commutative algebra: the *Castelnuovo-Mumford regularity*. We recall the definition of this invariant and its basic properties before illustrating the link with the solving degree.

Let $R = k[x_1, \ldots, x_n]$ be a polynomial ring in n variables over a field k and let I be a homogeneous ideal of R. For any integer j we recall that R_j denotes the k-vector space of homogeneous elements of R of degree j.

Choose a minimal system of generators $f_1, \ldots, f_{\beta_0}$ of I. We recall that, since I is homogeneous, the number β_0 and the degrees $d_i = \deg f_i$ are uniquely determined. We fix an epimorphism $\varphi : R^{\beta_0} \to I$ sending the canonical basis $\{e_1, \ldots, e_{\beta_0}\}$ of the free module R^{β_0} to $\{f_1, \ldots, f_{\beta_0}\}$. The map φ is in general not homogeneous of degree 0, so we introduce degree shifts on R: For any integer d, we denote by $R(-d)$ the R-module R, whose j-th homogeneous component is

$R(-d)_j = R_{-d+j}$. For example, the variables $x_1, \ldots, x_n$ have degree 2 in $R(-1)$, and degree 0 in $R(1)$. The map

$$\varphi : \bigoplus_{j=1}^{\beta_0} R(-d_j) \to I$$

is homogeneous of degree 0, that is $\deg(\varphi(f)) = \deg f$ for every f.

Now consider the submodule $\ker \varphi \subseteq \bigoplus_{j=1}^{\beta_0} R(-d_j)$. It is again finitely generated and graded, and is called (first) syzygy module of I. We choose a minimal system of generators of $\ker \varphi$ and we continue similarly defining an epimorphism from a free R-module (with appropriate shifts) to $\ker \varphi$ and so on.

Hilbert's Syzygy Theorem guarantees that this procedure terminates after a finite number of steps. Thus, we obtain a *minimal graded free resolution* of I:

$$0 \to F_p \to \cdots \to F_1 \to F_0 \xrightarrow{\varphi} I \to 0,$$

where the F_i are free R-modules of the form

$$F_i = \bigoplus_{j=0}^{\beta_i} R(-d_{i,j})$$

for appropriate shifts $d_{i,j} \in \mathbb{Z}$. By regrouping the shifts, we may write the free R-modules of the minimal free resolution of I as

$$F_i = \bigoplus_{j \in \mathbb{Z}} R(-j)^{\beta_{i,j}}.$$

The numbers $\beta_{i,j} = \beta_{i,j}(I)$ are the *(graded) Betti numbers* of I.

Definition 10. *The* Castelnuovo-Mumford regularity *of I is*

$$\mathrm{reg}(I) = \max\{j - i : \ \beta_{i,j}(I) \neq 0\}.$$

If $\mathcal{F}$ is a homogeneous system of generators of I, we set also $\mathrm{reg}(\mathcal{F}) = \mathrm{reg}(I)$.

Example 9. We consider the ideal $I = (x^2, xy, xz, y^3)$ in $R = k[x, y, z]$. A minimal free resolution of I is given by

$$0 \to R(-4) \xrightarrow{\varphi_2} R(-3)^3 \oplus R(-4) \xrightarrow{\varphi_1} R(-2)^3 \oplus R(-3) \xrightarrow{\varphi_0} I \to 0,$$

with R-linear maps given by the following matrices

$$\varphi_0 = (x^2, xy, xz, y^3), \quad \varphi_1 = \begin{pmatrix} -y & -z & 0 & 0 \\ x & 0 & -z & -y^2 \\ 0 & x & y & 0 \\ 0 & 0 & 0 & x \end{pmatrix}, \quad \varphi_2 = \begin{pmatrix} z \\ -y \\ x \\ 0 \end{pmatrix}.$$

So the non-zero Betti numbers of I are $\beta_{0,2} = 3$, $\beta_{0,3} = 1$, $\beta_{1,3} = 3$, $\beta_{1,4} = 1$, $\beta_{2,4} = 1$, and the Castelnuovo-Mumford regularity is $\mathrm{reg}(I) = 3$.

For more on regularity and its properties, the interested reader may consult [Eis94, Chapter 20] or [Cha07]. In the sequel we only mention the facts that are relevant for our purposes.

Remark 10. In many texts in commutative algebra or algebraic geometry it is assumed that the field k is algebraically closed or infinite. However, the definition of regularity makes perfect sense over a finite field. The construction of a minimal free resolution that we illustrated can be carried out over a finite field. Moreover, it shows that the Castelnuovo-Mumford regularity is preserved under field extensions. In particular, if I is an ideal in a polynomial ring $R = \mathbb{F}_q[x_1, \ldots, x_n]$ over a finite field $\mathbb{F}_q$ and J is its extension to the polynomial ring $S = \overline{\mathbb{F}}_q[x_1, \ldots, x_n]$ over the algebraic closure of $\mathbb{F}_q$, then $\mathrm{reg}_R(I) = \mathrm{reg}_S(J)$.

The next theorem is due to Bayer and Stillman. It relates the regularity of a homogeneous ideal to the regularity of its *DRL* initial ideal. Combined with our Theorem 7, it will allow us to bound the solving degree of any system.

Theorem 8 ([BS87], Theorem 2.4 and Proposition 2.9). *Let J be a homogeneous ideal in $k[x_1, \ldots, x_n]$. Assume that J is in generic coordinates over $\overline{k}$, then*

$$\mathrm{reg}(J) = \mathrm{reg}(\mathrm{in}_{DRL}(J)).$$

Remark 11. Let J be a homogeneous ideal in generic coordinates. If k has characteristic zero, then we have $\mathrm{reg}(\mathrm{in}_{DRL}(J)) = \max. \mathrm{GB}. \deg_{DRL}(J)$, as shown in [BS87]. If k has positive characteristic, one still has that $\max. \mathrm{GB}. \deg_{DRL}(J) \leq \mathrm{reg}(\mathrm{in}_{DRL}(J))$ and the inequality is often an equality. In fact this was the case in all the examples that we computed while working on this paper. Nevertheless, in positive characteristic one can find examples of ideals J in generic coordinates for which the inequality is strict. E.g. $J = (x^p, y^p) \subseteq \overline{\mathbb{F}}_p[x, y]$ is in generic coordinates, $\max. \mathrm{GB}. \deg_{DRL}(J) = p$, and $\mathrm{reg}(J) = 2p - 1$.

Combining Theorem 7 and Theorem 8, one obtains bounds on the solving degree. Our bounds assume that the ideal generated by the (homogenized) system is in generic coordinates. Notice that this assumption is likely to be satisfied for systems of equations coming from multivariate cryptography, at least over a field of sufficiently large cardinality. In fact, multivariate schemes are often constructed by applying a generic change of coordinates (and a generic linear transformation) to the set of polynomials which constitutes the private key.

For the sake of clarity, we give a homogeneous and a non-homogeneous version of the result. Since the proofs are very similar, and in fact more complicated in the non-homogeneous case, we only give the proof in the latter case.

Theorem 9. *Let $\mathcal{F} \subseteq R$ be a system of homogeneous polynomials and assume that $(\mathcal{F})$ is in generic coordinates over $\overline{k}$. Then*

$$\mathrm{solv}. \deg_{DRL}(\mathcal{F}) \leq \mathrm{reg}(\mathcal{F}).$$

The following result allows us to bound the complexity of computing a Gröbner basis of a system of equations by establishing a connection with the Castelnuovo-Mumford regularity of the homogenization of the system.

Theorem 10. *Let $\mathcal{F} = \{f_1, \ldots, f_r\} \subseteq R$ be a system of polynomials, which is not homogeneous. Let $\mathcal{F}^h = \{f_1^h, \ldots, f_r^h\} \subseteq S = R[t]$ and assume that the ideal $(\mathcal{F}^h)$ is in generic coordinates over $\overline{k}$. Then*

$$\mathrm{solv.\,deg}_{DRL}(\mathcal{F}) \leq \mathrm{reg}(\mathcal{F}^h).$$

Proof. For a homogeneous ideal J in R or S, $\mathrm{max.\,GB.\,deg}_{DRL}(J)$ and $\mathrm{reg}(J)$ are invariant under field extension. So we may extend all ideals to the algebraic closure $\overline{k}$ of k. By Theorem 7 and Theorem 8 we have the chain of equalities and inequalities

$$\mathrm{solv.\,deg}_{DRL}(\mathcal{F}) = \mathrm{solv.\,deg}_{DRL}(\mathcal{F}^h)$$
$$= \mathrm{max.\,GB.\,deg}_{DRL}(\mathcal{F}^h) \leq \mathrm{reg}(\mathrm{in}_{DRL}(\mathcal{F}^h)) = \mathrm{reg}(\mathcal{F}^h).$$

$\square$

Remark 12. The upper bound in Theorem 9 and Theorem 10 is often an equality, since $\mathrm{max.\,GB.\,deg}_{DRL}(\mathcal{F}^h) = \mathrm{reg}(\mathrm{in}_{DRL}(\mathcal{F}^h))$ if k has characteristic zero and often even if it has positive characteristic (see Remark 11).

By combining Theorem 10 and classical results on the Castelnuovo-Mumford regularity (see e.g. [Cha07, Theorem 12.4]), one immediately obtains the following bound on the solving degree of systems which have finitely many solutions over $\overline{k}$. The bound is linear in both the number of variables and the degrees of the polynomials of the system.

Corollary 2 (Macaulay bound – [Laz83], Theorem 2). *Consider a system of equations $\mathcal{F} = \{f_1, \ldots, f_r\} \subseteq R$ with $d_i = \deg f_i$ and $d_1 \geq d_2 \geq \cdots \geq d_r$. Set $\ell = \min\{n + 1, r\}$. Assume that $|\mathcal{Z}_+(\mathcal{F}^h)| < \infty$ and that $(\mathcal{F}^h)$ is in generic coordinates over $\overline{k}$. Then*

$$\mathrm{solv.\,deg}_{DRL}(\mathcal{F}) \leq d_1 + \ldots + d_\ell - \ell + 1$$

and equality holds if $f_1, \ldots, f_r$ are a regular sequence. In particular, if $r > n$ and $d = d_1$, then

$$\mathrm{solv.\,deg}_{DRL}(\mathcal{F}) \leq (n + 1)(d - 1) + 1.$$

The condition that $(\mathcal{F}^h)$ is in generic coordinates is not always easy to verify. Nevertheless, if we add the field equations, or their fake Weil descent, to the generators of the ideal, then we can prove that the homogenized system is in generic coordinates.

Theorem 11. *Let $p > 0$ be a prime and let $q = p^e$, $e \geq 1$. Let k be a field of characteristic p and let $\mathcal{F} = \{f_1, \ldots, f_r\} \subseteq k[x_1, \ldots, x_n]$ be a system of polynomial equations. Set $d_i = \deg f_i$ with $d_1 \geq d_2 \geq \cdots \geq d_r$ and $\ell = \min\{n + 1, r\}$. Assume that one of the following holds:*

(i) $x_i^q - x_i \in \mathcal{F}$ for $i = 1, \ldots, n$, or

(ii) $x_1^q - x_2, \ldots, x_{n-1}^q - x_n, x_n^q - x_1 \in \mathcal{F}$.

Then the ideal $(\mathcal{F}^h) = (f_1^h, \ldots, f_r^h)$ is in generic coordinates over $\bar{k}$. In particular

$$\text{solv.} \deg_{DRL}(\mathcal{F}) \leq d_1 + \ldots + d_\ell - \ell + 1$$

and equality holds if $f_1, \ldots, f_r$ are a regular sequence. Moreover, if $r > n$ and $d = d_1$, then

$$\text{solv.} \deg_{DRL}(\mathcal{F}) \leq (n+1)(d-1) + 1.$$

Proof. According to [BS87, Theorem 2.4 and Definition 1.5], $J = (\mathcal{F}^h)$ is in generic coordinates over $\bar{k}$ if and only if t is not a zero divisor on the quotient $\bar{k}[x_1, \ldots, x_n, t]/J^{\text{sat}}$, where J^{sat} is the saturation of J with respect to the irrelevant maximal ideal $(x_1, \ldots, x_n, t)$. Substituting $t = 0$ in the equations of J one obtains the equations $x_1 = \ldots = x_n = 0$. Therefore the projective zero locus of J does not contain any point with $t = 0$. This means that $t \nmid 0$ modulo J^{sat}, hence proving that J is in generic coordinates. The second part of the statement then follows from Corollary 2.   □

Remark 13. From the proof of Theorem 11 one sees that a system is in generic coordinates whenever it contains equations of the form $x_i^{d_i} + p_i(x_1, \ldots, x_n)$ with $\deg(p_i) < d_i$, for $i = 1, \ldots, n$.

We may use the results established in this section to obtain bounds on the solving degree of the ABC encryption scheme. We assume that the systems have finite affine zero loci, which was the case for all the instances of the ABC cryptosystem that we computed.

Example 10. The system associated to the ABC cryptosystems [TDTD13, TXPD15] consists of $2n$ quadratic equations in n variables. Therefore by assuming that the system is in generic coordinates, or, if the ground field is $\mathbb{F}_2$, simply by adding the field equations to the system we obtain

$$\text{solv.} \deg(\mathcal{F}) \leq n + 2.$$

4 Solving Degree and Degree(s) of Regularity

In recent years, different invariants for measuring the complexity of solving a polynomial system of equations were introduced. In particular, the notion of *degree of regularity* gained importance and is widely used nowadays. In this section we discuss how the degree of regularity is related with the Castelnuovo-Mumford regularity.

In the literature we found several definitions of degree of regularity. However, they are mostly variations of the following two concepts:

1. the degree of regularity by Bardet, Faugère, and Salvy [Bar04, BFS04, BFS15];
2. the degree of regularity by Dubois and Gama, later studied by Ding, Schmidt, and Yang [DG10, DS13, DY13].

In this section we recall both definitions of degree of regularity and compare them with the Castelnuovo-Mumford regularity.

4.1   The Degree of Regularity by Bardet, Faugère, and Salvy

To the best of our knowledge, the degree of regularity appeared first in a paper by Bardet, Faugère, and Salvy [BFS04] and in Bardet's Ph.D. thesis [Bar04]. However, the idea of measuring the complexity of computing the Gröbner basis of a homogeneous ideal using its index of regularity can be traced back to Lazard's seminal work [Laz83]. Before giving the definition, we recall some concepts from commutative algebra.

Let $R = k[x_1, \ldots, x_n]$ be a polynomial ring over a field k, let I be a homogeneous ideal of R, and let $A = R/I$. For an integer $d \geq 0$, we recall that A_d denotes the homogeneous part of degree d of A. The function $HF_A(-) : \mathbb{N} \to \mathbb{N}$, $HF_A(d) = \dim_k A_d$ is called *Hilbert function* of A. It is well known that for large d, the Hilbert function of A is a polynomial in d called *Hilbert polynomial* and denoted by $HP_A(d)$. The generating series of HF_A is called *Hilbert series* of A. We denote it by $HS_A(z) = \sum_{d \in \mathbb{N}} HF_A(d) z^d$. A classical theorem by Hilbert and Serre says that the Hilbert series of A is a rational function, and more precisely has the form

$$HS_A(z) = \frac{h_A(z)}{(1-z)^\ell} \tag{3}$$

where $h_A(z)$ is a polynomial such that $h_A(1) \neq 0$, called *h-polynomial* of A.

Definition 11. *The* index of regularity *of I is the smallest integer $i_{\text{reg}}(I) \geq 0$ such that $HF_{R/I}(d) = HP_{R/I}(d)$ for all $d \geq i_{\text{reg}}(I)$. If $\mathcal{F}$ is a system of generators for I, we set also $i_{\text{reg}}(\mathcal{F}) = i_{\text{reg}}(I)$.*

The index of regularity can be read off the Hilbert series of the ideal, as shown in the next theorem.

Theorem 12 ([BH98], Proposition 4.1.12). *Let $I \subseteq R$ be a homogeneous ideal with Hilbert series as in* (3) *and let $\delta = \deg h_A$. Then $i_{\text{reg}}(I) = \delta - \ell + 1$.*

Let $I \subseteq R$ be a homogeneous ideal. Applying the Grothendieck-Serre's Formula [BH98, Theorem 4.4.3] to R/I one obtains

$$i_{\text{reg}}(I) \leq \text{reg}(I). \tag{4}$$

Moreover, if I is homogeneous and $I_d = R_d$ for $d \gg 0$, then $i_{\text{reg}}(I) = \text{reg}(I)$ by [Eis05, Corollary 4.15].

Definition 12. *Let $\mathcal{F} = \{f_1, \ldots, f_r\} \subseteq R$ be a system of equations and let $(\mathcal{F}^{\text{top}}) = (f_1^{\text{top}}, \ldots, f_r^{\text{top}})$ be the ideal of R generated by the homogeneous part of highest degree of $\mathcal{F}$. Assume that $(\mathcal{F}^{\text{top}})_d = R_d$ for $d \gg 0$. The* degree of regularity *of $\mathcal{F}$ is*

$$d_{\text{reg}}(\mathcal{F}) = i_{\text{reg}}(\mathcal{F}^{\text{top}}).$$

Remark 14. If $(\mathcal{F}^{\text{top}})_d = R_d$ for $d \gg 0$, then $|\mathcal{Z}(\mathcal{F})| < \infty$. The converse, however, does not hold in general. See Example 12 for an example where $\mathcal{F}$ has finitely many solutions over $\bar{k}$, but $(\mathcal{F}^{\text{top}})_d \neq R_d$ for all d.

The following is an easy consequence of the definitions.

Proposition 2. *Let $\mathcal{F} \subseteq R$ be a system of equations. Assume that $(\mathcal{F}^{\text{top}})_d = R_d$ for $d \gg 0$. Then*

$$d_{\text{reg}}(\mathcal{F}) = \text{reg}(\mathcal{F}^{\text{top}}).$$

If in addition $\mathcal{F}$ is homogeneous, then $\mathcal{F}^{\text{top}} = \mathcal{F}$ and

$$d_{\text{reg}}(\mathcal{F}) = \text{reg}(\mathcal{F}).$$

In the context of multivariate cryptosystems however, it is almost never the case that $\mathcal{F}$ is homogeneous and $(\mathcal{F})_d = R_d$ for $d \gg 0$. In fact, this is equivalent to saying that $\mathcal{Z}(I) = \{(0, \ldots, 0)\}$ by Remark 2.

For a system $\mathcal{F}$ such that $I = (\mathcal{F})$ has finite affine zero locus, we may interpret the condition $(\mathcal{F}^{\text{top}})_d = R_d$ for $d \gg 0$ as a *genericity* assumption. This assumption guarantees that the degree of regularity gives an upper bound on the maximum degree of a polynomial in a Gröbner basis of I, with respect to any degree-compatible term order.

Remark 15. Let τ be a degree-compatible term order and assume that $(\mathcal{F}^{\text{top}})_d = R_d$ for $d \gg 0$. Let $I = (\mathcal{F})$ and $J = (\mathcal{F}^{\text{top}})$. Then $HP_{R/J}(z) = 0$, hence $J_d = \text{in}_\tau(J)_d = R_d$ for $d \geq d_{\text{reg}}(\mathcal{F})$. The inclusion $\text{in}_\tau(J)_d \subseteq \text{in}_\tau(I)_d$ holds for any d, since τ is degree-compatible. So we obtain $\text{in}_\tau(I)_d = R_d$ for $d \geq d_{\text{reg}}(\mathcal{F})$. This implies that every element of the reduced Gröbner basis of I has degree at most $d_{\text{reg}}(\mathcal{F})$, that is

$$\max. \, \text{GB.} \deg_\tau(\mathcal{F}) \leq d_{\text{reg}}(\mathcal{F}). \tag{5}$$

Notice however that (5) does not yield a bound on the solving degree of $\mathcal{F}$, as we show in the next example.

Example 11. We consider the polynomial systems $\mathcal{F}$ obtained in [BG18] (see also [Bia17, Chapter 5]) for collecting relations for index calculus following the approach outlined by Gaudry in [Gau09]. For $n = 3$, they consist of three non-homogeneous equations f_1, f_2, f_3 of degree 3 in two variables. Computing 150'000 randomly generated examples of cryptographic size (3 different q's, 5 elliptic curves for each q, 10'000 random points per curve), we found that $(\mathcal{F}^{\text{top}})_d = R_d$ for $d \gg 0$ and

$$\text{solv.} \deg_{DRL}(\mathcal{F}) = \text{reg}(\mathcal{F}^h) = 5 > 4 = d_{\text{reg}}(\mathcal{F}) = i_{\text{reg}}(\mathcal{F}^{\text{top}}).$$

The computations were performed by G. Bianco with MAGMA [BCP97].

Notice moreover that there are systems $\mathcal{F}$ for which $|\mathcal{Z}(\mathcal{F})| < \infty$ and $(\mathcal{F}^{\text{top}})_d \neq R_d$ for all $d \geq 0$. Definition 12 and inequality (5) do not apply to such systems. This can happen also for polynomial systems arising in cryptography.

When this happens, one may be tempted to consider $i_{\text{reg}}(\mathcal{F}^{\text{top}})$ anyway, and use it to bound the solving degree of $\mathcal{F}$. Unfortunately this approach fails since

$i_{\mathrm{reg}}(\mathcal{F}^{\mathrm{top}})$ and solv. $\deg(\mathcal{F})$ might be far apart, as the next examples shows. On the other hand, the Castelnuovo-Mumford regularity of $\mathcal{F}^h$ still allows us to correctly bound the solving degree of $\mathcal{F}$.

Example 12. We consider the polynomial systems obtained in [GM15] for collecting relations for index calculus following the approach outlined by Gaudry in [Gau09]. For $n = 3$, they consist of three non-homogeneous equations f_1, f_2, f_3 in two variables, of degrees 7,7, and 8. Let $\mathcal{F} = \{f_1, f_2, f_3\}$, $\mathcal{F}^h = \{f_1^h, f_2^h, f_3^h\}$, and $\mathcal{F}^{\mathrm{top}} = \{f_1^{\mathrm{top}}, f_2^{\mathrm{top}}, f_3^{\mathrm{top}}\}$. For 150'000 randomly generated examples of cryptographic size (as in Example 11) we found that solv. $\deg_{DRL}(\mathcal{F}) = \mathrm{reg}(\mathcal{F}^h) = 15$, $(\mathcal{F}^{\mathrm{top}})_d \neq R_d$ for all $d \geq 0$, and $i_{\mathrm{reg}}(\mathcal{F}^{\mathrm{top}}) = 8$. The computations were performed by G. Bianco with MAGMA [BCP97].

Finally, given a polynomial system $\mathcal{F} = \{f_1, \dots, f_r\}$ there is a simple relation between the ideals $(\mathcal{F}^{\mathrm{top}}) \subseteq R$ and $(\mathcal{F}^h) \subseteq S$, namely

$$(\mathcal{F}^{\mathrm{top}})S + (t) = (\mathcal{F}^h) + (t). \tag{6}$$

Here $(\mathcal{F}^{\mathrm{top}})S$ denotes the extension of $(\mathcal{F}^{\mathrm{top}})$ to S, i.e., the ideal of S generated by $\mathcal{F}^{\mathrm{top}}$. Since $\mathcal{F}^{\mathrm{top}} \subseteq R$, $t \nmid 0$ modulo $(\mathcal{F}^{\mathrm{top}})S$. If $t \nmid 0$ modulo $(\mathcal{F}^h)$, then $(\mathcal{F}^h) = (\mathcal{F})^h$ is the homogenization of $(\mathcal{F})$ and $\mathrm{reg}(\mathcal{F}^h) = \mathrm{reg}(\mathcal{F}^{\mathrm{top}})$. Therefore, if $t \nmid 0$ modulo $\mathcal{F}^h$ and $(\mathcal{F}^{\mathrm{top}})_d = R_d$ for $d \gg 0$, then

$$d_{\mathrm{reg}}(\mathcal{F}) = \mathrm{reg}(\mathcal{F}^h)$$

by Proposition 2. However, one expects that in most cases $t \mid 0$ modulo $(\mathcal{F}^h)$. In fact, $(\mathcal{F}^h) = (\mathcal{F})^h$ only in very special cases, namely when $f_1, \dots, f_r$ are a Macaulay basis of $(\mathcal{F})$ with respect to the standard grading (see [KR05, Theorem 4.3.19]). Therefore (6) usually does not allow us to compare the regularity and the index of regularity of $\mathcal{F}^h$ and $\mathcal{F}^{\mathrm{top}}$. See also [BDDGMT20, Section 4.1] for a more detailed discussion.

4.2   The Degree of Regularity by Ding and Schmidt

The second notion of degree of regularity is more recent. To the extent of our knowledge it has been introduced by Dubois and Gama [DG10], and later has been used by several authors such as Ding, Schmidt, and Yang [DS13, DY13]. The definition we present here is taken from [DS13], and differs slightly from the original one of Dubois and Gama.

Let $\mathbb{F}_q$ and let $B = \mathbb{F}_q[x_1, \dots, x_n]/(x_1^q, \dots, x_n^q)$. Let $f_1, \dots, f_r \in B$ be homogeneous polynomials of degree 2. We fix a B-module homomorphism φ sending the canonical basis $e_1, \dots, e_r$ of B^r to $\{f_1, \dots, f_r\}$, that is for every $(b_1, \dots, b_r) \in B^r$ we have $\varphi(b_1, \dots, b_r) = \sum_{i=1}^r b_i f_i$. We denote by $\mathrm{Syz}(f_1, \dots, f_r)$ the first syzygy module of $f_1, \dots, f_r$, that is the kernel of φ. An element of $\mathrm{Syz}(f_1, \dots, f_r)$ is a syzygy of $f_1, \dots, f_r$. In other words, it is a vector of polynomials $(b_1, \dots, b_r) \in B^r$ such that $\sum_{i=1}^r b_i f_i = 0$.

An example of syzygy is given by the Koszul syzygies $f_i e_j - f_j e_i$, where $i \neq j$ or by the syzygies coming by the quotient structure of B, that is $f_i^{q-1} e_i$.

Here e_i denotes the i-th element of the canonical basis of B. These syzygies are called *trivial syzygies*, because they are always present and do not depend on the structure of $f_1, \ldots, f_r$, but rather on the ring structure of B. We define the module $\mathrm{Triv}(f_1, \ldots, f_r)$ of trivial syzygies of $f_1, \ldots, f_r$ as the submodule of $\mathrm{Syz}(f_1, \ldots, f_r)$ generated by $\{f_i e_j - f_j e_i : 1 \leq i < j \leq r\} \cup \{f_i^{q-1} e_i : 1 \leq i \leq r\}$.

For any $d \in \mathbb{N}$ we define the vector space $\mathrm{Syz}(\mathcal{F})_d = \mathrm{Syz}(\mathcal{F}) \cap B_d^r$ of syzygies of degree d. We define the vector subspace of trivial syzygies of degree d as $\mathrm{Triv}(\mathcal{F})_d = \mathrm{Triv}(\mathcal{F}) \cap B_d^r$. Clearly, we have $\mathrm{Triv}(\mathcal{F})_d \subseteq \mathrm{Syz}(\mathcal{F})_d$.

Definition 13. *Let $\mathcal{F} = \{f_1, \ldots, f_r\} \subseteq B$ be a system of polynomials of degree 2. The* degree of regularity *of $\mathcal{F}$ is*

$$\delta_{\mathrm{reg}}(\mathcal{F}) = \min\{d \geq 2 : \mathrm{Syz}(\mathcal{F}^{\mathrm{top}})_{d-2} / \mathrm{Triv}(\mathcal{F}^{\mathrm{top}})_{d-2} \neq 0\}.$$

Remark 16. Dubois and Gama [DG10] work in the ring $\mathbb{F}_q[x_1, \ldots, x_n]/(x_1^q - x_1, \ldots, x_n^q - x_n)$ and not in $B = \mathbb{F}_q[x_1, \ldots, x_n]/(x_1^q, \ldots, x_n^q)$.

The degree of regularity is the first degree where we have a linear combination of multiples of $f_1, \ldots, f_r$ which produces a non-trivial cancellation of their top degree parts. For this reason, some authors refer to it as *first fall degree*.

One may wonder whether the degree of regularity by Ding and Schmidt is close to the solving degree of a polynomial system of quadratic equations. Ding and Schmidt showed that this is not always the case. In fact, it is easy to produce examples, the so-called degenerate systems, for which the degree of regularity and the solving degree are far apart. For a detailed exposition on this problem and several examples we refer the reader to their paper [DS13].

We are not aware of any results relating $\delta_{\mathrm{reg}}(\mathcal{F})$ (Definition 13) and $d_{\mathrm{reg}}(\mathcal{F})$ (Definition 12). Despite the fact that they share the name, we do not see an immediate connection. A comparison between these two invariants is beyond the scope of this paper.

5 Solving Degree of Ideals of Minors and the MinRank Problem

The goal of this section is giving an example of how the results from Sect. 3, in combination with known commutative algebra results, allow us to prove estimates for the solving degree in a simple and synthetic way. We consider polynomial systems coming from the MinRank Problem. For more bounds on the complexity of the MinRank Problem, see [CG20].

The MinRank Problem can be stated as follows. Given an integer $t \geq 1$ and a set $\{M_1, \ldots, M_n\}$ of $s \times s$ matrices with entries in a field k, find a non-zero tuple $\lambda = (\lambda_1, \ldots, \lambda_n) \in k^n$ such that

$$\mathrm{rank}\left(\sum_{i=1}^{n} \lambda_i M_i\right) \leq t - 1. \tag{7}$$

This problem finds several applications in multivariate cryptography and in other areas of cryptography as well. For example, Goubin and Courtois [GC00] solved a MinRank Problem to attack Stepwise Triangular Systems, and Kipnis and Shamir [KS99] solved an instance of MinRank in their cryptanalysis of the HFE cryptosystem.

Consider the matrix $M = \sum_{i=1}^{n} x_i M_i$, whose entries are homogeneous linear forms in R. Condition (7) is equivalent to requiring that the minors of size $t \times t$ of M vanish. Therefore, every solution of the MinRank Problem corresponds to a non-zero point in the zero locus in k^n of the ideal $I_t(M)$ of t-minors of M. A similar algebraic formulation can be given for the Generalized MinRank Problem, which finds applications within coding theory, non-linear computational geometry, real geometry, and optimization. We refer the interested reader to [FSS13] for a discussion of the applications of the Generalized MinRank Problem and a list of references.

Problem 1 (Generalized MinRank Problem). Given a field k, an $r \times s$ matrix M whose entries are polynomials in $R = k[x_1, \ldots, x_n]$, and an integer $1 \leq t \leq \min\{r, s\}$, find a point in $k^n \setminus \{(0, \ldots, 0)\}$ at which the evaluation of M has rank at most $t - 1$.

The Generalized MinRank Problem can be solved by computing the zero locus of the ideal of t-minors $I_t(M)$. The minors of size $t \times t$ of the matrix M form an algebraic system of multivariate polynomials, which one can attempt to solve by computing a Gröbner basis. This motivates our interest in estimating the solving degree of this system for large classes of matrices.

Ideals of minors of a matrix with entries in a polynomial ring are called *determinantal ideals* and have been extensively studied in commutative algebra and algebraic geometry. Using Theorem 9, we can take advantage of the literature on the regularity of determinantal ideals to give bounds on the solving degree of systems of minors of certain large classes of matrices. For simplicity, we focus on homogeneous matrices.

Definition 14. *Let M be an $r \times s$ matrix with $r \leq s$, whose entries are elements of R. The matrix M is* homogeneous *if both its entries and its 2-minors are homogeneous polynomials.*

It is easy to see that the minors of any size of a homogeneous matrix are homogeneous polynomials. Moreover, observe that a matrix whose entries are homogeneous polynomials of the same degree is a homogeneous matrix, but there are homogeneous matrices whose entries have different degrees. After possibly exchanging some rows and columns, we may assume without loss of generality that the degrees of the entries of a homogeneous matrix increase from left to right and from top to bottom. With this notation, we can compute the solving degree of our first family of systems of minors. We refer the reader to [Eis94] for the definition of height of an ideal.

Theorem 13. *Let $M = (f_{ij})$ be an $r \times s$ homogeneous matrix with $r \leq s$, whose entries are elements of R, $n \geq s - r + 1$. Let $\mathcal{F}$ be the polynomial system of the* *minors of size r of M and assume that* height$(I_r(M)) = s - r + 1$. *Then the solving degree of $\mathcal{F}$ is upper bounded by*

$$\text{solv.} \deg(\mathcal{F}) \leq \deg(f_{1,1}) + \ldots + \deg(f_{m,m}) + \deg(f_{m,m+1}) + \ldots + \deg(f_{m,n}) - s + r.$$

If $\deg(f_{i,j}) = 1$ *for all* i, j, *then* solv.$\deg(\mathcal{F}) = r$.

Proof. Since the matrix M is homogeneous, the system of minors $\mathcal{F}$ consists of homogeneous polynomials. The regularity of the corresponding ideal $I_r(M) = (\mathcal{F})$ is

$$\text{reg}(I_r(M)) = \deg(f_{1,1}) + \ldots + \deg(f_{r,r}) + \deg(f_{r,r+1}) + \ldots + \deg(f_{r,s}) - s + r.$$

The formula can be found in [BCG04, Proposition 2.4] and is derived from a classical result of Eagon and Northcott [EN62]. The bound on the solving degree now follows from Theorem 9. In particular, if $\deg(f_{i,j}) = 1$ for all i, j, then solv.$\deg(\mathcal{F}) \leq r$. Since $I_r(M)$ is generated in degree r, then solv.$\deg(\mathcal{F}) = r$. □

Notice that the assumption on the height is satisfied by a matrix M whose entries are generic homogeneous polynomials of fixed degrees. If $n = s - r + 1$, then $I_r(M)_d = R_d$ for $d \gg 0$, hence $d_{\text{reg}}(\mathcal{F}) = \text{reg}(\mathcal{F})$, where $\mathcal{F}$ is the set of maximal minors of M. Therefore, Theorem 13 recovers the results of [FSS10, FSS13] for $n = s - r + 1$ and $t = r$, and extends them to homogeneous matrices whose entries do not necessarily have the same degree.

We now restrict to systems of maximal minors of matrices of linear forms. The MinRank Problem associated to this class of matrices is a slight generalization of the classical MinRank Problem of (7). From the previous result it follows that, if the height of the ideal of maximal minors is as large as possible, then the solving degree of the corresponding system is as small as possible, namely r. We now give different assumptions which allows us to obtain the same estimate on the solving degree, for ideals of maximal minors whose height is not maximal. We are also able to bound the solving degree of the system of 2-minors.

Let R have a standard $\mathbb{Z}^v$-graded structure, i.e., the degree of every indeterminate of R is an element of the canonical basis $\{e_1, \ldots, e_v\}$ of $\mathbb{Z}^v$.

Definition 15. *Let $M = (f_{i,j})$ be an $r \times s$ matrix with entries in R, $r \leq s$. We say that M is column-graded if $s \leq v$, and $f_{i,j} = 0$ or it is homogeneous of degree $\deg(f_{i,j}) = e_j \in \mathbb{Z}^v$ for every i, j. We say that M is row-graded if $r \leq v$, and $f_{i,j} = 0$ or it is homogeneous of degree $\deg(f_{i,j}) = e_i \in \mathbb{Z}^v$ for every i, j.*

Informally, a matrix is row-graded if the entries of each row are homogeneous linear forms in a different set of variables. Similarly for a column-graded matrix.

Theorem 14. *Let M be an $r \times s$ row-graded or column-graded matrix with entries in R. Assume that $r \leq s$ and $I_r(M) \neq 0$. Then:*

– *if $\mathcal{F}$ is the system of maximal minors of M then* solv.$\deg(\mathcal{F}) = r$,

– *if $\mathcal{F}$ is the system of 2-minors of M then* solv. $\deg(\mathcal{F}) \leq s$ *in the column-graded case, and* solv. $\deg(\mathcal{F}) \leq r$ *in the row-graded case.*

Proof. It is shown in [CDG15,CDG20] that $\mathrm{reg}(I_r(M)) = r$, $\mathrm{reg}(I_2(M)) \leq s$ in the column-graded case, and $\mathrm{reg}(I_2(M)) \leq r$ in the row-graded case. The bounds on the solving degree now follow from Theorem 9. □

Acknowledgements. The authors are grateful to Albrecht Petzoldt for help with MAGMA computations, to Wouter Castryck for pointing out some typos in an earlier version of this paper, and to Marc Chardin, Teo Mora, Christophe Petit, and Pierre-Jean Spaenlehauer for useful discussions on the material of this paper. This work was made possible thanks to funding from Armasuisse.

References

[Bar04] Bardet, M.: Étude des systémes algébriques surdéterminés. Applications aux codes correcteurs et á la cryptographie, Ph.D. thesis, Université Paris 6 (2004)

[BFS04] Bardet, M., Faugère, J.-C., Salvy, B.: On the complexity of Gröbner basis computation of semi-regular overdetermined algebraic equations. In: ICPPSS International Conference on Polynomial System Solving (2004)

[BFS15] Bardet, M., Faugère, J.-C., Salvy, B.: On the complexity of the F_5 Gröbner basis algorithm. J. Symb. Comput. **70**, 49–70 (2015)

[BS87] Bayer, D., Stillman, M.: A criterion for detecting m-regularity. Invent. Math. **87**(1), 1–11 (1987). https://doi.org/10.1007/BF01389151

[Bia17] Bianco, G.: Trace-zero subgroups of elliptic and twisted Edwards curves: a study for cryptographic applications, Ph.D. thesis (2017). https://doi.org/10.35662/unine-thesis-2631

[BG18] Bianco, G., Gorla, E.: Index calculus in trace-zero subgroups and generalized summation polynomials, preprint (2018)

[BDDGMT20] Bigdeli, M., De Negri, E., Dizdarevic, M.M., Gorla, E., Minko, R., Tsakou, S.: Semi-regular sequences and other random systems of equations, preprint (2020)

[BCP97] Bosma, W., Cannon, J., Playoust, C.: The Magma algebra system. I. The user language. J. Symb. Comput. **24**, 235–265 (1997)

[BH98] Bruns, W., Herzog, J.: Cohen-Macaulay rings. Revised ed. Cambridge Studies in Advanced Mathematics, vol. 39. Cambridge University Press, Cambridge (1998)

[BCG04] Budur, N., Casanellas, M., Gorla, E.: Hilbert functions of irreducible arithmetically Gorenstein schemes. J. Algebra **272**(1), 292–310 (2004)

[CG20] Caminata, A., Gorla, E.: The complexity of MinRank. In: Cojocaru, A., Ionica, S., Lorenzo Garcia, E. (eds.) Women in Numbers Europe III: Research Directions in Number Theory. Springer (to appear)

[Cha07] Chardin, M.: Some results and questions on Castelnuovo-Mumford regularity. In: Syzygies and Hilbert Functions. Lecture Notes in Pure and Applied Mathematics, vol. 254, pp. 1–40 (2007)

[CDG15] Conca, A., De Negri, E., Gorla, E.: Universal Gröbner bases for maximal minors. In: International Mathematics Research Notices, IMRN 2015, no. 11, pp. 3245–3262 (2015)

[CDG20] Conca, A., De Negri, E., Gorla, E.: Universal Gröbner bases and Cartwright-Sturmfels ideals. International Mathematics Research Notices, IMRN 2020, no. 7, pp. 1979–1991 (2020)

[CKPS00] Courtois, N., Klimov, A., Patarin, J., Shamir, A.: Efficient algorithms for solving overdefined systems of multivariate polynomial equations. In: Preneel, B. (ed.) EUROCRYPT 2000. LNCS, vol. 1807, pp. 392–407. Springer, Heidelberg (2000). https://doi.org/10.1007/3-540-45539-6_27

[CLO07] Cox, D., Little, J., O'Shea, D.: Ideals, Varieties, and Algorithms. An Introduction to Computational Algebraic Geometry and Commutative Algebra. Springer, Cham (2015). https://doi.org/10.1007/978-0-387-35651-8

[DBMMW08] Ding, J., Buchmann, J., Mohamed, M.S.E., Moahmed, W.S.A.E., Weinmann, R.-P.: MutantXL. In: Proceedings of the 1st international conference on Symbolic Computation and Cryptography (SCC08), Beijing, China, LMIB, pp. 16–22 (2008)

[DS13] Ding, J., Schmidt, D.: Solving degree and degree of regularity for polynomial systems over a finite fields. In: Fischlin, M., Katzenbeisser, S. (eds.) Number Theory and Cryptography. LNCS, vol. 8260, pp. 34–49. Springer, Heidelberg (2013). https://doi.org/10.1007/978-3-642-42001-6_4

[DY13] Ding, J., Yang, B.-Y.: Degree of regularity for HFEv and HFEv-. In: Gaborit, P. (ed.) PQCrypto 2013. LNCS, vol. 7932, pp. 52–66. Springer, Heidelberg (2013). https://doi.org/10.1007/978-3-642-38616-9_4

[DG10] Dubois, V., Gama, N.: The degree of regularity of HFE systems. In: Abe, M. (ed.) ASIACRYPT 2010. LNCS, vol. 6477, pp. 557–576. Springer, Heidelberg (2010). https://doi.org/10.1007/978-3-642-17373-8_32

[Eis94] Eisenbud, D.: Commutative Algebra. With a View Toward Algebraic Geometry. Graduate Texts in Mathematics, vol. 150. Springer, New York (1994). https://doi.org/10.1007/978-1-4612-5350-1

[Eis05] Eisenbud, D.: The Geometry of Syzygies A Second Course in Algebraic Geometry and Commutative Algebra. Graduate Texts in Mathematics, vol. 229. Springer, New York (2005). https://doi.org/10.1007/b137572

[EN62] Eagon, J.A., Northcott, D.G.: Ideals defined by matrices and a certain complex associated with them. In: Proceedings of the Royal Society of London. Series A, Mathematical and Physical Sciences, vol. 269, no. 1337, pp. 188–204 (1962)

[Fau99] Faugère, J.-C.: A new efficient algorithm for computing Gröbner bases (F4). J. Pure Appl. Algebra **139**, 61–88 (1999)

[Fau02] Faugère, J.-C.: A new efficient algorithm for computing Gröbner bases without reduction to zero (F5). In: Proceedings of the 2002 International Symposium on Symbolic and Algebraic Computation, ISSAC 2002, New York, NY, USA, pp. 75–83 (2002)

[FGLM93] Faugère, J.-C., Gianni, P.M., Lazard, D., Mora, T.: Efficient computation of zero-dimensional Gröbner bases by change of ordering. J. Symb. Comput. **16**(4), 329–344 (1993)

[FSS10] Faugère, J.-C., El Din, M.S., Spaenlehauer, P.-J.: Computing loci of rank defects of linear matrices using Gröbner bases and applications to cryptology. In: Proceedings of the 2010 International Symposium on Symbolic and Algebraic Computation, ISSAC 2010, Munich, Germany, pp. 257–264 (2010)

[FSS13] Faugère, J.-C., Din, M.S.E., Spaenlehauer, P.-J.: On the complexity of the generalized MinRank problem. J. Symb. Comput. **55**, 30–58 (2013)

[Gal74] Galligo, A.: A propos du théorème de préparation de weierstrass. In: Norguet, F. (ed.) Fonctions de Plusieurs Variables Complexes. LNM, vol. 409, pp. 543–579. Springer, Heidelberg (1974). https://doi.org/10. 1007/BFb0068121

[Gau09] Gaudry, P.: Index calculus for abelian varieties of small dimension and the elliptic curve discrete logarithm problem. J. Symb. Comput. **44**(12), 1690–1702 (2009)

[GC00] Goubin, L., Courtois, N.T.: Cryptanalysis of the TTM cryptosystem. In: Okamoto, T. (ed.) ASIACRYPT 2000. LNCS, vol. 1976, pp. 44–57. Springer, Heidelberg (2000). https://doi.org/10.1007/3-540-44448-3_4

[GM15] Gorla, E., Massierer, M.: Index calculus in the trace zero variety. Adv. Math. Commun. **9**(4), 515–539 (2015)

[KS99] Kipnis, A., Shamir, A.: Cryptanalysis of the HFE public key cryptosystem by relinearization. In: Wiener, M. (ed.) CRYPTO 1999. LNCS, vol. 1666, pp. 19–30. Springer, Heidelberg (1999). https://doi.org/10.1007/ 3-540-48405-1_2

[KR00] Kreuzer, M., Robbiano, L.: Computational Commutative Algebra 1. Springer, Heidelberg (2000). https://doi.org/10.1007/978-3-540-70628-1

[KR05] Kreuzer, M., Robbiano, L.: Computational Commutative Algebra 2. Springer, Heidelberg (2005). https://doi.org/10.1007/3-540-28296-3

[KR16] Kreuzer, M., Robbiano, L.: Computational Linear and Commutative Algebra. Springer, Heidelberg (2016). https://doi.org/10.1007/978-3-319-43601-2

[Laz83] Lazard, D.: Gröbner bases, Gaussian elimination and resolution of systems of algebraic equations. In: Computer Algebra (London, 1983). Lecture Notes in Computer Science, vol. 162, pp. 146–156. Springer, Berlin (1983)

[NIST] National Institute of Standards, Post-Quantum Cryptography, Round 3 Submissions. https://csrc.nist.gov/projects/post-quantum-cryptography/round-3-submissions

[TDTD13] Tao, C., Diene, A., Tang, S., Ding, J.: Simple matrix scheme for encryption. In: Gaborit, P. (ed.) PQCrypto 2013. LNCS, vol. 7932, pp. 231–242. Springer, Heidelberg (2013). https://doi.org/10.1007/978-3-642-38616-9_16

[TXPD15] Tao, C., Xiang, H., Petzoldt, A., Ding, J.: Simple matrix - a multivariate public key cryptosystem (MPKC) for encryption. Finite Fields Appl. **35**, 352–368 (2015)

Linearized Polynomials and Their Adjoints, and Some Connections to Linear Sets and Semifields

Gary McGuire and John Sheekey[✉]

UCD School of Mathematics and Statistics, University College Dublin,
Dublin, Ireland
{gary.mcguire,john.sheekey}@ucd.ie

For a q-linearized polynomial function L on a finite field, we give a new short proof of a known result, that $L(x)/x$ and $L^*(x)/x$ have the same image, where $L^*(x)$ denotes the adjoint of L. We give some consequences for semifields, recovering results first proved by Lavrauw and Sheekey. We also give a characterization of planar functions.

1 Introduction

Throughout this paper we let p be a prime number, let $q = p^r$ and let $\mathbb{F}_{q^n}$ denote a finite field with q^n elements, where n is a positive integer.

Any function $\mathbb{F}_{q^n} \longrightarrow \mathbb{F}_{q^n}$ can be expressed uniquely as a polynomial function (with coefficients in $\mathbb{F}_{q^n}$) of degree less than q^n. This is because there are $(q^n)^{q^n}$ such polynomials, they are distinct as functions, and this is also the total number of functions. We call this polynomial the reduced form of the function.

A polynomial in $\mathbb{F}_{q^n}[x]$ is called a permutation polynomial (PP) if its reduced form induces a bijective function $\mathbb{F}_{q^n} \longrightarrow \mathbb{F}_{q^n}$.

Thinking of $\mathbb{F}_{q^n}$ as an n-dimensional vector space over $\mathbb{F}_q$, a polynomial of the form

$$a_0 x + a_1 x^q + a_2 x^{q^2} + \cdots + a_{n-1} x^{q^{n-1}} \tag{1}$$

with $a_i \in \mathbb{F}_{q^n}$ induces an $\mathbb{F}_q$-linear function $\mathbb{F}_{q^n} \longrightarrow \mathbb{F}_{q^n}$. Conversely, any $\mathbb{F}_q$-linear function $\mathbb{F}_{q^n} \longrightarrow \mathbb{F}_{q^n}$ can be written in this form, because there are $(q^n)^n$ such polynomials, they are distinct as functions, and this is also the total number of $\mathbb{F}_q$-linear functions. A polynomial of the form (1) is called a q-linearized polynomial. This is already in reduced form. In this paper, when we use the term q-linearized polynomial, we mean the function $\mathbb{F}_{q^n} \longrightarrow \mathbb{F}_{q^n}$ that is induced by the polynomial.

Let Tr denote the absolute trace map $\mathbb{F}_{q^n} \longrightarrow \mathbb{F}_p$ defined by

$$\mathrm{Tr}(x) = x + x^p + x^{p^2} + \cdots + x^{p^{rn-1}}.$$

Let tr denote the relative trace map $\mathbb{F}_{q^n} \longrightarrow \mathbb{F}_q$ defined by

$$tr(x) = x + x^q + x^{q^2} + \cdots + x^{q^{n-1}}.$$

© Springer Nature Switzerland AG 2021
J. C. Bajard and A. Topuzoğlu (Eds.): WAIFI 2020, LNCS 12542, pp. 37–41, 2021.
https://doi.org/10.1007/978-3-030-68869-1_2

The *adjoint* of $L(x) = a_0 x + a_1 x^q + a_2 x^{q^2} + \cdots + a_{n-1} x^{q^{n-1}}$ is defined to be

$$L^*(x) = a_0 x + a_1^{q^{n-1}} x^{q^{n-1}} + a_2^{q^{n-2}} x^{q^{n-2}} + \cdots + a_{n-1}^q x^q.$$

The adjoint has the property that $tr(L(u)v) = tr(uL^*(v))$ for all $u, v \in \mathbb{F}_{q^n}$. This property implies that $\mathrm{Tr}(L(u)v) = \mathrm{Tr}(uL^*(v))$ for all $u, v \in \mathbb{F}_{q^n}$.

We introduce some notation. Let

$$V(L) = \left\{ -a \in \mathbb{F}_{q^n} : L(x) + ax \text{ is a PP} \right\}$$

and let

$$I(L) = \left\{ \frac{L(z)}{z} : z \in \mathbb{F}_{q^n}, z \neq 0 \right\}.$$

The following theorem was first proved in Lemma 2.6 of [2].

Theorem 1. *Let $L(x)$ be a q-linearized polynomial. Then $I(L) = I(L^*)$ and $V(L) = V(L^*)$.*

In this paper we will provide a new proof of this fact. In addition to giving an alternative viewpoint on this result, this approach may be of use towards studying the following problem.

Open Question. Let $L(x)$ be a q-linearized polynomial. For what other q-linearized polynomials $M(x)$ does it hold that $I(L) = I(M)$ and $V(L) = V(M)$?

This question has been addressed in [3]; in particular it has been shown that for $n \leq 5$, and $L(x)$ *not* a monomial, then $I(L) = I(M)$ if and only if $L(x) = M(\lambda x)/\lambda$ or $L^*(x) = M(\lambda x)/\lambda$ for some $\lambda \in \mathbb{F}_{q^n}^\times$. If $L(x) = x^{q^i}$ and $M(x) = x^{q^j}$ then $I(L) = I(M)$ if and only if $(i, n) = (j, n)$. The general case remains an open problem.

Motivation for this question stems from the study of *linear sets*, which are sets of points on a projective line $\mathrm{PG}(1, q^n)$. The set $U_L = \{(x, L(x)) : x \in \mathbb{F}_{q^n}^\times\}$ defines a set $\mathcal{L}_L$ of points on the projective line $\mathrm{PG}(1, q^n)$ in a natural way. Then it is straightforward to see that $\mathcal{L}_L = \mathcal{L}_M$ if and only if $I(L) = I(M)$. This problem, which has been studied in [4,5], has applications in the study of MRD codes, as well as for semifields, which we will see in Sect. 3.

2   Alternative Proof of Main Theorem

Let ζ be a primitive complex p-th root of unity. The additive characters of $\mathbb{F}_{q^n}$ may be written

$$\chi_\alpha(x) = \zeta^{\mathrm{Tr}(\alpha x)},$$

one character for each $\alpha \in \mathbb{F}_{q^n}$.

The following is a well known characterization of PPs based on additive characters (Theorem 7.7 in [8]).

Theorem 2. *A polynomial $P(x) \in \mathbb{F}_{q^n}[x]$ is a permutation polynomial if and only if*

$$\sum_{x \in \mathbb{F}_{q^n}} \chi(P(x)) = 0$$

for every nontrivial additive character χ of $\mathbb{F}_{q^n}$.

We will use the following well known fact (Theorem 7.9 in [8]).

Lemma 1. *If $L(x) \in \mathbb{F}_{q^n}[x]$ is a q-linearized polynomial, then $L(x)$ is a PP on $\mathbb{F}_{q^n}$ if and only if the only solution in $\mathbb{F}_{q^n}$ of $L(x) = 0$ is $x = 0$.*

We use this characterisation in order to provide a new proof of the Main Theorem.

Theorem 3. *Let $L(x)$ be a q-linearized polynomial. Then $I(L) = I(L^*)$ and $V(L) = V(L^*)$.*

Proof. We will show that both $I(L)$ and $I(L^*)$ are equal to the complement of $V(L)$. In part (i) we show that $-a \in I(L)$ if and only if $L(x) + ax$ is not a PP, and in part (ii) we will show that $-a \in I(L^*)$ if and only if $L(x) + ax$ is not a PP.

(i) Note that $L(x) + ax$ maps 0 to 0, and so $L(x) + ax$ is a PP if and only if $-a \notin Im(L(x)/x)$ by Lemma 1. This proves that

$$I(L) = \left\{ -a \in \mathbb{F}_{q^n} : L(x) + ax \text{ is not a PP} \right\}$$

which shows that $I(L)$ is equal to the complement of $V(L)$.

(ii) By Theorem 2, $L(x) + ax$ is a PP if and only if

$$\sum_{x \in \mathbb{F}_{q^n}} \chi(L(x) + ax) = 0$$

for all nontrivial additive characters χ, or equivalently, if and only if

$$\sum_{x \in \mathbb{F}_{q^n}} \zeta^{\text{Tr}(\alpha(L(x)+ax))} = 0$$

for all nonzero $\alpha \in \mathbb{F}_q$. But

$$\sum_{x \in \mathbb{F}_{q^n}} \zeta^{\text{Tr}(\alpha(L(x)+ax))} = \sum_{x \in \mathbb{F}_{q^n}} \zeta^{\text{Tr}(L^*(\alpha)x+\alpha ax)} = \sum_{x \in \mathbb{F}_{q^n}} \zeta^{\text{Tr}((L^*(\alpha)+\alpha a)x)}$$

which is 0 if and only if $L^*(\alpha) + \alpha a \neq 0$. In other words, $L(x) + ax$ is a PP if and only if $L^*(\alpha) + \alpha a \neq 0$ for all nonzero $\alpha \in \mathbb{F}_{q^n}$. Thus $L(x) + ax$ is a PP if and only if $-a \notin Im(L^*(x)/x)$. This proves that $I(L^*)$ is equal to the complement of $V(L)$.

We have shown that both $I(L)$ and $I(L^*)$ are equal to the complement of $V(L)$, and it follows that $I(L) = I(L^*)$. Applying this to L^* instead of L shows that both $I(L)$ and $I(L^*)$ are equal to the complement of $V(L^*)$. Therefore $V(L) = V(L^*)$, and $L(x) + ax$ is a PP if and only if $L^*(x) + ax$ is a PP.

3   Application to Semifields

We now present an alternative proof of a result of Lavrauw and Sheekey [6].

A finite *semifield* is a nonassociative division algebra of finite dimension over $\mathbb{F}_q$. There are many constructions for semifields, many of which use q-linearized polynomials. In [6] a particular class of semifields were studied, namely those of *BEL-rank two*. These are those semifields whose multiplication can be written in the form

$$x \circ y = xL(y) - M(x)y$$

for some q-linearized polynomials $L(x)$ and $M(x)$. As noted and studied in [7,9], the condition for the pair (L, M) to defines a semifield is equivalent to the condition $I(L) \cap I(M) = \emptyset$, and equivalent to the condition that the sets of points $\mathcal{L}_L$ and $\mathcal{L}_M$ in $\mathrm{PG}(1, q^n)$ are disjoint. In [6] it was shown that if the pair (L, M) define a semifield, then so do the pairs (L^*, M), (L, M^*), and (L^*, M^*) (as well as the obvious fact that (M, L) also defines a semifield, the dual or opposite semifield). The proof of this was an application of the *switching* operation defined in [1]. In fact we can now see that this is an immediate consequence of the main theorem.

Corollary 1. *Let $L(x)$ and $M(x)$ be q-linearized polynomials. Suppose $I(L)$ and $I(M)$ are disjoint, so that $xL(y) - M(x)y$ defines a semifield multiplication law. Then*

1. *$xL^*(y) - M(x)y$ defines a semifield,*
2. *$xL^*(y) - M^*(x)y$ defines a semifield,*
3. *$xL(y) - M^*(x)y$ defines a semifield.*

Proof. If $I(L) \cap I(M) = \emptyset$ then $x * y = xL(y) - M(x)y$ defines a semifield multiplication law. By Theorem 3 we have $I(L) = I(L^*)$ and $I(M) = I(M^*)$. Since $I(L) \cap I(M) = \emptyset$ we also get $I(L^*) \cap I(M) = \emptyset$ and $I(L^*) \cap I(M^*) = \emptyset$ and $I(L) \cap I(M^*) = \emptyset$. The result follows.

Note that the main theorem is in fact stronger than the result of [6], in which it was shown that if $I(L)$ and $I(M)$ are disjoint, then (for example) $I(L^*)$ and $I(M)$ are disjoint, which does not necessarily imply that $I(L) = I(L^*)$.

4   A Criterion for Planarity

Assume q is odd. A function $f : \mathbb{F}_{q^n} \longrightarrow \mathbb{F}_{q^n}$ is said to be *planar* if the functions $x \mapsto f(x + a) - f(x)$ are bijective for all nonzero $a \in \mathbb{F}_{q^n}$. The term PN (perfect nonlinear) is also used instead of the word 'planar'.

Sometimes a polynomial $xL(x)$ will be planar, where $L(x)$ is a q-linearized polynomial. For example, x^2 is planar. We present a criterion for the planarity of $xL(x)$.

Theorem 4. *Let $L(x)$ be a q-linearized polynomial. The polynomial $xL(x)$ is planar if and only if $L^*(bx) + bL(x)$ is a PP for all nonzero $b \in \mathbb{F}_{q^n}$.*

Proof. First,

$$xL(x) \text{ is PN} \iff (x+u)L(x+u) - xL(x) \text{ is a PP for all nonzero } u$$
$$\iff uL(x) + xL(u) + uL(u) \text{ is a PP for all nonzero } u$$
$$\iff uL(x) + xL(u) \text{ is a PP for all nonzero } u.$$

By Theorem 2, $uL(x) + xL(u)$ is a PP if and only if

$$\sum_{x \in \mathbb{F}_{q^n}} \zeta^{\mathrm{Tr}(b(uL(x)+xL(u)))} = 0$$

for all nonzero $b \in \mathbb{F}_{q^n}$. However

$$\sum_{x \in \mathbb{F}_{q^n}} \zeta^{\mathrm{Tr}(buL(x)+bxL(u))} = \sum_{x \in \mathbb{F}_{q^n}} \zeta^{\mathrm{Tr}(L^*(bu)x+bxL(u))}$$

so $uL(x) + xL(u)$ is a PP if and only if $L^*(bu) + bL(u) \neq 0$ for all nonzero b. By Lemma 1 we are done.

References

1. Ball, S., Ebert, G., Lavrauw, M.: A geometric construction of finite semifields. J. Algebra **311**, 117–129 (2007)
2. Bartoli, D., Giulietti, M., Marino, G., Polverino, O.: Maximum scattered linear sets and complete caps in Galois spaces. Combinatorica **38**, 255–278 (2018)
3. Csajbók, B., Marino, G., Polverino, O.: A Carlitz type result for linearized polynomials. Ars Math. Contemp. **16**(2), 585–608 (2019)
4. Csajbók, B., Marino, G., Polverino, O.: Classes and equivalence of linear sets in PG(1, q^n). J. Comb. Theory Ser. A **157**, 402–426 (2018)
5. Csajbók, B., Zanella, C.: On the equivalence of linear sets. Des. Codes Cryptogr. **81**, 269–281 (2016)
6. Lavrauw, M., Sheekey, J.: The BEL-rank of finite semifields. Des. Codes Cryptogr. **84**, 345–358 (2017)
7. Sheekey, J., Van de Voorde, G.: Rank-metric codes, linear sets, and their duality. Des. Codes Cryptogr. **88**, 655–675 (2020)
8. Lidl, R., Niederreiter, H.: Finite Fields. Addison-Wesley (1983)
9. Zini, G., Zullo, F.: On the intersection problem for linear sets in the projective line. arXiv:2004.09441

Efficient, Actively Secure MPC
with a Dishonest Majority: A Survey

Emmanuela Orsini$^{(\boxtimes)}$ (iD)

imec-COSIC, KU Leuven, Leuven, Belgium
emmanuela.orsini@kuleuven.be

Abstract. The last ten years have seen a tremendous growth in the interest and practicality of secure multiparty computation (MPC) and its possible applications. Secure MPC is indeed a very hot research topic and recent advances in the field have already been translated into commercial products world-wide. A major pillar in this advance has been in the case of active security with a dishonest majority, mainly due to the SPDZ-line of work protocols. This survey gives an overview of these protocols, with a focus of the original SPDZ paper (Damgård et al. CRYPTO 2012) and its subsequent optimizations.

1 Introduction

Secure Multiparty Computation (MPC) allows a set of parties to compute a joint function on their inputs while maintaining *privacy*, meaning that the output of the computation should not reveal anything but the output itself.

The concept of secure computation was introduced by Andrew Yao [71] who presented a two-party protocol for Boolean circuits based on the idea of *garbled circuits*. Yao's protocol is a constant-round protocol, where one party, the garbler, generates an encrypted version of the circuit that is securely evaluated by the other party, the evaluator. After forty years this protocol still remains the basis for many efficient MPC implementations.

After Yao's garbled-circuit based protocol was proposed, several multiparty protocols appeared both for Boolean and arithmetic circuits, including those given by Goldreich, Micali and Wigderson (GMW) [37], Ben Or, Goldwasser and Wigderson (BGW) [15], Chaum, Crepeau and Damgård (CCD) [23]. All of these protocols have a number of rounds linear in the depth of the circuit to be evaluated and consist in evaluating the circuit gate-by-gate using a secret-sharing of the data. In 1990, Beaver, Micali and Rogaway presented the BMR protocol [14] generalizing Yao's approach to the multiparty setting. The BMR protocol runs in a constant number of rounds, while achieving security against dishonest majority. Almost all known secure MPC protocols rely on techniques described in these fundamental works.

Secure multiparty computation should guarantee a number of security properties other than privacy, even in the presence of some adversarial entity that controls a subset of the parties, usually referred to as *corrupt* parties. The most

© Springer Nature Switzerland AG 2021
J. C. Bajard and A. Topuzoğlu (Eds.): WAIFI 2020, LNCS 12542, pp. 42–71, 2021.
https://doi.org/10.1007/978-3-030-68869-1_3

significant of these properties are: a) *correctness*, meaning that each party should receive a correct output; b) *independence of the inputs*, i.e. corrupt parties' inputs should be independent of honest parties' inputs; c) *guaranteed output delivery*, namely honest parties should always be able to receive their outputs; d) *fairness*, i.e. corrupt parties should receive their outputs if and only if honest parties do. Note that fairness is a weaker requirement than guaranteed output delivery; indeed guaranteed output delivery implies fairness, but the opposite is not always true [25].

Secure multiparty computation comes in different flavors according to diverse corruption strategies, security requirements, model of computation, communication channels, etc. However, not all the possible combinations of these properties and settings are possible. One major distinction is between protocols that rely on the existence of a *honest majority*, i.e. less than a half of the total number of parties is corrupt, and protocols that can be proven secure even with no honest majority. In the first case it is possible to describe unconditionally secure protocols, whereas to deal with a dishonest majority it is necessary to restrict to computational security that holds under some cryptographic assumption. Note that assuming a honest majority is sometimes a strong requirement, and it is pointless in the important case of two-party computation.

Another crucial division is determined by the type of corruptions that the protocol can support. There are three main adversary models that are usually considered: 1) *semi-honest/passive adversary*, that follows the protocol specifications but tries to gain more information than what is allowed; 2) *malicious/active adversary*, that can arbitrarily deviate from the protocol in order to break the inputs' privacy and/or the outputs' correctness; 3) *covert adversary* that may behave maliciously, but with a fixed probability to be spotted. While actively secure protocols are always able to detect malicious corruptions (but not necessarily the identity of corrupt parties), in covert secure protocols a cheating party might not be detected with a certain non-negligible probability. Semi-honest protocols offer a rather weak security guarantee, but they are much more efficient than maliciously secure protocols. Interestingly, covert security can be thought as a compromise between the other two more standard models, as it can offer more efficiency than active security and stronger guarantees than semi-honest one.

Besides this efficiency issue and the need of cryptographic assumptions, Cleve in [24] showed that in the very desirable setting of active security and dishonest majority it is impossible to obtain protocols for secure computation that provide fairness and guaranteed output delivery. Consequently, many secure MPC protocols with these strong security properties simply abort if a cheating is observed, realizing the weaker notion of *security with abort*. In particular, this means that, either the protocol succeeds and every party receives its outputs, or the protocol aborts, and this can happen even after the adversary has learnt the output of the computation, which could be a serious issue in some applications. One of the main drawbacks is that these protocols are vulnerable to *denial-of-service attacks* where corrupt parties can force the protocol to abort so that honest parties never learn the output of the computation.

This motivates the study of secure computation protocols with *identifiable abort* (ID-MPC) [9,10,25,27,44]. In this setting, if some malicious behaviour is detected or the adversary abort, the honest parties will agree upon the identity of at least one corrupt party. Even though this notion of security remains strictly weaker of fairness or guaranteed output delivery, it is very useful in practice as it discourages corrupt parties to behave maliciously, because upon abort at least one of them would be detected and maybe excluded from future computations.

1.1   Actively-Secure MPC with Dishonest Majority

The last decade has seen a huge progress in the practicality of secure computations. Although it seems fairly natural to imagine efficient protocols with restricted security against semi-honest adversaries and/or assuming an honest majority, surprisingly a major advance has been in the dishonest majority case with active corruptions with the SPDZ line of works [16,29,30].

MPC in the Correlated Randomness Model. A theoretically interesting and practically effective way to obtain efficiency in secure computation is by designing protocols with a randomness distribution phase, which is independent of the inputs to the function being computed, and sometimes also to the function itself. During this phase, parties receive randomness that are correlated from a pre-determined joint distribution. Using these random strings in the actual computation, it is possible to circumvent impossibility results such as impossibility of unconditional security in the plain model. Practically, one way to instantiate this model is through *MPC with pre-processing*.

Secure MPC protocols in this model restrict all the expensive operations to a pre-processing phase that can be both function and input independent. If this is the case we talk of *universal pre-processing* and if the pre-processing is only input independent, then we talk of *dedicated pre-processing*.

The randomness generated in the pre-processing stage is consumed by a lightweight non-cryptographic *online phase* that performs the actual circuit evaluation. Typically, the main goal of the pre-processing (or, *offline* phase) in MPC protocols is to produce randomness that enables an efficient, both in terms of communication and computation, evaluation of multiplication gates. In 1991, Beaver [11] introduced a neat trick that permits efficient secure evaluation of circuits by randomizing the inputs to each multiplication gate using a pre-processed *random multiplication triple*. Using Beaver's trick, online evaluation turns out to be very efficient, involving only information-theoretic techniques, and creating triples becomes the main bottleneck.

From Passive to Active Security. Even though the pre-processing model allows to perform most of the work in a offline phase, leading to a very efficient online computation, this does not reduce the price to pay to have actively secure protocols compared to passively-secure ones. A typical example is the GMW protocol which needs expensive generic zero-knowledge (ZK) proofs to achieve active security. In 2008, Ishai, Prabhakaran and Sahai in [45], described a novel technique, also known as IPS compiler, for actively secure MPC for Boolean circuits and constant number of parties, having asymptotic constant overhead over passively secure protocols. The IPS compiler is based on Oblivious Transfer (OT)[1] and hence can also be expressed in the correlated randomness paradigm as OT can be pre-processed as shown by Beaver [12]. In [55], Lindell et al. presented a protocol based on the IPS compiler that converts semi-honest protocols in the dishonest majority setting into covertly secure ones. Later, Genkin et al. [34] proposed an MPC protocol for arbitrary number of parties based on Oblivious Linear Evaluation (OLE) for large fields with constant communication overhead. This technique was extended in [35] to obtain active security for Boolean circuits. Recently, Hazay, Venkitasubramaniam and Weiss [41], have proposed a more efficient compiler from passive to active that works over arbitrary fields and arbitrary number of parties. Almost all these works make black-box use of the underlying cryptographic primitives, OT, OLE, etc., and are mainly concerned with asymptotic complexity.

Concrete Efficiency. A different line of works, more focused on *concrete* efficiency, started with the paper by Damgård and Orlandi [29] in 2010. To generate triples, the pre-processing phase utilizes an additively homomorphic encryption scheme, plus a "sacrifice" technique and homomorphic commitments to accomplish active security. This protocol uses commitments also during the circuit evaluation, still limiting the online phase to computational security. This issue was solved shortly after by Bendlin, Damgård, Orlandi and Zakarias [16]. In this protocol, often called BDOZ, the homomorphic commitment scheme is replaced by a pairwise *information-theoretic* Message Authentication Code (MAC). A further optimization was introduced by Damgård, Pastro, Smart and Zakarias in 2012 with the SPDZ protocol [30]. In SPDZ, the pairwise MAC used in BDOZ was simplified to a "global" MAC so that each party only stores a single field element for each MAC value instead of $n - 1$ (where n is the number of parties). This leads to an information-theoretic online phase which is, roughly, only two times less efficient than the passive variant of the protocol. A second efficiency improvement provided by SPDZ is in the pre-processing phase, and comes from replacing the additive homomorphic encryption scheme with a somewhat homomorphic scheme. In particular, SPDZ uses the lattice-based scheme by Brakersy, Gentry and Vaikuntanathan (BGV) [21], making extensive use of the packing technique of Smart and Vercauteren [67], which allows the manipulation of several plaintexts at once using SIMD (Single Instruction Multiple Data) operations.

---

[1] OT is a fundamental cryptographic primitive [63,69]. In its classical formulation, a (one-out-of-two) oblivious transfer is a two-party protocol between a sender P_S and a receiver P_R: P_S inputs two messages x_0, x_1, P_R inputs a bit b, and the goal is for the receiver to learn x_b and nothing more, whilst the sender learns no information about b.

After SPDZ, there has been a very large body of work mainly aiming to improve the SPDZ pre-processing phase, and in particular the triples generation step. In the next sections we will briefly describe some of these results, however we stress that what we are going to present is far from being exhaustive, and many interesting results are not going to be covered or even mentioned, due to space limitation. The aim of this paper is to introduce the main ideas and high level techniques used in protocols that are closely related to SPDZ, rather than giving a detailed and complete description of all the protocols dealing with active security and arbitrary number of corruptions.

1.2 Instantiating the Preprocessing and Alternative Approaches

Concurrently to SPDZ, at CRYPTO 2012, another practical secure MPC protocol was presented by Nielsen et al. [58], usually referred to as TinyOT. It is a very efficient two-party protocol for secure computation of Boolean circuit, hence, in some sense, it can be considered complementary to SPDZ that on the contrary achieves better performances over large fields and allows arbitrary number of parties. TinyOT-online phase is very similar to SPDZ-online phase, except for the use of pair-wise MACs à la BDOZ. On the other hand, the offline phase differs significantly as it is based on oblivious transfer. TinyOT was later generalized to the multiparty case by Larraia et al. [54], and to work on arithmetic circuit in MASCOT [50]. The most efficient versions of the offline phase of SPDZ, TopGear [6,51], and MASCOT still represent the state-of-the-art of linear secret-shared based MPC for arithmetic and binary circuits, respectively. We provide a more detailed comparison between these two approaches in Sect. 5.

1.3 Is MPC Any Good in Practice? [59]

Secure multiparty computation has been studied since the mid 1980s and back then the research on this field was mainly focused on feasibility results. Now, after almost 40 years, MPC is a rather mature technology that has rapidly progressed, especially in the last decade, from a notion of theoretical interest only into a technology that is starting to being commercialized.

In some aspects the extraordinary advance in the practicality of secure computation has been surprising, and it can be considered to be a consequence of a combination of algorithmic, technological and computational progress. Practically, if we consider malicious secure two party computation, the first implementation of Pinkas et al. [62] reports roughly 1114 sec for the evaluation of AES-128, i.e. a Boolean circuit of roughly 30000 gates (6400 AND gates and the rest XOR gates). Recent protocols [48,68] require roughly 10 ms for the same circuit. Possibly, recent improvements, for example in OT-extension protocols, will further improve these running times.

A number of online implementations are available, and SCALE-MAMBA [1] and MP-SPDZ [47] are the ones most closely related to SPDZ. We refer to [40], for a more detailed discussion about these and other available frameworks. Even if we only focus on the dishonest majority setting, we want to remark that significant advances have also been made in the practicality of MPC protocols in different settings.

The current state of affairs is that secure MPC is able to compute relatively simple functions very efficiently, but it fails when we try to scale to more involved computations or involving a large number of parties. One of the main problem is communication, especially for linear secret-sharing (LSSS) based protocols like SPDZ or TinyOT. A major progress would consist in designing protocols that have both low bandwidth, like LSSS-based protocols, and small number of rounds, like GC-based protocols.

Despite the extraordinary progresses in the last years, there is still a long way to go before we can assert that secure multiparty computation is practically efficient in every scenario, for example for huge data sets or for Internet-like settings. Research in MPC is very active, and range from fundamental research, to implementation, to products deployments. It a very fast-moving field and, considering recent improvements, one would expect to see more breakthroughs in the area, with secure computation taking a leading role in most practical privacy-preserving solutions.

1.4 This Survey

The aim of this work is to give an overview of the techniques used in concretely efficient MPC protocols with active security and dishonest majority, giving a high level description of the main building blocks of SPDZ and providing a limited literary review of the main related works. Other than describing the SPDZ protocol in its most recent and efficient version, we also provide a rough description of the alternative approaches and, maybe more importantly, references to the relative papers.

We start with basic notation and preliminaries in Sect. 2. We explain how data is authenticated via information theoretic MACs in Sect. 3. Assuming a trusted functionality $\mathcal{F}_{\mathsf{Prep}}$ for the pre-processing phase, we describe the SPDZ online protocol in Sect. 4, and finally, in Sect. 5, we show how SPDZ implements $\mathcal{F}_{\mathsf{Prep}}$. In this section we also describe an alternative approach to the pre-processing implementation using oblivious transfer.

2 Preliminaries

We let κ (resp. s) denote the computational (resp. statistical) security parameter. We say that a function $\mu : \mathbb{N} \to \mathbb{N}$ is negligible if for every positive polynomial[2] $p(\cdot)$, and all sufficiently large κ, it holds that $\mu(\kappa) < 1/p(\kappa)$. We use the abbreviation PPT to denote probabilistic polynomial-time. Let $\mathbb{F}$ denote a finite field, we consider protocols that allow to evaluate circuits C_f representing functions $f : \mathbb{F}^{n_{\mathsf{in}}} \to \mathbb{F}^{n_{\mathsf{out}}}$ with n_{in} inputs and n_{out} outputs. To ease the reading, we drop the dependence on f, when it is clear from the context.

---

[2] Given a set S, a positive polynomial on S is such that $p(x) > 0$ for every $x \in S$.

We use lower case letters to denote finite field elements and bold lower case letters for vectors in $\mathbb{F}^\kappa$, for any finite field $\mathbb{F}$. If $\mathbf{x}, \mathbf{y}$ are vectors over $\mathbb{F}$, then $\mathbf{x} * \mathbf{y}$ denotes the component-wise products of the vectors. If A is a (probabilistic) algorithm then we denote by $a \leftarrow A$ the assignment of the output of A where the probability distribution is over the random tape of A and we denote by by $s \xleftarrow{\$} S$ the uniform sampling of s from a set S. We also use the notation $[d]$ as shorthand for the set of integers $\{1, \ldots, d\}$.

Security Model. Protocols described in this paper work with n parties from the set $\mathcal{P} = \{P_1, \ldots, P_n\}$, and we consider security against malicious, static adversaries, i.e. corruption may only take place before the protocols start, corrupting up to $n - 1$ parties.

All the protocols described in this paper can be proved to be secure in the universal composition (UC) framework of Canetti [22]. Even though we omit these proofs here and provide the references to the relevant papers only, we will maintain some of the terminology used in the UC framework. For example we are going to use ideal functionalities in most of the protocols described in this survey. The reader who is not familiar with this notation and security model, and is not interested in understanding it, can simply imagine these functionalities as trusted entities that are called to securely perform some specific tasks.

Loosely speaking, protocols that aim to achieve security in the UC model are defined in three steps. First, the protocol and its execution in the presence of an adversary are formalized, this represents the real-life model which we also call the *real world*. Next, an *ideal functionality* for executing the task is defined; its role is to act as a trusted party by separately receiving the input of each party, both honest and corrupt, and honestly computing the result of the protocol internally and returning the output assigned to each party. In this ideal process, also called *ideal world*, the parties do not communicate with one another but instead solely rely on the ideal functionality to provide them with their output. Finally, we say that the protocol in question UC-realizes the ideal functionality if running the protocol is equivalent, or *indistinguishable*, from emulating the ideal functionality. When we say that a protocol Π securely implements an ideal functionality $\mathcal{F}$ with computational (resp. statistical) security parameter κ (resp. λ), our theorems guarantee that the advantage in distinguishing the real and ideal executions is in $O(2^{-\kappa})$ (resp. $O(2^{-s})$).

Communication Model. We assume all parties are connected via authenticated communication channels, as well as secure point-to-point channels and a broadcast channel. In practice, since we are considering security with abort, broadcast can be implemented with point-to-point channels requiring only two rounds of communication as follows [38]: 1) The party that needs to broadcast a value sends this to all parties; 2) All the receiving parties send the value they received to all other parties. It can be proven that either all the parties output the same value or the protocol aborts. This broadcast is also called *broadcast with abort*. It requires $O(n^2)$ communication per broadcast. SPDZ [30] (Appendix $A.3$) describes how to optimize it in the case there are many broadcasts to perform, like in MPC protocols. Roughly, in all the broadcast instances parties maintain a running hash of all values sent and received, and these are checked later, at the end of the protocol. With this optimization the amortized cost per broadcast value is $O(n)$.

Secret Sharing Scheme. We only consider computation on values that are additively secret shared among parties, i.e. each shared value $x \in \mathbb{F}$ is represented as

$$\langle x \rangle = (x^{(1)}, \ldots, x^{(n)}),$$

where each party $P_i \in \mathcal{P}$ holds a random share $x^{(i)}$ and $x = \sum_{i \in [n]} x^{(i)}$. In this way, by setting all but a single share to be a random value in $\mathbb{F}$, we have that any subset of $n - 1$ parties cannot recover the secret value x. We give a more formal definition below.

Definition 1 (Additive secret-sharing scheme). *Let $\mathbb{F}$ be a finite field and $n \in \mathbb{N}$ a positive integer. We define an additive secret sharing scheme $\mathbb{S} = (\mathsf{Share}, \mathsf{Recover})$ such that:*

- $\mathsf{Share}(x, n)$: *on input a secret x and an integer n, first it generates shares $(x^{(1)}, \ldots, x^{(n-1)})$ uniformly at random from $\mathbb{F}$ and define $x^{(n)} = x - \sum_{i=1}^{n-1} x^{(i)}$; then it outputs $(x^{(1)}, \ldots, x^{(n)})$, where $x^{(i)}$ is the share of party P_i*
- $\mathsf{Recover}(\mathsf{x}^{(1)}, \ldots, \mathsf{x}^{(n)})$: *given all the shares $x^{(i)}, i \in [n]$, parties compute $x = \sum_{i=1}^{n} x^{(i)}$.*

Trivially, this secret sharing scheme is linear, therefore linear operations can be performed locally without interactions among parties, as described below.

- Addition of secret-shared values: $\langle x \rangle + \langle y \rangle = (x^{(1)} + y^{(1)}, \ldots, x^{(n)} + y^{(n)}) = \langle x + y \rangle$
- Addition by a public value a: $a + \langle x \rangle = (a + x^{(1)}, \ldots, x^{(n)}) = \langle a + x \rangle$
- Multiplication by a public value a: $a \cdot \langle x \rangle = (a \cdot x^{(1)}, \ldots, a \cdot x^{(n)}) = \langle a \cdot x \rangle$

Statistical Distance. Let E be a finite set, Ω be a probability space and $X, Y : \Omega \to E$ be random variables. The *statistical distance* between X, Y is defined as:

$$\Delta(X, Y) = \frac{1}{2} \sum_{x \in E} \left| \Pr_X(X = x) - \Pr_Y(Y = x) \right|$$

We recall the following result from [2].

Lemma 1 (Smudging Lemma). *Let B_1 and B_2 be positive integers, let $e \in [-B_1, B_2]$ be a fixed integer, and let E_1, E_2 be independent random variables uniformly distributed in $[-B_1, B_2]$. Define the two stochastic variables $X_1 = E_1 + e$ and $X_2 = E_2$. Then, it holds that:*

$$\Delta(E_1, E_2) < B_1 / B_2.$$

This lemma allows to "smudge out" small differences between distributions adding large noise. It will be used many times in the protocols we describe in the next sections, often implicitly.

2.1 Threshold (L-leveled) Homomorphic Encryption

We briefly recall the definition of threshold (L-leveled) homomorphic encryption (THE) [2,17] scheme. It is similar to a standard (leveled) homomorphic encryption scheme, but with different key-generation and decryption algorithms. The scheme is parametrized by security parameters (κ, s), the number of levels L, the amount of packing of plaintext elements which can be made into a single ciphertext N, and by a linear secret scheme $\mathbb{S}$. We instantiate $\mathbb{S}$ with an additive secret sharing scheme as describe before, and hence we give a less general definition of a THE scheme. Informally, a threshold L-leveled HE scheme supports homomorphic evaluation of any circuit C consisting of addition and multiplication gates and of multiplicative depth at most L, with the provision for distributed (threshold) decryption.

Definition 2. *An L-leveled public key homomorphic encryption scheme with message space* $\mathcal{M} = \mathbb{F}^N$, *is a tuple of PPT algorithms* THE = (KeyGen, Enc, Eval, PartDec, DistDec), *satisfying the following specifications:*

$(\mathsf{pk}, \mathsf{sk}^{(1)}, \ldots, \mathsf{sk}^{(n)}) \leftarrow \mathsf{KeyGen}(1^\kappa, 1^s, n)$: *taking as input the security parameters* κ *and* s, *and the number of parties* n, *it outputs a public key* pk, *and secret key additive shares* $(\mathsf{sk}^{(1)}, \ldots, \mathsf{sk}^{(n)})$;

$\mathsf{ct} \leftarrow \mathsf{Enc}(m; \mathsf{pk})$: *it takes a plaintext* $m \in \mathcal{M}$ *and public key* pk, *and output a ciphertext* ct;

$\hat{\mathsf{ct}} \leftarrow \mathsf{Eval}(C, \mathsf{ct}_1, \ldots, \mathsf{ct}_t)$: *it takes as input a circuit* $C : \mathbb{F}^t \to \mathbb{F}$, *with multiplicative depth at most* L *and* t *ciphertexts* $\mathsf{ct}_1, \ldots, \mathsf{ct}_t$, *and outputs an evaluation ciphertext* $\hat{\mathsf{ct}}$;

$(p^{(1)}, \ldots, p^{(n)}) \leftarrow \mathsf{PartDec}(\mathsf{ct}; \mathsf{sk}^{(1)}, \ldots, \mathsf{sk}^{(n)}$: *given a ciphertext* ct *and a secret key share* $\mathsf{ct}^{(i)}$, *it outputs a partial decryption* $p^{(i)}$ *to party* P_i, *for every* $i \in [n]$;

$\hat{m} \leftarrow \mathsf{DistDec}(p^{(1)}, \ldots, p^{(n)}; \mathsf{pk})$: *it takes as input the public key and all the partial decryption outputs, and outputs a plaintext* $\hat{m}$.

The scheme needs to satisfy correctness, semantic security and simulation security as described in [17]. Here we omit the formal definitions of these properties. While the first two are relatively standard, the latter essentially says that no information about the key shares and plaintext should be leaked by the decryption algorithms other than what is already implied by the result of homomorphic operations.

In SPDZ, a THE scheme is used to generate random triples in the preprocessing phase, therefore we need a very simple THE supporting only one homomorphic operation, i.e. $L = 1$. Concretely, the THE scheme is instantiated with the scheme by Brakersky, Gentry and Vaikuntanathan (BGV) [21], based on the Ring Learning with Error assumption [56,64], and supporting packing operations [67] that permits to handle many plaintexts in a single ciphertext. We omit the description of BGV in this survey.

2.2   UC Commitments

Functionality $\mathcal{F}_{\mathsf{Commit}}$

Commit: On input $(\mathtt{Commit}, m, i, \tau_m)$ from P_i, store (m, i, τ_m). τ_m is a handle for the commitment. and output (i, τ_m) to all parties.

Open: On input $(\mathtt{Open}, i, \tau_m)$ by P_i, output (m, i, τ_m) to all parties.
  If instead $(\mathtt{NoOpen}, i, \tau_m)$ is given by the adversary, and P_i is corrupt, the functionality outputs $(\perp, i, \tau_m)$ to all parties.

Fig. 1. Commitments functionality.

A commitment scheme allows a commiter holding a secret value m to send a commitment c of m to a verifier, and later on to "open" this commitment to reveal m. More formally, a commitment scheme is defined by three algorithms.

- Setup(1^κ) : given as input the security parameter, it generates the global parameters that will be implicitly used by the other algorithms;
- $(c, w) \leftarrow$ Commit(m) : given a message m it produces a commitment c on m and the opening information w;
- $m \leftarrow$ Open(c, w) : it decommits c using w and outputs either the message m or $\perp$ if the opening fails.

The scheme has to be both *binding*, i.e. the opening should successfully open to one value only, and *hiding* which means that the commitment c should not reveal any information about m. These two properties can be achieved in a perfect, statistical or computational way. A UC-secure commitment must be both *extractable* (meaning that it is possible to extract the value that a corrupted party commits to) and *equivocable* (meaning that it is possible to generate commitments that can be opened to any value). In this survey we will use an ideal functionality $\mathcal{F}_{\mathsf{Commit}}$ as described in Fig. 1.

This ideal functionality can be implemented assuming a random oracle, by defining $c = \mathsf{H}(m, i, r)$, where H is a random oracle, $r \leftarrow \{0, 1\}^\kappa$ and $w = (c, r)$.

Using this hash-based commitment we can also efficiently implement a standard coin flipping functionality $\mathcal{F}_{\mathsf{Rand}}$. We refer to [28] for more details.

2.3   Zero Knowledge Proofs

A zero-knowledge (ZK) proof [39] is an interactive protocol between a prover P and a verifier V that allows the prover to demonstrate that a statement is true without revealing any further information about the proof beyond the fact that the statement is true.

An NP-relation $\mathcal{R}(x, w)$ is an efficiently decidable binary relation $\mathcal{R}(x, w)$ that is polynomially bounded, i.e. if $\mathcal{R}(x, w)$ is satisfied, then $|w| \leq \mathsf{poly}(|w|)$.

Any NP-relation defines a language $\mathcal{L} = \{x : \exists w, \mathcal{R}(x, w) = 1\}$. Usually w is called a *witness* for the statement $x \in \mathcal{L}$.

A ZK proof protocol for the NP relation $\mathcal{R}(x, w)$, with common input x and additional input w for P, satisfies three properties that we can informally describe as follows:

- Completeness: if $x \in \mathcal{L}$, and P knows a proof of this, she/he will succeed in convincing V;
- Soundness: if the statement is false, no prover can convince the verifier of the truth of the statement except with probability ϵ, where ϵ is the *soundness error* of the protocol;
- Zero-knowledge: the interaction between P and V yields nothing beyond the fact that the statement is true. This is equivalent to require the existence of a simulator that can produce an honest-looking transcript for the protocol, without knowing anything about the statement.

2.4 Oblivious Transfer

Oblivious transfer is a fundamental cryptographic primitive originally introduced by Rabin [63] and Wiesner [69]. Subsequent works by [33,52] showed oblivious transfer to be a very powerful primitive. In particular, Kilian [52] showed that OT is complete for secure multi-party computation. Many MPC protocols have been constructed based on OT, including the GMW protocol, Yao's garbled circuits and the IPS compiler that we have already mentioned before.

In its classical formulation, a (one-out-of-two) oblivious transfer is a two-party protocol between a sender P_S and a receiver P_R: the sender inputs two messages x_0, x_1, a receiver inputs a bit b, and the goal is for the receiver to learn x_b and nothing more, whilst the sender learns no information about b. In Fig. 2 we describe the ideal functionality for oblivious transfer on bit strings of length k, meaning that sender's inputs x_0, x_1 are elements in $\{0, 1\}^k$. Given the inputs from P_S and P_R the functionality outputs the string x_b corresponding to receiver's input b.

Functionality $\mathcal{F}_{\mathsf{OT}}^k$

Running between a sender P_S and a receiver P_R, it operates as follows.

- P_S inputs $(x_0, x_1) \in \{0, 1\}^k \times \{0, 1\}^k$ and P_R inputs b.
- The functionality outputs x_b to P_R.

Fig. 2. Functionality for one-out-of-two oblivious transfers on k-bit strings.

Oblivious Transfer Extension. Although oblivious transfer is a fundamental building block for many cryptographic constructions, it used to be considered an expensive primitive. Indeed, Impagliazzo and Rudich [42] showed a black-box separation result that is strong evidence that OT is impossible without the use of expensive public-key cryptography. However, thanks to recent, and somehow surprisingly, advances in the field, we can fairly claim that in practice OT is no longer an expensive primitive.

Beaver in [13] first showed that OT can be "extended", i.e. starting from few OTs one could generate a large amount of additional OTs using only cheap symmetric primitives. Albeit elegant, Beaver's protocol is highly impractical. The first efficient OT-extension protocol was described by Ishai, Kilian, Nissim and Petrank [43] in the passive setting. Subsequent works, secure against both passive [3,53] and active [4,49,60] adversary, all follow the IKNP blueprint. These protocols are computationally very efficient and allow to create more than 10 million of OTs in 1 sec. The main bottleneck remains communication. A different approach, that uses LPN-based (LPN stands for Learning Parity with Noise[3]) PCG (i.e. Pseudorandom Correlation Generators see Sect. 5.3) for OT-extension, outperforms previous solutions in terms of communications in low bandwidth network, but at price of high computational overhead [18,66]. A very recent protocol by Yang et al. [70] achieves impressive performances both in terms of communication and computation requiring only 21 nanoseconds (resp. 22 nanoseconds) for generating one (correlated) OT in a 50 Mbps network with passive (resp. active) security. Correlated OT (COT) is a slightly different variant of OT, but sufficient for many practical MPC protocols. We will define COT in Sect. 5.2.

3 Data Representation

While an additive secret sharing scheme is sufficient to guarantee privacy and hence security in the weak model of semi-honest security, we need extra caution in presence of an active adversary in order to prevent corrupt parties to inject incorrect values to the protocol that could lead to erroneous results or information leakage.

As we mentioned in the introduction, SPDZ protocols achieve active security by authenticating each shared value with an information-theoretic MAC. This can be done either in a pairwise manner [16,58], or in a global manner [30, 54]. Both of these variants can be applied, yet implying significant practical differences in the total amount of data each party needs to store, in the ZK proofs and in the way MACs are checked. We describe both these variants below.

BDOZ-style MAC: Each value $x \in \mathbb{F}$ is authenticated and additively secret shared among parties in $\mathcal{P}$ in such a way that each party P_i holds a share $x^{(i)}$ and $n-1$ pairwise MACs

$$\mathsf{m}_x^{(ij)} = \mathsf{k}_x^{(ji)} + x^{(i)} \cdot \Delta^{(j)},$$

for each $j \neq i$. This notation implies that P_i holds $x^{(i)}$ and $\{\mathsf{m}_x^{(ij)}\}_j$, and each other party P_j holds a local key $\mathsf{k}_x^{(ji)}$, i.e. depending on the value $x^{(i)}$, and a

---

[3] Roughly, the LPN assumption says that given a random linear code C, a noisy random codeword of C is pseudo-random.

global key $\Delta^{(j)}$ fixed for the entire computation. The values $m_x^{(ij)}, k_x^{(ji)}, \Delta^{(j)}$ are either elements of $\mathbb{F}$, or elements of an extension field $\mathbb{E}$ of $\mathbb{F}$. Typically, if $\log_2 |\mathbb{F}| \geq \kappa$, then $\mathbb{E} = \mathbb{F}$. We will use the following notation to represent this type of authenticated values:

$$[x]_B^j = \left(\langle x \rangle, (m_x^{(1j)}, \ldots, m_x^{(nj)}), (k_x^{(j1)}, \ldots, k_x^{(jn)}), \Delta^j \right),$$

to denote each party authenticating their share of $\langle x \rangle$ towards party P_j, and

$$[x]_B = \left([x]_B^1, \ldots, [x]_B^n \right),$$

for the global representation. It is easy to see that parties can locally perform linear operations on authenticated data.

Addition of pairwise authenticated secret-shared values.

$$\begin{aligned} [x]_B + [y]_B &= \left([x]_B^1 + [y]_B^1, \ldots, [x]_B^n + [y]_B^n \right) = \left([x+y]_B^1, \ldots, [x+y]_B^n \right) \\ &= [x+y]_B, \end{aligned}$$

since, for each i, it holds:

$$\begin{aligned} [x]_B^i + [y]_B^i =& \left(\langle x \rangle + \langle y \rangle, (m_x^{(1j)} + m_y^{(1j)}, \ldots, m_x^{(nj)} + m_y^{(nj)}), \right. \\ & \left. (k_x^{(j1)} + k_y^{(j1)}, \ldots, k_x^{(jn)} + k_y^{(jn)}), \Delta^j \right) \\ =& \left(\langle x+y \rangle, (m_{x+y}^{(1j)}, \ldots, m_{x+y}^{(nj)}), (k_{x+y}^{(j1)}, \ldots, k_{x+y}^{(jn)}), \Delta^j \right) \end{aligned}$$

Addition by a public value. Given a publicly known value $a \in \mathbb{F}$,

$$a + [x]_B = [a + x]_B,$$

where $\langle a+x \rangle$ is obtained as described in the introduction. All the MAC values and keys remain the same, except for $k_{a+x}^{(j1)} = k_x^{(j1)} - a \cdot \Delta^{(j)}$.

Multiplication by a public value. As before, given a public $a \in \mathbb{F}$, $a \cdot [x]_B = [a \cdot x]_B$, obtained by multiplying each share, MAC and local key by a.

SPDZ-style MAC: Each value $x \in \mathbb{F}$ is additively secret shared and authenticated as follows.

$$[x]_S = \left(\langle x \rangle, \langle m_x \rangle, \langle \Delta \rangle \right),$$

where $m_x = \sum_i m_x^{(i)} = x \cdot \Delta$ and $\Delta = \sum_i \Delta^{(i)}$ is the MAC key, that is hence unknown to the parties. As for $[\cdot]_B$, we assume m_x and Δ to be elements of $\mathbb{F}$ or $\mathbb{E}$, such that $\mathbb{F} \subset \mathbb{E}$. Again, due to the linear relation between authenticated values and MAC, linear operations can be carried out locally.

Addition of pairwise authenticated secret-shared values. $[x]_S + [y]_S = \left(\langle x \rangle + \langle y \rangle, \langle m_x \rangle + \langle m_y \rangle, \langle \Delta \rangle \right) = \left(\langle x+y \rangle, \langle m_x + m_y \rangle, \langle \Delta \rangle \right) = [x+y]_S$.
In a similar way we can perform addition and multiplication by a public value. Note that given a public value a, the MAC value on a is defined by each party setting $m_a^{(i)} = a \cdot \Delta^{(i)}$, to obtain a valid authenticated share.

Conversion to $[\cdot]_S$. It is possible to locally convert the BDOZ representation to the SPDZ representation [54]. As we will see in Sect. 5, this conversion is particularly useful in the case we want to use a two-party primitive, like oblivious transfer, to generate authenticated values and random triples. This allows a more efficient memory usage and exploits a less expensive, global MAC check procedure in the online evaluation. Given a BDOZ-style authenticated value $[x]_B$, parties already hold $\langle x \rangle$ and additive shares of Δ, so to obtain a SPDZ-style representation, it is enough to generate shares $\mathsf{m}_x^{(i)}$ of $\mathsf{m}_x = x \cdot \Delta$. This is done without any interaction by parties combining their pairwise MACs and keys as follows:

$$\mathsf{m}_x^{(i)} = \sum_{j \neq i} (\mathsf{m}_x^{(ij)} - \mathsf{k}_x^{(ij)}) + x^{(i)} \cdot \Delta^{(i)}.$$

Indeed the following relations hold:

$$\mathsf{m}_x = \sum_i \mathsf{m}_x^{(i)} = \sum_i x^{(i)} \cdot \Delta^{(i)} + \sum_i \sum_{j \neq i} (\mathsf{m}_x^{(ij)} - \mathsf{k}_x^{(ij)})$$
$$= \sum_i x^{(i)} \cdot \Delta^{(i)} + \sum_i \sum_{j \neq i} x^{(i)} \cdot \Delta^{(j)}$$
$$= x \cdot \Delta.$$

3.1 Checking MACs

During the online evaluation of the circuit, parties need to communicate or, more precisely, they need to be able to *reveal* secret shared values $[x]$. This is done by sending over all the private shares $x^{(i)}$ of $\langle x \rangle$:

Open: On input (Open, x) from every party, each P_i broadcasts $x^{(i)}$, recovers $x = \sum_i x^{(i)}$ and stores x.

Moreover, we need to prevent corrupt parties to disclose incorrect values, therefore we need a way to check MACs on opened values. This can be done in different ways. Note that since it is always possible to convert from $[\cdot]_B$ to $[\cdot]_S$, and the latter allows a more efficient check, we only consider the case of SPDZ-style MAC.

An obvious way to check if a reconstructed value is correct is by revealing shares $\mathsf{m}_x^{(i)}$ and $\Delta^{(i)}$, for each $i \in [n]$, along with $x^{(i)}$. Clearly, we can perform this check only once because after the MAC key Δ is revealed all parties can forge new MACs and introduce incorrect values, so a new MAC key should be generated (along with new pre-processed material).

To overcome this problem in original SPDZ protocol, parties wait until the end of the computation to reveal the MACs and the MAC key. Only when the circuit evaluation is completed, parties check the MACs on opened values, and if the check passes the final result of the computation is opened.

However, this approach limits the use of Δ to a single evaluation, preventing reactive computations without generating fresh MAC keys and pre-processed randomness. The following two procedures, firstly described in [28], allow to check a single MAC and a batch of MACs, respectively, on opened values *without disclosing the global MAC key*. At a high level, given an opened value $\tilde{x}$ and authenticated value $[x]$, the goal is to check whether $\tilde{x} = x$ by checking the MAC relation $\mathsf{m}_x = \tilde{x} \cdot \Delta$. To this end parties broadcast $\sigma^{(i)} = \mathsf{m}_x^{(i)} - \tilde{x} \cdot \Delta^{(i)}$ and then check that $\sum_i \sigma^{(i)} = 0$. Note that the shares $\mathsf{m}_x^{(i)}$ are uniformly random, so sending $\sigma^{(i)}$ does not leak any private information, in particular about $\Delta^{(i)}$.

MAC Check. On input $(\mathsf{CheckMAC}, \tilde{x})$ from all parties:

1. Each party P_i computes $\sigma^{(i)} = \mathsf{m}_x^{(i)} - \tilde{x} \cdot \Delta^{(i)}$
2. P_i calls the functionality $\mathcal{F}_{\mathsf{Commit}}$ on command $(\mathsf{Commit}, \sigma^{(i)}, i, \tau^{(i)})$ to broadcast $(i, \tau^{(i)})$
3. All parties call $\mathcal{F}_{\mathsf{Commit}}$ with command $(\mathsf{Open}, i, \tau^{(i)})$, obtaining $\sigma^{(i)}$, for all $i \in [n]$
4. Parties check if $\sigma^{(1)} + \cdots + \sigma^{(n)} = 0$. If the check passes, accept x as a correct authenticated value, otherwise output $\perp$ and abort.

Batch MAC Check. On input $(\mathsf{CheckMAC}, \tilde{x}_1, \ldots \tilde{x}_t)$ from all parties:

1. Parties use $\mathcal{F}_{\mathsf{Rand}}$ to sample a random vector $\mathbf{r} \leftarrow \mathbb{F}^t$
2. Each party locally computes $\tilde{x} = \sum_{j=1}^t r_j \cdot \tilde{x}_j$
3. Each party P_i computes $\langle \tilde{\mathsf{m}} \rangle \leftarrow \sum_{i=1}^t r_j \cdot \langle \mathsf{m}_{\tilde{x}_j} \rangle$ and $\langle \sigma \rangle = \langle \mathsf{m} \rangle - \tilde{x} \cdot \langle \Delta \rangle$
4. Use the (single) **MAC Check** procedure described above to check the MAC relation on the value $\tilde{x}$, with MAC $\tilde{\mathsf{m}}$ and Δ.

Remark 1. These procedures use the $\mathcal{F}_{\mathsf{Commit}}$ (Fig. 1) functionality so that a corrupt party is not able to cheat in the broadcast of its share of σ, for example using information on shares sent by honest parties. In the same spirit, during the **Batch MAC Check**, the sampling of the vector $\mathbf{r}$, used to generate random linear combinations of opened values and MACs, is performed by $\mathcal{F}_{\mathsf{Rand}}$ to ensure that corrupt parties are not able to influence it in the attempt of passing the check with incorrect shares.

We omit the security proof of the MAC Check procedure, however, the intuition is that if corrupt parties send incorrect shares, such that the opened value is $x + \delta$, for some adversarial chosen δ, then to pass the check it should hold that

$$\sum_{i \in [n]} \sigma^{(i)} = (x + \delta) \cdot \Delta - \sum_i \mathsf{m}_x^{(i)} = (x + \delta) \cdot \Delta - \mathsf{m}_x = 0.$$

This means that the adversary should be able to "correct" the corrupt parties' MAC share by the value $\delta \cdot \Delta$, which in turns implies to guess the global key Δ. Hence the probability of passing the check is $1/|\mathbb{E}|$, which is negligible when the field is large. As a consequence, to carry out computation over small fields we need to take a large enough extension field $\mathbb{E}$ and embed the whole computation in that field, generating a significant communication overhead.

Another issue with this technique is that it does not work over other rings, for example over the modular rings $\mathbb{Z}/2^k\mathbb{Z}$. The reason is that these rings contain zero divisors and hence guessing $\delta \cdot \Delta \mod 2^k$ is much easier than over fields. We discuss in the next sections how to overcome these difficulties.

4   SPDZ Online Evaluation

In this section we describe the circuit evaluation phase (or, online phase) of SPDZ, assuming a trusted setup, $\mathcal{F}_{\mathsf{Prep}}$, that generates the correlated randomness used in the actual circuit computation. Recall that we assume that the circuit being evaluated is an arithmetic circuit over the finite field $\mathbb{F}$.

The main question is "What do we need for $\mathcal{F}_{\mathsf{Prep}}$?" The minimal requirements are the following. 1) Random authenticated values, $(r, [r]_S)$, that are used as masks to create authenticated sharings of the inputs. The value r is secret shared, but known to the input party; 2) Random authenticated triples, $([a]_S, [b]_S, [c]_S)$, $c = a \cdot b$, used to multiply two shared values.

During the online protocol the circuit is evaluated gate by gate on shared values and using the linearity of the $[\cdot]_S$-representation. To share an input x_i, party P_i takes a pre-processed random value $[r]_S$ and broadcast the value $x_i - r$. Since r is uniformly random in $\mathbb{F}$ and unknown to all other parties, it acts as a one-time pad to perfectly hide x_i. All parties can then locally compute $[r]_S + (x_i - r)$ to obtain $[x_i]_S$.

Multiplication of two shared values $[x]_S$ and $[y]_S$ uses Beaver's trick. Using a multiplication triple $[a]_S, [b]_S, [c]_S$, first parties open and recover the values $\epsilon = x - a$ and $\rho = y - b$. Again, the triple values perfectly mask the inputs x and y, and the opened values appear uniformly random to corrupt parties. Given ϵ and ρ, a sharing of the product $x \cdot y$ can be locally computed by all parties using the triple as follows:

$$[x \cdot y]_S = [c]_S + \epsilon \cdot [b]_S + \rho \cdot [a]_S + \epsilon \cdot \rho.$$

When the circuit evaluation is completed, parties check the MACs on all the values revealed during the input and non-linear operations. If the check passes, they open and recover the output, otherwise the protocol aborts.

Online protocol
 Initialize. Parties call $\mathcal{F}_{\mathsf{Prep}}$ to get the shares $\Delta^{(i)}$ of the MAC key, multiplication triples $([a]_S, [b]_S, [c]_S)$ and mask values $(r_i, [r_i]_S)$ as needed for the function under evaluation. If $\mathcal{F}_{\mathsf{Prep}}$ aborts then the parties output $\perp$ and abort.
 Input. To share an input x_i, party P_i takes an available mask value $(r_i, [r_i]_S)$ and does the following:
  1. Broadcast $\epsilon \leftarrow x_i - r_i$.
  2. The parties compute $[x_i]_S$ as $[r_i] + \epsilon$.
 Add. On input $([x]_S, [y]_S)$, locally compute $[x + y]_S \leftarrow [x]_S + [y]_S$.
 Multiply. On input $([x]_S, [y]_S)$, the parties do the following:
  1. Take one multiplication triple $([a]_S, [b]_S, [c]_S)$, compute $[\epsilon]_S \leftarrow [x]_S - [a]_S$, $[\rho]_S \leftarrow [y]_S - [b]_S$. Open those values and run MAC Check.
  2. Use Beaver's trick described above.
 Output. To output a share $[y]_S$, do the following:
  1. Run MAC Check with input all opened values so far. If it fails, output $\perp$ and abort.
  2. Open and MAC Check $[y]_S$. If the check fails, output $\perp$ and abort, otherwise accept y as a valid output.

5   SPDZ Pre-processing

Here we show different ways of implementing the pre-processing phase. We recall, once again, that the main (basic) tasks of this step is to produce the following type of random authenticated values:

Input mask. $([r]_S, P_i)$, with the value r known by P_i
Triples. $([a]_S, [b]_S, [c]_S)$, where $c = a \cdot b$

Of course it is possible to pre-process different types of correlated randomness, such as random bits, squares, etc, that can help to improve the efficiency of certain online operations. However, explaining this kind of optimization is out of the scope of this work.

5.1   Pre-processing Using Threshold Homomorphic Encryption

As mentioned before, SPDZ offline protocol is based on a 1-leveled threshold homomorphic encryption scheme (introduced in Sect. 2.1), supporting $O(n)$ additions and one homomorphic multiplication, instantiated with BGV. Let us assume that $\mathcal{M} = \mathbb{F}^N$, where N is the packing parameter. This allows to produce many correlated random values in parallel. The original SPDZ paper, and several subsequent related works, assume a trusted setup $\mathcal{F}_{\text{KEYGEN}}$ for the KeyGen algorithm. We recall that this algorithm securely provides to the parties the BGV public key pk and a sharing $\langle \text{sk} \rangle$ of the secret key sk. [28] describes a covertly secure protocol that achieves this task, and only recently Rotaru et al. [65] have introduced a protocol that implements $\mathcal{F}_{\text{KEYGEN}}$ with active security. This protocol is based on oblivious transfer and, specifically, on the MASCOT protocol [50]. The interested reader can find the implementation of the so-called "SPDZ setup functionality" in [65], here we make use of ideal functionality $\mathcal{F}_{\text{KEYGEN}}$ in the description of the pre-processing protocol.

Other than $\mathcal{F}_{\text{KEYGEN}}$, we also assume another ideal functionality, $\mathcal{F}_{\text{DISTDEC}}$, that extends the standard BGV decryption algorithm to securely allow distributed decryption inside SPDZ. Now we give an overview of the pre-processing protocol, and later we provide and discuss it in greater detail.

High Level Description. The passive version of the pre-processing protocol works as follows. Let us assume that the parties have $(\text{pk}, \langle \text{sk} \rangle)$ and $(\langle \Delta \rangle, \text{ct}_\Delta)$, where ct_Δ is an encryption of the MAC key Δ using BGV.

– To create an input mask $([r]_S, P_i)$, each party P_i samples a random value r and creates a random sharing $\langle r \rangle$, that is P_i sends the relative share $r^{(j)}$ to P_j, for each $j \neq i$. Parties then locally compute the ciphertexts ct_{r^j} using the common public key pk and broadcast them. Using the homomorphic properties of BGV, parties can locally compute ct_r and $\text{ct}_r \cdot \text{ct}_\Delta = \text{ct}_{m_r}$, i.e. encryptions of the mask r and its MAC. Using a distributed decryption algorithm with $\langle \text{sk} \rangle$, each party obtains a share of m_r. Note that this step requires interaction. The output of this simple procedure is used in the Input step of the online evaluation to mask the actual input value, as described in Sect. 4.

– A similar technique is used to produce triples. Each party samples random shares $a^{(i)}, b^{(i)}$ and broadcast the corresponding ciphertexts $\mathsf{ct}_{a^i}, \mathsf{ct}_{b^i}$. Parties can compute $\mathsf{ct}_c, \mathsf{ct}_{m_a}, \mathsf{ct}_{m_b}$ as before and using the distributed decryption, the MAC sharing $\langle m_a \rangle$ and $\langle m_b \rangle$. Since we allow only one homomorphic multiplication, to produce $\langle m_c \rangle$ parties first decrypt ct_c, and, with $\langle c \rangle$, they produce a fresh encryption $\tilde{\mathsf{ct}}_c$ of c that can then be multiplied by ct_Δ.

Unfortunately, this simple protocol is not sufficient against active corruptions. Indeed, corrupt parties have the freedom to generate incorrect ciphertexts containing maliciously chosen noise or unknown plaintexts, that would result either in selective failure attacks or information leakage during distributed decryption. To solve this problem SPDZ uses zero-knowledge proofs of plaintext knowledge for every sent ciphertext, to prove that it is correctly generated. A second issue arises in the distributed decryption itself. During this interactive procedure an adversary might add errors both to triples and MAC values. While correctness of triples is checked through an additional check, called "sacrifice" (that we will describe later), errors on MACs have no impact on protocol security as potential errors cause the MAC Check to fail except with negligible probability.

SPDZ Zero-Knowledge Proofs. As mentioned before, to achieve active security SPDZ uses zero-knowledge proof of plaintext knowledge in order to prove that the ciphertexts used to generate pre-processed randomness are correctly generated. While it would be very convenient, in terms of efficiency, to avoid these expensive proofs all together, they seem to be quite unavoidable if we do not want to occur in decryption failures and information leakage both in the pre-processing and, more importantly, in the online computation. Zero-knowledge proofs constitute the main bottleneck in SPDZ implementations, both in terms of communication and runtime. For this reason a consistent amount of work have been devoted to the optimization of those proofs [5–8, 31, 51]. Here we informally describe the main idea of these proofs and explain why they are so expensive. For the details the reader may refer to [6, 7].

Roughly, in SPDZ ZK proofs, each party P_i, acting as a prover P, has to prove knowledge of a short preimage x of a linearly homomorphic function f such that $f(x) = y$ and $\|x\| \leq B$, for some bound B. Here f is the BGV encryption function and y is the ciphertext that P_i has to prove being correctly generated. More formally, we need a zero-knowledge proof of knowledge for the relation

$$\mathcal{R}_{ZK} = \{(x,y) \mid y = f(x) \land \|x\| \leq B\}.$$

This kind of proofs usually consist of a standard Σ-protocol:

1. P samples a random r such that $\|r\| \leq \tau \cdot B$, for τ sufficiently large (see below), and sends $f(r) = a$ to the verifier V;
2. V samples a random challenge $e \in \{0, 1\}$ and sends it to P;
3. P replies with $z = r + e \cdot x$.

Finally, the verifier checks whether $f(z) = a + e \cdot y$ and that $\|z\| \leq \tau \cdot B$.

It is evident that the bound proven above is not tight. Indeed a sufficiently large τ is necessary to make the distribution of z statistically independent of x and hence provide (honest-verifier) zero-knowledge. Also, we can extract the witness x (and get special soundness) from two correct transcripts $(a, e, z), (a, e+1, z')$ that a cheating prover can provide, by $f(z - z') = y$, so that $\|z - z'\| \leq 2 \cdot \tau \cdot B$. The term $2 \cdot \tau \cdot B$ is known as *soundness slack* and quantifies the difference between the bound used by an honest prover and what we can force a cheating prover to do.

In short, this approach has two main drawbacks. Firstly, it needs to be repeated many times to reach a sufficiently small soundness. Secondly, a large soundness slack implies in SPDZ larger parameters in the underlying BGV cryptosystem, with consequences in terms of computation and also communication as these ciphertexts need to be sent to all parties in the protocol. As described in [6], the slack can be removed by a modulus switch operation after the ZK proof is executed. Loosely speaking, a modulus switching operation is a noise management technique, introduced by Brakerski et al. [21], that transforms a ciphertext over a certain modulo into a ciphertext defined over a smaller modulo.

A common solution to the first issue is to use standard amortized techniques [26], and prove several statements at once. Even if on one hand the amortization reduces the soundness from $1/2$ to 2^{-t}, where t is the number of instances we are proving, on the other hand it introduces even more slack.

Different alternatives to this general approach have been proposed.

In [28] a cut-and-choose based check is described to replace the zero-knowledge proofs. This method needs a large number of additional ciphertexts and it seems to require too much memory to be practical.

In [51], Keller, Pastro and Rotaru revisited the original SPDZ ZK proofs by noting that in the pre-processing it is not required that each ciphertext $\mathfrak{ct}_{x_i}$ was correctly generated, but rather than the sum of those is "correct". This is because only this sum is going to be used in the distributed decryption, and not the single shares. In this way it is possible to replace the per-party proof with a global proof, improving the overall computational complexity, as each party needs to check only a single proof instead of $n - 1$, but not the overall communication.

In a recent work [6], Baum, Cozzo and Smart improve the soundness of the global proof introduced in [51]. This work implies a reduction in the amount of amortization required to achieve the desired soundness and also smaller slack. With this technique it is only possible to prove the validity of ciphertexts $2 \cdot \mathfrak{ct}_x$, and not of $\mathfrak{ct}_x$, but in SPDZ this can be mitigated by slightly modifying some of the shares in the MPC protocol.

SPDZ Sacrifice. To ensure triples correctness essentially all SPDZ-style protocols use a standard sacrifice technique that checks a pair of triples such that one can be then used securely. While the original SPDZ protocol used two independent random triples $([a]_S, [b]_S, [c]_S)$ and $([a']_S, [b']_S, [c']_S)$, checking one against the other, in MASCOT it was noticed that the check also works with "correlated" triples $([a]_S, [b]_S, [c]_S)$ and $([a']_S, [b]_S, [c']_S)$, i.e. with the same b (or equivalently same a). In this way we have a cheaper check requiring less authenticated randomness and also less opening, and hence less communication. It proceeds as follows.

Sacrifice: Given two correlated triples $([a]_S, [b]_S, [c]_S)$ and $([a']_S, [b]_S, [c']_S)$:
1. Parties call the ideal functionality $\mathcal{F}_{\mathsf{Rand}}$ to obtain a random $r \in \mathbb{F}$
2. Parties open the value $\rho = r \cdot [a] - [a']$
3. Parties compute $r \cdot [c] - [c'] - \rho \cdot [b]$, and check whether it is equal to zero. If not, the protocol outputs $\bot$ and aborts.

If the triples are correct:

$$r \cdot c - c' - \rho \cdot b = r \cdot (a \cdot b) - (a' \cdot b) - b \cdot (r \cdot a - a') = 0.$$

If the triples are incorrect, that is $(a, b, c + \delta)$ and $(a', b, c' + \delta')$, where δ, δ' are chosen by the adversary:

$$r \cdot (c + \delta) - (c' + \delta') - \rho \cdot b =$$
$$r \cdot (a \cdot b + \delta) - (a' \cdot b + \delta') - b \cdot (r \cdot a - a') = r \cdot \delta - \delta',$$

which is zero with probability $1/|\mathbb{F}|$.

Putting Everything Together. We can finally show the SPDZ offline protocol. We make the following assumptions:

- A global zero-knowledge protocol Π_{gZKPoK} as we have described above (for details we refer to [6,51,61]).
- A key generation functionality $\mathcal{F}_{\mathrm{KEYGEN}}$ that distributes $(\mathsf{pk}, \langle \mathsf{sk} \rangle)$ among the parties (in SPDZ instantiation these keys are BGV encryption and decryption keys).
- A distributed decryption functionality $\mathcal{F}_{\mathrm{DISTDEC}}$ that, given a correctly generated ciphertext, output a sharing of the decryption output.

When we implement $\mathcal{F}_{\mathrm{DISTDEC}}$ in SPDZ using BGV, we provide this ideal functionality of two commands **DDM** and **DDT**, that we describe below.

Distributed decryption MACs (DDM): It takes as input a valid ciphertext $\mathfrak{ct}_m$ and the BGV keys $(\mathsf{pk}, \mathsf{sk})$.
1. Decrypt $\mathfrak{ct}_m$ and send the output of the decryption m to the adversary.
2. Wait for an input from the adversary. If receive Abort, send Abort to the parties and halt, otherwise on receiving $m' = m + \delta$, send $\langle m' \rangle$ to the parties.

Distributed decryption triples (DDT): It takes as input a valid ciphertext $\mathfrak{ct}_m$ and the BGV keys $(\mathsf{pk}, \mathsf{sk})$.
1. Do as **DDM** in steps 1. and 2.
2. Compute a fresh encryption $\mathfrak{ct}_{m'}$ of m' and send it to the parties.

DDT is essentially the Reshare protocol given in [28, 30]. When we implement this functionality with BGV, the protocol requires a masking ciphertext, and hence a ZK proof, that is used in the distributed decryption. Other than the decryption sharing $\langle m' \rangle$, it also produces a fresh encryption $\mathsf{ct}_{m'}$ of the output of the decryption. On the other hand, in the protocol implementing DDM [51], a large "plaintext" mask is introduced directly in the decryption procedure, and there is no need of ZK proof for this mask. Therefore, this latter protocol is cheaper than DDT, but it can only be used if the result of the decryption does not need to be re-encrypted. In particular, it can only be used for MACs generations and not for generating c and m_c, and for this reason it only outputs decryption shares and not a fresh encryption of it as DDT does. Finally, note that both of the commands allow the corrupt parties to add some error to the outputs. This does not break the security of the protocol as these errors will be detected by MAC Check failures.

Pre-processing protocol. Parties receive pk from $\mathcal{F}_{\text{KeyGen}}$.

   Initialize. Parties create a ciphertext ct_Δ encrypting the MAC key Δ:
   1. Each party P_i samples a random $\Delta^{(i)}$. Set $\Delta = \sum_i \Delta^{(i)}$
   2. Each $P_i, i \in [n]$, computes and broadcasts ct_{Δ^i}
   3. Parties run Π_{gZKPoK} to check that ct_Δ is valid

   Input. On input (Input, P_i) from all parties:
   1. P_i samples $r \leftarrow \mathbb{F}$, creates $\langle r \rangle$ and sends $r^{(j)}$ to $P_j, j \neq i$
   2. Each party P_i creates ct_{r^i} and broadcasts this value
   3. Parties run Π_{gZKPoK} and compute $\mathsf{ct}_{r \cdot \Delta}$
   4. Parties call $\mathcal{F}_{\text{DistDec}}$ on command **DDM** receiving $\langle m_r \rangle$
   5. Parties run MAC Check, if it fails, the protocol Abort

   Triples. On input (Triple) from all parties:
   1. Each P_i samples random shares $a^{(i)}, b^{(i)} \leftarrow \mathbb{F}$, computes $\mathsf{ct}_{a^i}, \mathsf{ct}_{b^i}$ and broadcasts these values
   2. Parties run Π_{gZKPoK} to check the validity of ct_a and ct_b
   3. Parties compute $\mathsf{ct}_a \cdot \mathsf{ct}_b = \mathsf{ct}_c$
   4. Parties call $\mathcal{F}_{\text{DistDec}}$ on command **DDT** obtaining $\langle c \rangle$ and a fresh encryption of c, $\tilde{\mathsf{ct}}_c$
   5. Parties obtain $\langle m_a \rangle, \langle m_b \rangle, \langle m_c \rangle$ calling $\mathcal{F}_{\text{DistDec}}$ on command **DDM** on inputs $\mathsf{ct}_a, \mathsf{ct}_b, \tilde{\mathsf{ct}}_c$.
   6. Parties repeat steps 2–5 with value a', obtaining $\langle c' \rangle$, such that $c' = a' \cdot b$ and $\langle m_{a'} \rangle, \langle m_{c'} \rangle$.
   7. Parties run the **Sacrifice** check on input $\big((a, b, c), (a', b, c')\big)$. If the check fails, the protocol Abort
   8. Parties run MAC Check, if it fails, the protocol Abort

5.2   Pre-processing Using Oblivious Transfer

Here we describe how to generate random authenticated values and triples using oblivious transfer instead of homomorphic encryption. In order to do this we need some more notation.

We define the 'gadget' vector $\mathbf{g}$ consisting of the powers of two (in $\mathbb{F}_p$) or powers of X (in extension fields $\mathbb{F}_{p^k}$), so that

$$\mathbf{g} = (1, g, g^2, \ldots, g^{k-1}) \in \mathbb{F}^k,$$

where, as said before, $g = 2$ in $\mathbb{F}_p$ and $g = X$ in $\mathbb{F}_{p^k}$. Let $\mathbf{g}^{-1} : \mathbb{F} \to \{0,1\}^k$ be the 'bit decomposition' function that maps $x \in \mathbb{F}$ to a bit vector $\mathbf{x}_B = \mathbf{g}^{-1}(x) \in \{0,1\}^k$, such that $\mathbf{x}_B$ can be mapped back to $\mathbb{F}$ by taking the inner product $\langle \mathbf{g}, \mathbf{g}^{-1}(x) \rangle = x$. This tool permits to switch between field elements and vectors of bits whilst remaining independent of the underlying finite field.

Passively-Secure Multiplication Using OT. We are now ready to show how to use OT to produce a secret sharing of an arithmetic product. In a standard one-out-of-two OT, the sender inputs two messages $x_0, x_1 \in \mathbb{F}$, and the receiver inputs a bit b, receiving $x_b = x_0 + b \cdot (x_1 - x_0)$. Setting $a = x_1 - x_0$, we obtain

$$x_b - x_0 = b \cdot a,$$

where $x_b, x_0, a \in \mathbb{F}$ and $b \in \{0,1\}$. The value a is called *correlation*, and the corresponding OT functionality, *correlated OT* (Fig. 3).

Functionality $\mathcal{F}_{\mathsf{COT}}$

Running between a sender P_S and a receiver P_R, it operates as follows.

- P_S inputs $(x_0, x_0 + a) \in \mathbb{F} \times \mathbb{F}$ and P_R inputs b.
- The functionality outputs $x_b = x_0 + b \cdot a$ to P_R.

Fig. 3. Functionality for one-out-of-two oblivious transfers on k-bit strings.

We can then combine k correlated OTs into one arithmetic OT, as follows. Parties P_S and P_R input $(x_i, x_i + a)$, for some fixed correlation $a \in \mathbb{F}$, and $(b_1, \ldots, b_k)$, such that $(b_1, \ldots, b_k) = \mathbf{g}^{-1}(b)$, $b \in \mathbb{F}$, respectively. The receiver then obtains $y_i = x_i + b_i \cdot a$, $i \in [k]$. By setting $q = \langle \mathbf{g}, \mathbf{y} \rangle$, with $\mathbf{y} = (y_1, \ldots, y_k)$ and $t = \langle \mathbf{g}, \mathbf{x} \rangle$, with $\mathbf{x} = (x_1, \ldots, x_k)$, we obtain $q = t + b \cdot a$, where the sender holds $q, a \in \mathbb{F}$ and the receiver holds $t, b \in \mathbb{F}$. We have thus transformed oblivious transfer into a secret sharing of the product of both parties' inputs in $\mathbb{F}$.[4] Using this building block, constructing a passively secure protocol for secret-shared multiplication triples is straightforward by simply running the protocol between every pair of parties and summing the shares.

---

[4] This generalisation of oblivious transfer is also referred to as *oblivious linear function evaluation* (OLE) [57].

Efficient Authentication Using Correlated OT. As for the case of homomorphic encryption, also in oblivious transfer based pre-processing protocols we can use the same approach to create triples and MACs, because the relation between authenticated values and MAC keys is the same as the multiplication triple relation. The main difference is that in an authentication procedure, the global MAC key is fixed, so while in triples generation we need to use a fresh correlation for each triple, the correlation remains the same for all the values we need to authenticate. More precisely, the MAC generation for an additively secret shared value $x \in \mathbb{F}$ proceeds as follows.

1. Each party P_i samples a random share $\Delta^{(i)}$ of the global MAC key
2. Each pair of parties, (P_i, P_j), run k $\mathcal{F}_{\mathsf{COT}}$ on input $x^{(i)}$, $\Delta^{(j)}$, respectively, obtaining
$$q^{(j,i)} = t^{(i,j)} + x^{(i)} \cdot \Delta^{(j)}.$$

3. After all the $n(n-1)$ executions, each party P_i locally combines their results to generate the MAC share

$$m^{(i)} = x^{(i)} \cdot \Delta^{(i)} + \sum_{j \neq i} \left(q^{(i,j)} - t^{(i,j)} \right).$$

Essentially, using $\mathcal{F}_{\mathsf{COT}}$, i.e. a 2-party functionality, we naturally obtain a BDOZ-style authentication that can be locally converted to SPDZ-style MACs as we described in Sect. 3.

OT-Based Pre-processing with Active Security. It is clear, from previous description of the authentication and triple generation protocols, that an adversary could easily cheat, for example by inputting inconsistent values in one of the $\mathcal{F}_{\mathsf{COT}}$ instances, or using different MAC key shares with different parties.

Here we discuss separately how to achieve active security of the MACs generation and triples generation protocols.

For the MAC generation, it turns out that the passively secure protocol is almost enough. This is because during the authentication, the correlation Δ is fixed at the beginning, so the adversary does not have much possibility to deviate from protocol instructions later on. However, even after the correlation has been fixed, the adversary is still able to create wrong MACs which contain errors depending on the global key. It was proved in [50], that to obtain active security it is enough to run a MAC Check opening a random linear combination of authentication values just after their generation. This somehow fixes the global key and ensures correctness of subsequent checks. Note that during the MAC checks an adversary is still able to pass the check even in the presence of some errors by guessing some bits of Δ, however if the guess is incorrect the protocol aborts. So the only thing we have to make sure is that the global MAC key still has sufficient entropy to prevent cheating in MAC checks, even if a few bits have been guessed.

For triple generation achieving active security is more involved, since we do not have a fixed correlation, and hence a linear combination on which running a check. Note that we need to ensure both correctness and privacy of the triples. Correctness is easily verified with a pairwise, standard sacrifice technique. This check, however, raises the possibility of selective failure attacks, so that if for example the adversary cheats in just a single bit, and the check passes, then this bit of the triple is leaked to the adversary. To prevent this, a simple variant of privacy amplification is used. First we generate several leaky triples, from which a single, random triple is extracted by taking random combinations [50].

SHE vs OT - Comparison. Here we compare the efficiency of OT-based and HE-based pre-processing, reporting the figures provided by [51]. The values in Table 1 confirm the complementarity of these two approaches, even if some recent improvements in OT-extension protocols could greatly improve the efficiency of protocols relying on oblivious transfer. We can see that MASCOT is more efficient over binary extension fields and LowGear over prime fields of odd characteristics. In the multiparty case, essentially when the number of parties is larger than ~ 7, HighGear will become more efficient than LowGear [51].

Table 1. Triple generation for prime and binary fields with two-party and 64 bits of statistical security [51].

Protocol	Triples/sec	Network	Field
MASCOT	5100	1 Gbit/s	Prime field $\log_2 |\mathbb{F}| = 128$
	214	50 Mbit/s	Prime field $\log_2 |\mathbb{F}| = 128$
	5100	1 Gbit/s	Binary field $\mathbb{F}_{2^{128}}$
LowGear	**30000**	1 Gbit/s	Prime field $\log_2 |\mathbb{F}| = 128$
	3200	50 Mbit/s	Prime field $\log_2 |\mathbb{F}| = 128$
	117	1 Gbit/s	Binary field $\mathbb{F}_{2^{128}}$
HighGear	5600	1 Gbit/s	Prime field $\log_2 |\mathbb{F}| = 128$
	1300	50 Mbit/s	Prime field $\log_2 |\mathbb{F}| = 128$
	67	1 Gbit/s	Binary field $\mathbb{F}_{2^{128}}$

Pre-processing with OLE. We can naturally instantiate the pre-processing with OLE (Oblivious Linear-function Evaluation) instead of OT. As we said previously, OLE is an arithmetic generalization of OT to larger fields. More formally, it is a two-party functionality where the sender P_S inputs two values $a, b \in \mathbb{F}$ and the receiver P_R inputs a value $x \in \mathbb{F}$ obtaining $y = x + a \cdot b$. OLE can be constructed from several assumptions and public-key based constructions, like OT (as seen before), homomorphic encryption, noisy encodings [36,46], etc. An efficient arithmetic implementation of OLE can potentially lead to a very efficient pre-processing phase, as it will avoid running $\mathcal{F}_{\mathsf{COT}}$ for each bit of the binary representation of the values involved in the computation. A two-party protocol based on OLE is described by Döttling et al. [32], that can be considered as a natural generalization of TinyOT to the arithmetic setting, however this work does not give an implementation of the protocols described, so the actual efficiency of this approach is not completely clear.

5.3   Silent Pre-processing via PCG

In a recent line of work Boyle et al. [19,20] show how to generate correlated randomness that can be used as pre-processsd material in MPC protocols using *pseudorandom correlation generators* (PCGs). A PCG is a deterministic function that allows to extend short seeds to long instances of a desired correlation, i.e. OT, OLE, triples etc. Using a PCG we can have a so-called "silent pre-processing". After a setup that consists of generation and distribution of the seeds, the expansion is local, i.e. does not require communication, and hence the term "silent".

This is a very promising approach as it allows to reduce significatively the communication and memory usage, even if it still require, in some useful case like generation of authenticated triples, a quite expensive setup.

Acknowledgements. I would like to thank the organizers of WAIFI 2020 for inviting me to give a talk there. I am also grateful to Axel Mertens and Nigel Smart for helpful comments. This work has been supported in part by ERC Advanced Grant ERC-2015-AdG-IMPaCT and by the FWO under an Odysseus project GOH9718N.

References

1. Aly, A., et al.: Scale - mamba v1.9: documentation
2. Asharov, G., Jain, A., López-Alt, A., Tromer, E., Vaikuntanathan, V., Wichs, D.: Multiparty computation with low communication, computation and interaction via threshold FHE. In: Pointcheval, D., Johansson, T. (eds.) EUROCRYPT 2012. LNCS, vol. 7237, pp. 483–501. Springer, Heidelberg (2012). https://doi.org/10.1007/978-3-642-29011-4_29
3. Asharov, G., Lindell, Y., Schneider, T., Zohner, M.: More efficient oblivious transfer and extensions for faster secure computation. In: Sadeghi, A.R., Gligor, V.D., Yung, M. (eds.) ACM CCS 2013, pp. 535–548. ACM Press, November 2013. https://doi.org/10.1145/2508859.2516738
4. Asharov, G., Lindell, Y., Schneider, T., Zohner, M.: More efficient oblivious transfer extensions with security for malicious adversaries. In: Oswald, E., Fischlin, M. (eds.) EUROCRYPT 2015. LNCS, vol. 9056, pp. 673–701. Springer, Heidelberg (2015). https://doi.org/10.1007/978-3-662-46800-5_26
5. Baum, C., Bootle, J., Cerulli, A., del Pino, R., Groth, J., Lyubashevsky, V.: Sublinear lattice-based zero-knowledge arguments for arithmetic circuits. In: Shacham, H., Boldyreva, A. (eds.) CRYPTO 2018. LNCS, vol. 10992, pp. 669–699. Springer, Cham (2018). https://doi.org/10.1007/978-3-319-96881-0_23
6. Baum, C., Cozzo, D., Smart, N.P.: Using TopGear in overdrive: a more efficient ZKPoK for SPDZ. In: Paterson, K.G., Stebila, D. (eds.) SAC 2019. LNCS, vol. 11959, pp. 274–302. Springer, Cham (2020). https://doi.org/10.1007/978-3-030-38471-5_12

7. Baum, C., Damgård, I., Larsen, K.G., Nielsen, M.: How to prove knowledge of small secrets. In: Robshaw, M., Katz, J. (eds.) CRYPTO 2016. LNCS, vol. 9816, pp. 478–498. Springer, Heidelberg (2016). https://doi.org/10.1007/978-3-662-53015-3_17

8. Baum, C., Damgård, I., Toft, T., Zakarias, R.: Better preprocessing for secure multiparty computation. In: Manulis, M., Sadeghi, A.-R., Schneider, S. (eds.) ACNS 2016. LNCS, vol. 9696, pp. 327–345. Springer, Cham (2016). https://doi.org/10.1007/978-3-319-39555-5_18

9. Baum, C., Orsini, E., Scholl, P.: Efficient secure multiparty computation with identifiable abort. In: Hirt, M., Smith, A. (eds.) TCC 2016. LNCS, vol. 9985, pp. 461–490. Springer, Heidelberg (2016). https://doi.org/10.1007/978-3-662-53641-4_18

10. Baum, C., Orsini, E., Scholl, P., Soria-Vazquez, E.: Efficient constant-round MPC with identifiable abort and public verifiability. In: Micciancio, D., Ristenpart, T. (eds.) CRYPTO 2020. LNCS, vol. 12171, pp. 562–592. Springer, Cham (2020). https://doi.org/10.1007/978-3-030-56880-1_20

11. Beaver, D.: Efficient multiparty protocols using circuit randomization. In: Feigenbaum, J. (ed.) CRYPTO 1991. LNCS, vol. 576, pp. 420–432. Springer, Heidelberg (1992). https://doi.org/10.1007/3-540-46766-1_34

12. Beaver, D.: Precomputing oblivious transfer. In: Coppersmith, D. (ed.) CRYPTO 1995. LNCS, vol. 963, pp. 97–109. Springer, Heidelberg (1995). https://doi.org/10.1007/3-540-44750-4_8

13. Beaver, D.: Correlated pseudorandomness and the complexity of private computations. In: 28th ACM STOC, pp. 479–488. ACM Press, May 1996. https://doi.org/10.1145/237814.237996

14. Beaver, D., Micali, S., Rogaway, P.: The round complexity of secure protocols (extended abstract). In: 22nd ACM STOC, pp. 503–513. ACM Press, May 1990. https://doi.org/10.1145/100216.100287

15. Ben-Or, M., Goldwasser, S., Wigderson, A.: Completeness theorems for non-cryptographic fault-tolerant distributed computation (extended abstract). In: 20th ACM STOC, pp. 1–10. ACM Press, May 1988. https://doi.org/10.1145/62212.62213

16. Bendlin, R., Damgård, I., Orlandi, C., Zakarias, S.: Semi-homomorphic encryption and multiparty computation. In: Paterson, K.G. (ed.) EUROCRYPT 2011. LNCS, vol. 6632, pp. 169–188. Springer, Heidelberg (2011). https://doi.org/10.1007/978-3-642-20465-4_11

17. Boneh, D., et al.: Threshold cryptosystems from threshold fully homomorphic encryption. In: Shacham, H., Boldyreva, A. (eds.) CRYPTO 2018. LNCS, vol. 10991, pp. 565–596. Springer, Cham (2018). https://doi.org/10.1007/978-3-319-96884-1_19

18. Boyle, E., et al.: Efficient two-round OT extension and silent non-interactive secure computation. In: Cavallaro, L., Kinder, J., Wang, X., Katz, J. (eds.) ACM CCS 2019, pp. 291–308. ACM Press, November 2019. https://doi.org/10.1145/3319535.3354255

19. Boyle, E., Couteau, G., Gilboa, N., Ishai, Y., Kohl, L., Scholl, P.: Efficient pseudorandom correlation generators: silent OT extension and more. In: Boldyreva, A., Micciancio, D. (eds.) CRYPTO 2019. LNCS, vol. 11694, pp. 489–518. Springer, Cham (2019). https://doi.org/10.1007/978-3-030-26954-8_16

20. Boyle, E., Couteau, G., Gilboa, N., Ishai, Y., Kohl, L., Scholl, P.: Efficient pseudorandom correlation generators from ring-LPN. In: Micciancio, D., Ristenpart, T. (eds.) CRYPTO 2020. LNCS, vol. 12171, pp. 387–416. Springer, Cham (2020). https://doi.org/10.1007/978-3-030-56880-1_14

21. Brakerski, Z., Gentry, C., Vaikuntanathan, V.: (Leveled) fully homomorphic encryption without bootstrapping. In: Goldwasser, S. (ed.) ITCS 2012, pp. 309–325. ACM, January 2012. https://doi.org/10.1145/2090236.2090262
22. Canetti, R.: Universally composable security: a new paradigm for cryptographic protocols. In: 42nd FOCS, pp. 136–145. IEEE Computer Society Press, October 2001. https://doi.org/10.1109/SFCS.2001.959888
23. Chaum, D., Crépeau, C., Damgård, I.: Multiparty unconditionally secure protocols (extended abstract). In: 20th ACM STOC, pp. 11–19. ACM Press, May 1988. https://doi.org/10.1145/62212.62214
24. Cleve, R.: Limits on the security of coin flips when half the processors are faulty (extended abstract). In: 18th ACM STOC, pp. 364–369. ACM Press, May 1986. https://doi.org/10.1145/12130.12168
25. Cohen, R., Lindell, Y.: Fairness versus guaranteed output delivery in secure multiparty computation. In: Sarkar, P., Iwata, T. (eds.) ASIACRYPT 2014. LNCS, vol. 8874, pp. 466–485. Springer, Heidelberg (2014). https://doi.org/10.1007/978-3-662-45608-8_25
26. Cramer, R., Damgård, I.: On the amortized complexity of zero-knowledge protocols. In: Halevi, S. (ed.) CRYPTO 2009. LNCS, vol. 5677, pp. 177–191. Springer, Heidelberg (2009). https://doi.org/10.1007/978-3-642-03356-8_11
27. Cunningham, R., Fuller, B., Yakoubov, S.: Catching MPC cheaters: identification and openability. In: Shikata, J. (ed.) ICITS 2017. LNCS, vol. 10681, pp. 110–134. Springer, Cham (2017). https://doi.org/10.1007/978-3-319-72089-0_7
28. Damgård, I., Keller, M., Larraia, E., Pastro, V., Scholl, P., Smart, N.P.: Practical covertly secure MPC for dishonest majority – or: breaking the SPDZ limits. In: Crampton, J., Jajodia, S., Mayes, K. (eds.) ESORICS 2013. LNCS, vol. 8134, pp. 1–18. Springer, Heidelberg (2013). https://doi.org/10.1007/978-3-642-40203-6_1
29. Damgård, I., Orlandi, C.: Multiparty computation for dishonest majority: from passive to active security at low cost. In: Rabin, T. (ed.) CRYPTO 2010. LNCS, vol. 6223, pp. 558–576. Springer, Heidelberg (2010). https://doi.org/10.1007/978-3-642-14623-7_30
30. Damgård, I., Pastro, V., Smart, N., Zakarias, S.: Multiparty computation from somewhat homomorphic encryption. In: Safavi-Naini, R., Canetti, R. (eds.) CRYPTO 2012. LNCS, vol. 7417, pp. 643–662. Springer, Heidelberg (2012). https://doi.org/10.1007/978-3-642-32009-5_38
31. del Pino, R., Lyubashevsky, V.: Amortization with fewer equations for proving knowledge of small secrets. In: Katz, J., Shacham, H. (eds.) CRYPTO 2017. LNCS, vol. 10403, pp. 365–394. Springer, Cham (2017). https://doi.org/10.1007/978-3-319-63697-9_13
32. Döttling, N., Ghosh, S., Nielsen, J.B., Nilges, T., Trifiletti, R.: TinyOLE: efficient actively secure two-party computation from oblivious linear function evaluation. In: Thuraisingham, B.M., Evans, D., Malkin, T., Xu, D. (eds.) ACM CCS 2017, pp. 2263–2276. ACM Press, October/November 2017. https://doi.org/10.1145/3133956.3134024
33. Even, S., Goldreich, O., Lempel, A.: A randomized protocol for signing contracts. Commun. ACM **28**(6), 637–647 (1985)
34. Genkin, D., Ishai, Y., Prabhakaran, M., Sahai, A., Tromer, E.: Circuits resilient to additive attacks with applications to secure computation. In: Shmoys, D.B. (ed.) 46th ACM STOC, pp. 495–504. ACM Press, May/Jun 2014. https://doi.org/10.1145/2591796.2591861

35. Genkin, D., Ishai, Y., Weiss, M.: Binary AMD circuits from secure multiparty computation. In: Hirt, M., Smith, A. (eds.) TCC 2016. LNCS, vol. 9985, pp. 336–366. Springer, Heidelberg (2016). https://doi.org/10.1007/978-3-662-53641-4_14

36. Ghosh, S., Nielsen, J.B., Nilges, T.: Maliciously secure oblivious linear function evaluation with constant overhead. In: Takagi, T., Peyrin, T. (eds.) ASIACRYPT 2017. LNCS, vol. 10624, pp. 629–659. Springer, Cham (2017). https://doi.org/10.1007/978-3-319-70694-8_22

37. Goldreich, O., Micali, S., Wigderson, A.: How to play any mental game or A completeness theorem for protocols with honest majority. In: Aho, A. (ed.) 19th ACM STOC, pp. 218–229. ACM Press, May 1987. https://doi.org/10.1145/28395.28420

38. Goldwasser, S., Lindell, Y.: Secure multi-party computation without agreement. J. Cryptol. **18**(3), 247–287 (2005). https://doi.org/10.1007/s00145-005-0319-z

39. Goldwasser, S., Micali, S., Rackoff, C.: The knowledge complexity of interactive proof systems. SIAM J. Comput. **18**(1), 186–208 (1989)

40. Hastings, M., Hemenway, B., Noble, D., Zdancewic, S.: SoK: general purpose compilers for secure multi-party computation. In: 2019 IEEE Symposium on Security and Privacy, pp. 1220–1237. IEEE Computer Society Press, May 2019. https://doi.org/10.1109/SP.2019.00028

41. Hazay, C., Venkitasubramaniam, M., Weiss, M.: The price of active security in cryptographic protocols. In: Canteaut, A., Ishai, Y. (eds.) EUROCRYPT 2020. LNCS, vol. 12106, pp. 184–215. Springer, Cham (2020). https://doi.org/10.1007/978-3-030-45724-2_7

42. Impagliazzo, R., Rudich, S.: Limits on the provable consequences of one-way permutations. In: 21st ACM STOC, pp. 44–61. ACM Press, May 1989. https://doi.org/10.1145/73007.73012

43. Ishai, Y., Kilian, J., Nissim, K., Petrank, E.: Extending oblivious transfers efficiently. In: Boneh, D. (ed.) CRYPTO 2003. LNCS, vol. 2729, pp. 145–161. Springer, Heidelberg (2003). https://doi.org/10.1007/978-3-540-45146-4_9

44. Ishai, Y., Ostrovsky, R., Zikas, V.: Secure multi-party computation with identifiable abort. In: Garay, J.A., Gennaro, R. (eds.) CRYPTO 2014. LNCS, vol. 8617, pp. 369–386. Springer, Heidelberg (2014). https://doi.org/10.1007/978-3-662-44381-1_21

45. Ishai, Y., Prabhakaran, M., Sahai, A.: Founding cryptography on oblivious transfer – efficiently. In: Wagner, D. (ed.) CRYPTO 2008. LNCS, vol. 5157, pp. 572–591. Springer, Heidelberg (2008). https://doi.org/10.1007/978-3-540-85174-5_32

46. Ishai, Y., Prabhakaran, M., Sahai, A.: Secure arithmetic computation with no honest majority. In: Reingold, O. (ed.) TCC 2009. LNCS, vol. 5444, pp. 294–314. Springer, Heidelberg (2009). https://doi.org/10.1007/978-3-642-00457-5_18

47. Keller, M.: MP-SPDZ: a versatile framework for multi-party computation. IACR Cryptology ePrint Archive **2020**, 521 (2020)

48. Keller, M., Orsini, E., Rotaru, D., Scholl, P., Soria-Vazquez, E., Vivek, S.: Faster secure multi-party computation of AES and DES using lookup tables. In: Gollmann, D., Miyaji, A., Kikuchi, H. (eds.) ACNS 2017. LNCS, vol. 10355, pp. 229–249. Springer, Cham (2017). https://doi.org/10.1007/978-3-319-61204-1_12

49. Keller, M., Orsini, E., Scholl, P.: Actively secure OT extension with optimal overhead. In: Gennaro, R., Robshaw, M. (eds.) CRYPTO 2015. LNCS, vol. 9215, pp. 724–741. Springer, Heidelberg (2015). https://doi.org/10.1007/978-3-662-47989-6_35

50. Keller, M., Orsini, E., Scholl, P.: MASCOT: faster malicious arithmetic secure computation with oblivious transfer. In: Weippl, E.R., Katzenbeisser, S., Kruegel, C., Myers, A.C., Halevi, S. (eds.) ACM CCS 2016, pp. 830–842. ACM Press, October 2016. https://doi.org/10.1145/2976749.2978357

51. Keller, M., Pastro, V., Rotaru, D.: Overdrive: making SPDZ great again. In: Nielsen, J.B., Rijmen, V. (eds.) EUROCRYPT 2018. LNCS, vol. 10822, pp. 158–189. Springer, Cham (2018). https://doi.org/10.1007/978-3-319-78372-7_6

52. Kilian, J.: Founding cryptography on oblivious transfer. In: 20th ACM STOC, pp. 20–31. ACM Press, May 1988. https://doi.org/10.1145/62212.62215

53. Kolesnikov, V., Kumaresan, R.: Improved OT extension for transferring short secrets. In: Canetti, R., Garay, J.A. (eds.) CRYPTO 2013. LNCS, vol. 8043, pp. 54–70. Springer, Heidelberg (2013). https://doi.org/10.1007/978-3-642-40084-1_4

54. Larraia, E., Orsini, E., Smart, N.P.: Dishonest majority multi-party computation for binary circuits. In: Garay, J.A., Gennaro, R. (eds.) CRYPTO 2014. LNCS, vol. 8617, pp. 495–512. Springer, Heidelberg (2014). https://doi.org/10.1007/978-3-662-44381-1_28

55. Lindell, Y., Pinkas, B., Oxman, E.: The IPS compiler: optimizations, variants and concrete efficiency. Cryptology ePrint Archive, Report 2011/435 (2011). http://eprint.iacr.org/2011/435

56. Lyubashevsky, V., Peikert, C., Regev, O.: A toolkit for Ring-LWE cryptography. In: Johansson, T., Nguyen, P.Q. (eds.) EUROCRYPT 2013. LNCS, vol. 7881, pp. 35–54. Springer, Heidelberg (2013). https://doi.org/10.1007/978-3-642-38348-9_3

57. Naor, M., Pinkas, B.: Oblivious transfer and polynomial evaluation. In: 31st ACM STOC, pp. 245–254. ACM Press, May 1999. https://doi.org/10.1145/301250.301312

58. Nielsen, J.B., Nordholt, P.S., Orlandi, C., Burra, S.S.: A new approach to practical active-secure two-party computation. In: Safavi-Naini, R., Canetti, R. (eds.) CRYPTO 2012. LNCS, vol. 7417, pp. 681–700. Springer, Heidelberg (2012). https://doi.org/10.1007/978-3-642-32009-5_40

59. Orlandi, C.: Is multiparty computation any good in practice? In: Proceedings of the IEEE International Conference on Acoustics, Speech, and Signal Processing, ICASSP 2011, Prague Congress Center, Prague, Czech Republic, 22–27 May 2011, pp. 5848–5851. IEEE (2011)

60. Orrù, M., Orsini, E., Scholl, P.: Actively secure 1-out-of-N OT extension with application to private set intersection. In: Handschuh, H. (ed.) CT-RSA 2017. LNCS, vol. 10159, pp. 381–396. Springer, Cham (2017). https://doi.org/10.1007/978-3-319-52153-4_22

61. Orsini, E., Smart, N.P., Vercauteren, F.: Overdrive2k: efficient secure MPC over $\mathbb{Z}_{2^k}$ from somewhat homomorphic encryption. In: Jarecki, S. (ed.) CT-RSA 2020. LNCS, vol. 12006, pp. 254–283. Springer, Cham (2020). https://doi.org/10.1007/978-3-030-40186-3_12

62. Pinkas, B., Schneider, T., Smart, N.P., Williams, S.C.: Secure two-party computation is practical. In: Matsui, M. (ed.) ASIACRYPT 2009. LNCS, vol. 5912, pp. 250–267. Springer, Heidelberg (2009). https://doi.org/10.1007/978-3-642-10366-7_15

63. Rabin, M.O.: How to exchange secrets with oblivious transfer (1981)

64. Regev, O.: On lattices, learning with errors, random linear codes, and cryptography. In: Gabow, H.N., Fagin, R. (eds.) 37th ACM STOC, pp. 84–93. ACM Press, May 2005. https://doi.org/10.1145/1060590.1060603

65. Rotaru, D., Smart, N.P., Tanguy, T., Vercauteren, F., Wood, T.: Actively secure setup for SPDZ. Cryptology ePrint Archive, Report 2019/1300 (2019). https://eprint.iacr.org/2019/1300

66. Schoppmann, P., Gascón, A., Reichert, L., Raykova, M.: Distributed vector-OLE: improved constructions and implementation. In: Cavallaro, L., Kinder, J., Wang, X., Katz, J. (eds.) ACM CCS 2019, pp. 1055–1072. ACM Press, November 2019. https://doi.org/10.1145/3319535.3363228

67. Smart, N.P., Vercauteren, F.: Fully homomorphic SIMD operations. Des. Codes Cryptogr. **71**(1), 57–81 (2014)

68. Wang, X., Ranellucci, S., Katz, J.: Authenticated garbling and efficient maliciously secure two-party computation. In: Thuraisingham, B.M., Evans, D., Malkin, T., Xu, D. (eds.) ACM CCS 2017, pp. 21–37. ACM Press, October/November 2017). https://doi.org/10.1145/3133956.3134053

69. Wiesner, S.: Conjugate coding. ACM SIGACT News **15**(1), 78–88 (1983)

70. Yang, K., Weng, C., Lan, X., Zhang, J., Wang, X.: Ferret: fast extension for correlated OT with small communication. IACR Cryptology ePrint Archive **2020**, 924 (2020)

71. Yao, A.C.C.: How to generate and exchange secrets (extended abstract). In: 27th FOCS, pp. 162–167. IEEE Computer Society Press, October 1986. https://doi.org/10.1109/SFCS.1986.25

Finite Field Arithmetic

A HDL Generator for Flexible and Efficient Finite-Field Multipliers on FPGAs

Joël Cathébras[(⊠)] and Roselyne Chotin[iD]

Sorbonne Université, CNRS, LIP6, 75005 Paris, France
joel.cathebras@gmail.com, roselyne.chotin@lip6.fr

Abstract. In this paper we propose a HDL generator for finite-field multipliers on FPGAs. The generated multipliers are based on the CIOS variant of Montgomery multiplication. They are designed to exploit finely the DSPs available on most FPGAs, interleaving independent computations to maximize throughput and DSP's workload. Beside their throughput-efficiency, these operators can dynamically adapt to different finite-fields by changing both operand width and precomputed elements.

From this flexible and efficient operator base, our HDL generator allows the exploration of a wide range of configurations. This is a valuable asset for specialized circuit designers who wish to tune state-of-the-art IPs and explore design space for their applications.

Keywords: Finite-field multiplier · FPGA design · Design space exploration

1 Introduction

In hardware design, when considering specific fields of application, FPGA targets are particularly attractive today and found in many hardware acceleration solutions. A classical step in the development of specialized hardware is the exploration of design space to make architectural choices [1,13]. This exploration may be necessary both at system's level and at IP's level. Exploration tools that allow different IP configurations to be tested are therefore valuable assets for digital circuit designers. This is also true for cryptography which is more and more present in our digital applications.

Modern cryptography is often build upon finite-field arithmetic. As in classical arithmetic, multiplication is an expensive operation and optimizations of multipliers are often the subject of researches and explorations [9,11,12].

In 2017 and 2018 Gallin and Tisserand [5,7] proposed a FPGA implementation of a Finely-Pipelined Modular Multiplier (FPMM) based on the CIOS variant [8] of the Montgomery modular multiplication [10]. It makes fine use of hardware resources present in FPGAs while exploiting in depth the characteristics of the chosen algorithm. Their operator has a good throughput per area ratio

© Springer Nature Switzerland AG 2021
J. C. Bajard and A. Topuzoğlu (Eds.): WAIFI 2020, LNCS 12542, pp. 75–91, 2021.
https://doi.org/10.1007/978-3-030-68869-1_4

compared to the state of the art due to a pipeline interleaving several independent computations. Their approach is quite parametrizable but the developed generator [6] is restricted to the parameters that were suitable for embedded elliptic and hyperelliptic curve cryptography.

This paper presents our work of building up an extended generator for modular multiplier based on the FPMM's approach. Our main scientific contribution consists in the practical generalization of this operator. Another contribution is a new functionality: the ability to dynamically change, to a certain extent, the width of the finite-field elements handled by the operator. When enabled, this feature increases the flexibility of the original operator for dynamically reconfiguring the finite-field over which the multiplier is operating. This feature could be interesting in different contexts. For instance, a crypto-processor that implements several primitives requiring finite-fields of different widths (e.g. RSA, ECC, HECC, etc.). Another example may be the implementation of modulus switching homomorphic encryption schemes (e.g. BGV [2]), which leads to a regular decrease in the width of underlying finite-field arithmetic.

2 Preliminaries

2.1 Notations

Throughout this paper we will use the following notations. An element of a large finite field, as well as the prime number that defines it, are in upper case and bold (e.g. $\mathbf{A} \in \mathbb{Z}_{\mathbf{P}}$). The width in bits of these elements is noted N. The Montgomery constant is M-bit wide and noted as $\mathbf{R}$.

The radix considered for Montgomery multiplication algorithm is 2^{Ω}. The Ω-bit elements are just in upper-case, not bold. Hence, a finite field element is decomposed into s elements of width Ω-bit (i.e. $\mathbf{A} = \{A_0, ..., A_{s-1}\}$).

The width of the basic arithmetic considered in this paper is constrained by the hardware resources available on a FPGA. We note here ω the width of a DSP slice's input words. These "basic words" are written in lower-case, and k denotes the number of them needed to write a Ω-bit word (i.e $A = \{a_0, ..., a_{k-1}\}$).

2.2 DSP Slices

FPGAs are chips made of a grid of configurable basic hardware blocks, along with a configurable interconnection network. Within these basic hardware blocks are elementary resources allowing among other things: combinatorial logic (e.g. Look-Up-Tables), data storage (e.g. Flip-Flops), and clock generation for synchronous circuits [4]. With the growth of size and performance required for circuits to be programmed, more complex hardware blocks have been added to the bestiary (e.g. BRAM, DSP, μP core, ...). In this work we are particularly interested in DSP slices.

A DSP is a basic hardware block that embedded a small multiplier, accumulators and cascading capabilities. They were historically designed for digital

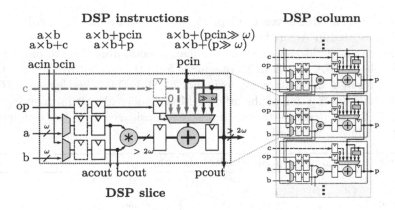

Fig. 1. DSP slice's characteristics used for our finite-fields multipliers. (Color figure online)

signal processing but could be efficiently used for finely-tuned arithmetic. They can achieve interesting running frequencies (up 700 MHz for last FPGAs).

Figure 1 presents the main DSP characteristics that have been exploited for our contributions. DSP are usually grouped in columns of half a dozen to a few hundred. This grouping facilitates cascading of operations and propagation of intermediate results within columns.

A single slice can be configured to perform one or several instructions multiplexed in time. In the latter, an operation code op is used to specify which instruction is issued. In this paper we are interested in three types of instructions : $a \times b$ (red) for a simple ω-bit multiplication, $a \times b + p$ (green) for a ω-bit multiplication followed by an accumulation, and $a \times b + (p \gg \omega)$ (blue) where the accumulated value is previously right-shifted by ω bits. For DSP slices that do not have this right-shift capability (like DSP48A slices in some Xilinx FPGA families), it is still possible to do so with external wires to the DSP. This makes use of the c port designed for $a \times b + c$ instructions (brown). It may require extra cycles in cascading operations to achieve the maximal running frequencies.

3   Previous Works on Finely-Pipelined Modular Multiplier

The main ideas behind the Finely-Pipelined Modular Multiplier (FPMM) are brought by Gallin and Tisserand's works [5,7]. They are introduced in this section but the details of the FPMM design comes in Sect. 4 along with our generalization of this approach.

Latency Optimized CIOS Algorithm Without Final Subtraction. The FPMM operator is based on the Coarsely Integrated Operand Scanning (CIOS) version of Montgomery multiplication [8], without final subtraction [14]. It is designed as a pipelined interleaving several independent computations. The resulting algorithm is presented in Algorithm 1.

As a remainder, Montgomery multiplication computes the product modulo P of two elements A and B in Montgomery form (i.e. scaled by $R = 2^M$), and return the result T in Montgomery form. For FPMM, P is taken less than $R/4$ to avoid the final subtraction of the original Montgomery algorithm (consequently $N \leq M - 2$). In practice, the operands are considered to be M-bit integers decomposed into $s > 1$ words of Ω-bit.

Algorithm 1: Latency optimized CIOS algorithm

Require: $P = \{P_0, ..., P_{s-1}\}$; $P' = -P^{-1} \bmod 2^\Omega$; $4P < R(= 2^M)$.
Input: $A = \{A_0, ..., A_{s-1}\}$; $B = \{B_0, ..., B_{s-1}\}$.
Output: $T = \{T_0, ..., T_{s-1}\}$ with $T = (AB \cdot R^{-1}) \bmod P$ and $0 \leq T < 2P$.

1 **begin**
2   **for** $i = 0$ *to* $s - 1$ **do**
3   **for** $j = 0$ *to* $s - 1$ **do**   /* L1 stage */
4 | $(D, U_j) \leftarrow A_i \times B_j + T_j + D$
5   **end**
6      $Q_i \leftarrow (V_0 \times P') \bmod 2^\Omega$ /* L2 stage: $V_0 = (A_i \times B_0 + T_0) \bmod 2^\Omega$ */
7   **for** $j = 0$ *to* $s - 1$ **do**   /* L3 stage */
8 | $(C, T_{j-1}) \leftarrow Q_i \times P_j + U_j + C$
9   **end**
10     $T_{s-1} \leftarrow T_{-1}^{(n)}$ /* i.e. $C + D$ */
11   **end**
12   **return** $T = \{T_0, ..., T_{s-1}\}$
13 **end**

The particularities of the proposed implementation are visible at line 6 and 10, and comes from the proposed pipeline. For line 6, the original CIOS algorithm computes Q_i from U_0, but also resets D for each new upper-loop's iteration. The equivalent behaviour is achieved with V_0 that is extracted from the L1 stage computation. For line 10, the authors have demonstrated in [5] that the summation of remaining upper-words from L1 and L3 stages (i.e. C and D) is actually propagated in the immediately successive computation. In different terms, the upper word T_{s-1} is actually the lower word in the immediately successive (and independent) computation $T_{-1}^{(n)}$. We do not go into the details of the demonstration and invite the reader to take a look at the original paper.

Interleaving Independent Computations. The latency of an outer-loop iteration is noted α. It is defined from the input of T in L1 stage to its retro-propagation at the end of L3 stage. This latency depends on the choice of Ω w.r.t. the decomposition of multiplications onto DSP slices. For typical applications α is larger than s, leaving room for interleaving $\sigma = \lceil \alpha/s \rceil$ independent computations in the outer-loop's pipeline, while increasing its latency by $l_T = \sigma \cdot s - \alpha$. Thus, increasing by $s \cdot l_T$ cycles a single modular multiplication. We note "slot" the space taken by a single computation in the pipeline.

Figure 2 presents the interleaving principle. When a slot is unused, a new computation can be requested. The storage of A and B in local memories and to start their sub-word's read routine takes l_{in} cycles. Then, each outer-loop

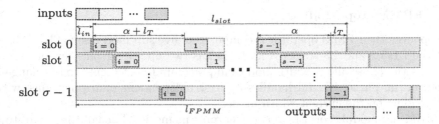

Fig. 2. Interleaving principle of the Finely-Pipelined Modular Multiplier.

iteration takes $\alpha + l_T$ cycles. While waiting for intermediate results, iterations of the other slots are performed.

Motivations for Extended Works. In [5], the authors explored FPMM implementations on different FPGAs from Xilinx. They varied several implementation parameters such as the number of slots σ or the type of memory used (BRAM or LUT based). However, their FPMM generator [6] is restricted to Ω being 2ω, which reduces FPMM's application ranges.

Therefore, our motivations are to generalize the FPMM principles to a wider range of configurations. In particular, we look to aim for larger finite fields, which require to choose $\Omega = k\omega$ with $k \geq 2$.

When studying the FPMM operator, we realized that it should be able to dynamically change the width of handled finite fields. Indeed, once parameter $\Omega = k\omega$ is chosen, FPMM's data-path is mainly fixed. The handled operands' width (M) drives s, σ and l_T parameters, which mainly impact control path. Thus, control path may be somehow duplicated for different width and a mode signal may select the current one.

4   FPMM's Model for HDL Generation

This section presents the generalized FPMM operator. It includes design modifications made to implement the multi-width feature.

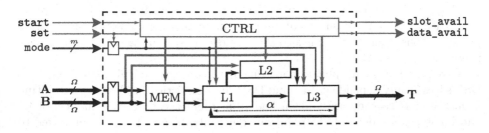

Fig. 3. FPMM top module.

4.1 FPMM Top Module

At top level, the FPMM operator is composed of five sub-modules: CTRL, MEM and the L1, L2 and L3 stages (Fig. 3). It has two operating phases: setup and run. A setup phase allows width mode and precomputed elements to be changed. When in run phase, FPMM handles up to σ independent multiplications, with σ depending on the current width mode.

MEM sub-module consists of two dual-port memories. Each of them can store up to $\max(s \cdot \sigma)$ Ω-bit words. Memory accesses are managed by CTRL while it is orchestrating the computations of the σ slots.

4.2 L1 and L3 Sub-modules

Regarding Algorithm 1, L1 and L3 stages are very close from each others. They both realize, in a i-indexed upper-loop, a j-indexed lower-loop of s iterations performing a multiply-accumulate operation of the form $(H_j, L_j) \leftarrow E_i \times F_j + G_j + H_{j-1}$. H_j and L_j are respectively the upper and lower resulting words, and E_i is constant for a whole lower loop.

Similarities of L1 and L3 Sub-modules. In practice, the multiply-accumulate operation is decomposed in three computation's sub-parts:

$$(H'_j, L'_j) \leftarrow E_i \times F_j \tag{1}$$

$$(c^{(1)}, V_j) \leftarrow L'_j + G_j \tag{2}$$

$$(c^{(2)}, L_j) \leftarrow V_j + H_{j-1} \tag{3}$$

The result L_j is the resulting lower word at the end of all sub-parts. The most significant word H_j is composed of H'_j and two carry bits $c^{(1)}$ and $c^{(2)}$ (i.e. $H_j = H'_j + c^{(1)} + c^{(2)}$).

To illustrate the following discussion on hardware implementation, we rely on Fig. 4. Equation 1 is broken down into ω-bit operations to be mapped onto DSP slices. It results in k^2 DSPs cascaded in space to fully pipeline the multiplication. Multiplexing in time with fewer DSPs would have reduced hardware utilization, but would have increased latency and impacted the whole operator's performances.

Due to this cascading, the sub-results of Eq. 1 are produced with some delay from each others. Each sub-result is used in further computation as soon as possible. Consequently, Eq. 2 is also broken down into k sub-additions with carry propagations (Fig. 4).

Given the Ω-bit multiplication algorithm, the mapping of sub-operations onto DSP slices is only dependent on k and DSP characteristics. All latencies resulting from the DSP cascade are known, and from them the hardware models of L1 and L3 stages are derived. To point out our contributions here: we managed to express hardware's models of each computation sub-parts depending on k and DSP characteristics. The automatic generation of HDL code is then possible from this modelisation.

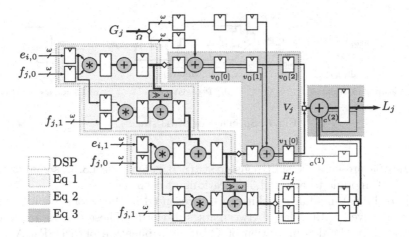

Fig. 4. Multiply-accumulate's pipeline in L1 and L3 sub-modules with $\Omega = 2\omega$.

L1 Stage Specificities. For the L1 stage, $(H_j, L_j, E_i, F_j, G_j)$ are identified with (D, U_j, A_i, B_j, T_j) from Algorithm 1. There are three specificities for L1 stage: handling the T_j's retro-propagated from L3 stage, reset D after a setup phase, and early propagation of V_0 to the L2 stage.

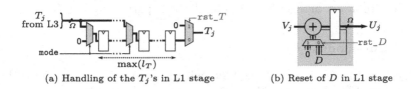

(a) Handling of the T_j's in L1 stage   (b) Reset of D in L1 stage

Fig. 5. Specificities of L1 stage.

Figure 5 illustrates design solutions for the first two specificities. In a multi-width context, different width mode may require different latencies l_T. The `mode` signal selects the shift register's depth according to current one (Fig. 5a). In addition, the T_j's are reset by the CTRL module (`rst_T` signal) whenever a new finite-field multiplication starts.

After a setup phase, the propagations of intermediate results through immediatly successive slots is mixed up. To restart a proper propagation, D is reset for the first slot going through the L1 stage after a setup phase (Fig. 5b).

Finally, the early propagation of V_0 is acheived by picking appropriately its sub-words from their delay lines. More details are given in Sect. 4.3.

L3 Stage Specificities. For the L3 stage, $(H_j, L_j, E_i, F_j, G_j)$ are identified with $(C, T_{j-1}, Q_i, P_j, U_j)$ from Algorithm 1. There are four specificities for the L3 stage. The first three are in the management of operands Q_i, P_j's and U_j's. The fourth one is the handling of outputs.

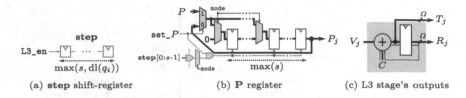

(a) **step** shift-register   (b) **P** register   (c) L3 stage's outputs

Fig. 6. Specificities of L3 stage.

Figure 6a presents the shift register **step**. It is fed by a L3_en signal coming from CTRL module to restart L3 operations at each new slot.

Due to L2 sub-module's architecture (presented in the next section), the Q_i input is received with delays between its sub-words. These delays are known and depend on the decomposition of Ω-bit multiplication onto DSPs. A write enable signal extracted from **step** (not shown) is generated for each sub-word to register them at appropriate time.

Figure 6b presents the storage module for **P**. It is composed of $\max(s)$ registers of Ω-bit that are reprogrammed whenever the prime **P** is changed (**set_P**). The depth used corresponds to the current width mode.

To synchronize the U_j's from L1 with Q_i input, an artificial latency l_U may be required depending on the FPMM configuration (not shown).

Finally, two L3 stage's outputs are differentiated (Fig. 6c): T_j feeds L1 stage for further iteration, and R_j feeds the final result port.

4.3   L2 Sub-module

L2 stage performs $Q_i = V_0 \times P' \mod 2^\Omega$, with V_0 coming from L1, and P' being precomputed. For convenience, V_0 and Q_i are noted V and Q.

Given Ω-bit multiplication's decomposition, result modulo 2^Ω requires only $k(k+1)/2$ sub-multiplications. Moreover, L2 stage has s cycles to reuse the hardware before the next slot's data arrive. Consequently, only $\left\lceil \frac{k(k+1)}{2s} \right\rceil$ DSPs are

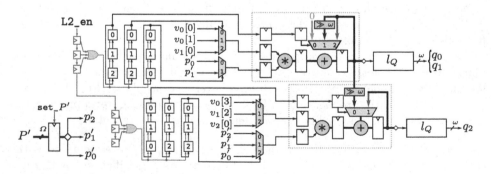

Fig. 7. Example of L2 stage for $\Omega = 3\omega$ and $s = 3$.

instantiated (minimum s in that case of multi-width mode). Figure 7 illustrates L2 stage for $k = 3$ and $s = 3$. The L2_en signal issued by CTRL restarts the sequence of operations each time a new slot arrives. Each DSP is configured with up to s different instructions, depending on the sub-multiplications it is taking care of.

As introduced earlier, V sub-words are extracted from L1 stage's data-path with appropriate delay. For instance, $v_i[d]$ is the d-th register delaying the result of the i-th sub-addition in Eq. 2's data-path (Fig. 4).

Outputs $(q_0, ..., q_{k-1})$ are progressively generated by appropriate DSPs, and stored in L3 stage as seen in L3 stage's specificities. Depending on the configuration, a latency l_Q may be required to synchronise L1 and L2's data-paths.

4.4 CTRL Sub-module

As FPMM's data-path handles slot-wise computations, CTRL module is ryhtmed over s cycles, depending on the current width mode (Fig. 8).

During a setup phase, CTRL updates precomputed values P' and $\mathbf{P}$ stored in L2 and L3 stages. It propagates set_P' (Fig. 7) and set_P (Fig. 6) appropriately (not shown for convenience).

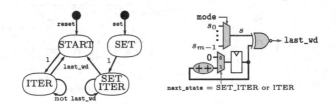

Fig. 8. Control FSM and slot cadencing generation.

During run phase, CTRL handles the succession of slots with the help of a control pipeline presented in Fig. 9. Whenever a last_wd is triggered, a verification is made of whether the next slot is free or already in use. This information is gathered from a shift register **slot_occ** presented latter.

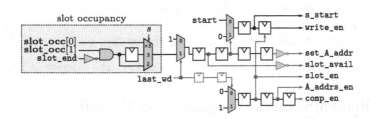

Fig. 9. Control pipeline orchestrated by last_wd signal.

From this, the FPMM signals any free slot (slot_avail) allowing a new computation to be required (start), in that case the storage of computation operands is issued (write_en). The other signals are used to handle slot and address managements.

Figure 10 presents the four shift registers for management of slots and addresses. They are all of depth $\max(\sigma)$, and the portion currently used depends on current width mode.

To the left, the shift register **slot_occ** memories the current slot occupancy. It is paired with **iter_count** that memories the current upper-loop's iteration of each slot (i-indexed). At a slot's last iteration the corresponding elements are reset in both shift registers (slot_end signal).

To the right, **base_addrs** stores for each slot the base address where operands are stored in MEM. It is reset with appropriate precomputed values during a setup phases (not shown here). **A_addrs** stores the address where A_i element is read for each slot current upper-loop's iteration. Read address for the B_j's is incremented at each cycle from the current slot's base address.

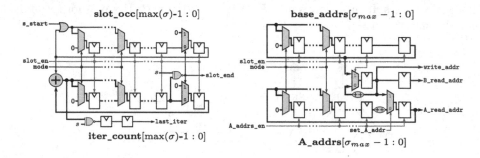

Fig. 10. Control shift registers for slot and address managements.

Finally, the different data path's control signals are delayed from the control signals shown in Fig. 9:

- rst_T: write_en delayed by $l_{MEM} + 4$ cycles.
- rst_D: s_start delayed by $1 + l_{MEM} + l_{L1}$ cycles, only for the first slot after a setup phase, there is no reset otherwise.
- L2_en: comp_en delayed by $l_{MEM} + 4$ cycles.
- L3_en: L2_en delayed by $l_Q + 4$ cycles.
- data_avail: (comp_en & last_iter) delayed by $l_{MEM} + 3 + \alpha$ cycles.

With l_{MEM} being the read latency of MEM, and l_{L1} the latency of L1 stage.

5   Implementation and Exploration Results

Some concrete implementation results are presented in this section. Each generated design has been tested with tens of random stimuli for each possible width mode. Regarding FPGA implementations, a first "place and route" while targeting the theoretically possible maximum frequency (limited by BRAM or DSP) was done for each design. If the timings are not met, performance optimizers from FPGA manufacturer are run. If a design still fails to meet propagation delays, we lower its targeted clock frequency and repeat the previous steps.

We named designs after their k parameter and the FPGA resource used to implement its internal memories - B for BRAM and D for LUT. The operand widths M for which the different designs are generated are made explicit in the different discussions.

5.1 Outlines for FPMM Design Space Exploration

This section gives the FPMM's practical limitations, as well as general observations on its performances and its FPGA utilization as a function of sizing parameters.

Parameter Ranges. The generator imposes by design s and σ to be greater than or equal to two. Thus, a given parameter k implies boundaries on operand widths (i.e. M). Given a FPGA target, the choice of k implies Ω and α, and M is restricted to the range from $\Omega + 1$ (for $s > 1$) to $(\alpha - 1) \cdot \Omega$ (for $\sigma > 1$).

Table 1. (M, s, σ)'s ranges and number of cascaded DSP depending on k. Considering DSP48E2 from Xilinx Ultrascale+, with $\omega = 17$ (unsigned) and internal right-shifting operation.

k	Ω	α	l_Q/l_U	M range	s range	σ range	Cascaded DSP
2	34	14	0/4	35–442	2–13	7–2	4
3	51	18	0/0	52–867	2–17	9–2	9
4	68	32	7/0	69–2108	2–31	16–2	16
5	85	50	16/0	86–4165	2–49	25–2	25
6	102	72	27/0	103–7242	2–71	36–2	36
7	119	98	40/0	120–11543	2–97	49–2	49
8	136	128	55/0	137–17272	2–127	64–2	64

In addition, an FPGA target provides a limit to the depth of DSP cascades, depending on the size of the DSP columns. Thus, k must be less than the square root of the deepest possible cascade. This limitation can be somehow circumvented by cascading across DSP columns, but our generator does not handle this limitation case at the moment. An example of parameter ranges for various k is given in Table 1.

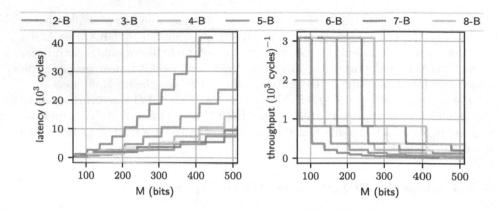

Fig. 11. Theoretical performance exploration for $M \in [64; 512]$.

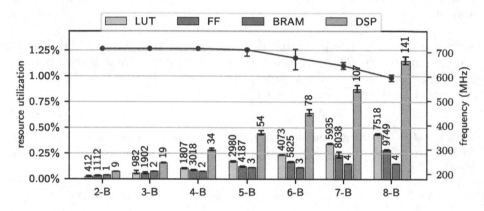

Fig. 12. Xilinx's Ultrascale+'s usage for $M \in [64; 512]$.

Theoretical Performances. Regarding latency, it takes $l_{FPMM} = l_{in} + (s-1)(\alpha + l_T) + \alpha$ cycles for a modular multiplication to be performed. For a given operand width M, a larger parameter k increases α but reduces s, thus an appropriate k may be found to reduce latency. Regarding throughput, the interleaving of independent computations allows the operator to output up to one operation every s^2 cycles[1]. A larger k always improves throughput by lowering s.

Figure 11 plots latency and throughput as a function of the operand width M for $k \in [2; 8]$. These are theoretical performances as it does not consider the running frequency achieved on a specific FPGA target. The stepped shapes are due to FPMM designs being actually dependent on (s, σ) pairs that are constants for M in $](s-1)\Omega; s\Omega]$. Nevertheless, these plots suggest that increasing k is a valid approach to improve both latency and throughput for larger M.

---

[1] When fully loaded with instructions, σ operations are outputed every $s^2\sigma$ cycles.

Resource Utilization. To give an overview of resource usage as a function of the parameter k, we implemented all designs possible with $M \in [64; 512]$ for each $k \in [2; 8]$. The targeted FPGA is an Ultrascale+ from Xilinx.

Figure 12 displays for each k averages of resource usage and running frequency, as well as deviations around these averages across designs (i.e. same k and different M). Designs are relatively small, and are able to reach important running frequency. The DSP are the limiting resources and their utilization is quadratic with the growth of k. The impact of M is rather insignificant on resource utilizations. For $k < 5$ the running frequency reach the upper limit imposed by BRAMs. The complexity of the design increasing for larger k, the maximum frequency achievable drops to 600 MHz (for $k = 8$).

5.2 Exploration's Examples for (H)ECC

In this section, we consider the use case of elliptic and hyper-elliptic curve cryptography to compare with previous works, and in particular [5] and [9]. For the sake of comparison, the FPGA target is now a Virtex-7 and the operand widths are 128-bit and 256-bit.

Table 2. Comparison of our generated FPMMs with [5] and [9].

M	Work	Name	CLB/LUT/FF	BRAM	DSP	Freq. MHz	Lat. ns	Thr. μs^{-1}
128	[9]	MA16	455/1182/1305	6	21	350	77.0	17.5
	[5]	F44B	325/545/725	2	9	528	141.8	33.1
	Our	2-B	305/406/1074	1	9	558	125.5	34.9
	Our	3-B	448/866/1785	2	20	396	156.4	44.0
	Our	4-B	839/1803/2823	2	37	373	193.2	93.2
	[5]	F44D	306/600/758	–	9	633	118.5	39.6
	Our	2-D	261/448/1218	–	9	539	129.8	33.7
	Our	3-D	497/921/1921	–	20	406	152.7	45.1
	Our	4-D	858/1902/2958	–	37	385	187.1	96.2
256	[9]	MA16	661/1770/2172	10	37	372	99.5	13.3
	[5]	F28B	296/556/743	2	9	528	270.3	8.3
	Our	2-B	481/578/1315	1	9	548	244.7	8.6
	Our	3-B	500/898/1732	2	19	414	280.3	11.5
	Our	4-B	860/1821/2829	2	35	367	370.9	22.9
	[5]	F28D	291/674/787	–	9	598	238.8	9.4
	Our	2-D	372/629/1401	–	9	527	254.3	8.2
	Our	3-D	466/956/1866	–	19	395	293.7	11.0
	Our	4-D	853/1913/2965	–	35	362	375.6	22.6

In Table 2 comparisons are regrouped according to M and the type of resources used to memorize input operands. Note that designs 2-[B/D] for

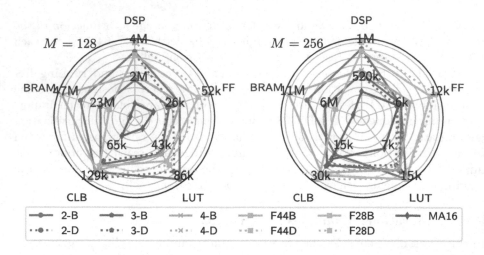

Fig. 13. TPAR comparisons on a Virtex-7 from Xilinx for $M \in \{128, 256\}$.

$M = 128$ (resp. $M = 256$) are actually our FPMM versions of F44[B/D] (resp. F28[B/D]), and are suppose to be very similar.

General Observations. A first observation is that increasing k makes the FPMM reach higher possible throughputs than the state of the art - more than twice the throughput of F44[D/B] and F28[B/D] with our 4-[B/D]. A second observation is that compared to the equivalent designs from [5] our FPMMs use more flip-flops and have a lower running frequency for designs with LUT based memories (e.g. F44D Vs 2-D). This is certainly due to specific optimizations for the case $k = 2$ that we given up to facilitate the generalization to larger k. A final observation concerns the operating frequency which seems to not be as stable with k growth as for Ultrascale+ target. One possible explanation could be the improvement of the configurable logic block carry-chains of the 7 Series for the Ultrascales (from 4-bit to 8-bit long). It may help to maintain the speed of the arithmetic as the operands' width grows. We did not investigate deeper this observation.

Throughput per Area Ratio. We then compare the different configurations through a typical metric that merges the information of throughput and hardware cost. This metric is called Throughput Per Area Ratio (TPAR) and gives a number of operations per second and per unit of utilized resources.

Figure 13 displays TPAR on Virtex-7 for 128 and 256 bits operators. According to the limiting resource (DSP) our best implementations are with $k = 2$. We believe that the the higher operating frequencies mainly explains this result.

From the flip-flops point of view, we find the consequence of the increased usage compared to the original works F44[D/B] and F28[D/B]. Nevertheless, all our different implementations remain relevant compared to Ma et al. [9] (MA16) approach. It has the closest performances to the FPMM's ones among the works to which [5] compared itself.

5.3 Multi-width FPMM

The multi-width feature allows an architect to implement a single operator to handle different operand widths. Among others, applications could be encryption circuits delivering several types of cryptographic primitives (e.g. RSA, DH, ECC, HECC,...), or different level of security.

To illustrate the gain brought by multi-width FPMM in such context, let's consider an application of elliptic and hyper elliptic curve cryptography with two different security levels, namely 128 and 256 bits. We further add the reasonable assumption that multiplication is the application's bottleneck and that the operator is working full time. We note p the fraction of time spent performing 128-bit multiplications and $(1 - p)$ the 256-bit one. We now compare a multi-width FPMM against two other implementation choices: a 256-bit single-width FPMM used for both operand widths, and two different single-width FPMMs, one for each operand width.

Table 3. Comparison between single-width and multi-width FPMMs for $k = 2$ and $M = 128, 256$ on Virtex-7 (V7) and Ultrascale+ (US+).

	Parameters		Single-width				Multi-width			
	k	$M/s/\sigma$	CLB/LUT/FF	BRAM	DSP	f	CLB/LUT/FF	BRAM	DSP	$f_\#$
V7	2	128/4/4	305/406/1074	1	9	557.7	372/547/1234	1	9	515.2
		256/8/2	481/578/1315	1	9	547.6				
US+	2	128/4/4	116/383/1074	1	9	714.3	135/532/1229	1	9	714.3
		256/8/2	152/552/1332	1	9	714.3				

Table 3 shows implementation results of single-widths and multi-width FPMM on Virtex-7 and Ultrascale+. Compared to the first implementation choice, the multi-width FPMM improves throughput of 128-bit operations. The application speedup is then derived from the fraction of time p spent on these small operands[2]. For instance, with p equals to 0.25, 0.5 and 0.75 the respective speedups are ×1.65, ×2.35 and ×3.06 on Virtex-7 and ×1.75, ×2.5 and ×3.25 on Ultrascale+. These improvements require no increase of resource utilization.

Compared to the second implementation choice, the multi-width FPMM does not improve performances on Ultrascale+, and is 6% slower on Virtex-7 due to the loss in running frequency. Nonetheless, it reduces by roughly half the utilization of FPGA's ressources.

In conclusion, the benefits of the multi-width option are case-critical when several operand widths must coexist in the same application.

6 Conclusion

This paper presented our realization of an extended FPMM generator. Its main scientific contributions are the generalization of the FPMM operator presented

---

[2] Speedup $= \dfrac{f_\# \left(\frac{p}{4^2} + \frac{1-p}{8^2} \right)}{\frac{f}{8^2}}$.

by Gallin and Tisserand [5] and the addition of a multi-width feature. The generator is provided as a python package under GPL3 licence [3]. The purpose is to propose a design space exploration tool for FPGA implementation of modular arithmetic.

Although the core principles of the FPMM are implemented, our generator is currently limited to Xillinx FPGA families. We have good reason to believe that extension to other types of FPGAs should not be a major problem. Indeed, the FPMM pre-requisites on DSP are typical characteristics of these arithmetic units, non-specific to Xilinx's ones.

A direct improvement of our generator would be the integration of design optimizations for $k = 2$ from [5]. Another area for improvement could be the mapping of Ω-bit multiplication onto DSPs. Karatsuba's algorithm would certainly reduce the number of DSPs for large k. Nevertheless, it remains to identify the impact on α and the differential in utilization of other hardware resources.

Ackowledgments. We would like to thank Arnaud Tisserand for our interesting exchanges and his encouragement to publish these results; as well as the anonymous reviewers for their pertinent and welcome remarks and suggestions.

References

1. Bossuet, L., Gogniat, G., Philippe, J.L.: Exploration de l'espace de conception des architectures reconfigurables. **25**(7), 921–946. https://doi.org/10.3166/tsi.25.921-946
2. Brakerski, Z., Gentry, C., Vaikuntanathan, V.: (Leveled) fully homomorphic encryption without bootstrapping. In: Proceedings of the 3rd Innovations in Theoretical Computer Science Conference on ITCS 2012. ACM Press (2012). https://doi.org/10.1145/2090236.2090262
3. Cathébras, J., Chotin, R.: Finely Pipelined Modular Multiplier (FPMM). https://gitlab.lip6.fr/roselyne/fpmm/. Accessed 12 Aug 2020
4. Deschamps, J.P., Sutter, G.D., Cantó, E.: Guide to FPGA Implementation of Arithmetic Functions. Springer, Netherlands (2012). https://doi.org/10.1007/978-94-007-2987-2
5. Gallin, G., Tisserand, A.: Generation of finely-pipelined GF(P) multipliers for flexible curve based cryptography on FPGAs, pp. 1–12. https://doi.org/10.1109/TC.2019.2920352
6. Gallin, G., Tisserand, A.: Hyper-Threaded Modular Multipliers (HTMM). https://sourcesup.renater.fr/www/htmm/. Accessed 25 Feb 2020
7. Gallin, G., Tisserand, A.: Hyper-threaded multiplier for HECC. In: 2017 51st Asilomar Conference on Signals, Systems, and Computers. IEEE (2017). https://doi.org/10.1109/acssc.2017.8335378
8. Koc, C.K., Acar, T., Kaliski, B.: Analyzing and comparing montgomery multiplication algorithms. IEEE Micro **16**(3), 26–33 (1996). https://doi.org/10.1109/40.502403
9. Ma, Y., Liu, Z., Pan, W., Jing, J.: A high-speed elliptic curve cryptographic processor for generic curves over GF(p). In: Lange, T., Lauter, K., Lisoněk, P. (eds.) SAC 2013. LNCS, vol. 8282, pp. 421–437. Springer, Heidelberg (2014). https://doi.org/10.1007/978-3-662-43414-7_21

10. Montgomery, P.L.: Modular multiplication without trial division. Math. Comput. **44**(170), 519 (1985). https://doi.org/10.1090/s0025-5718-1985-0777282-x
11. Morales-Sandoval, M., Diaz-Perez, A.: Scalable GF(P) montgomery multiplier based on a digit-digit computation approach. IET Comput. Digit. Tech. **10**(3), 102–109 (2016). https://doi.org/10.1049/iet-cdt.2015.0055
12. Mrabet, A., et al.: A scalable and systolic architectures of montgomery modular multiplication for public key cryptosystems based on DSPs. J. Hardw. Syst. Secur. **1**(3), 219–236 (2017). https://doi.org/10.1007/s41635-017-0018-x
13. Pimentel, A.D.: Exploring exploration: a tutorial introduction to embedded systems design space exploration. IEEE Des. Test **34**(1), 77–90 (2016). https://doi.org/10.1109/mdat.2016.2626445
14. Walter, C.: Montgomery exponentiation needs no final subtractions. Electron. Lett. **35**(21), 1831 (1999). https://doi.org/10.1049/el:19991230

Trisymmetric Multiplication Formulae
in Finite Fields

Hugues Randriambololona[1,2] and Édouard Rousseau[2,3(✉)]

[1] ANSSI, Paris, France
[2] Institut Polytechnique de Paris/Télécom Paris, Palaiseau, France
{randriam,erousseau}@telecom-paris.fr
[3] Université Paris-Saclay/UVSQ, Versailles, France

Abstract. Multiplication is an expensive arithmetic operation, therefore there has been extensive research to find Karatsuba-like formulae reducing the number of multiplications involved when computing a bilinear map. The minimal number of multiplications in such formulae is called the *bilinear complexity*, and it is also of theoretical interest to asymptotically understand it. Moreover, when the bilinear maps admit some kind of invariance, it is also desirable to find formulae keeping the same invariance. In this work, we study *trisymmetric, hypersymmetric*, and *Galois invariant* multiplication formulae over finite fields, and we give an algorithm to find such formulae. We also generalize the result that the bilinear complexity and symmetric bilinear complexity of the two-variable multiplication in an extension field are linear in the degree of the extension, to trisymmetric bilinear complexity, and to the complexity of t-variable multiplication for any $t \geq 3$.

1 Introduction

Given an algorithm that computes a polynomial map over a field $\mathbf{k}$ (or a family of such polynomial maps, with entries of length going to infinity), one is usually interested in the (asymptotic) cost of the algorithm. In order to understand this cost, one studies the *complexity* of the algorithm, *i.e.* the number of operations needed by the algorithm. We can for example count the number of bit operations, or the number of algebraic operations $(+, \times)$ in $\mathbf{k}$. The latter is called the *algebraic complexity* and in this model it is supposed that all algebraic operations have the same cost. Nevertheless, multiplication of two variable quantities in $\mathbf{k}$ is arguably more expensive than addition, or than multiplication of a variable by a fixed constant. In the context of the computation of bilinear maps, extensive work has been done to reduce the number of two-variable multiplications involved. Notable examples are Karatsuba's algorithm [11] and Strassen's algorithm [19]. Karatsuba's algorithm is based on the fact that the bilinear map associated to the product of two polynomials of degree 1

$$A = a_1 X + a_0 \text{ and } B = b_1 X + b_0$$

J. C. Bajard and A. Topuzoğlu (Eds.): WAIFI 2020, LNCS 12542, pp. 92–111, 2021.
https://doi.org/10.1007/978-3-030-68869-1_5

can be computed with three products $a_0 b_0, (a_0 + a_1)(b_0 + b_1), a_1 b_1$ instead of the four classic ones $a_0 b_0, a_0 b_1, a_1 b_0, a_1 b_1$. Strassen's algorithm exploits a similar idea in the case of 2×2 matrices: only 7 products are used instead of 8 in order to compute a matrix product. Both these algorithms have very practical consequences. The *bilinear complexity* $\mu(\Phi)$ of a bilinear map Φ over $\mathbf{k}$ represents the minimum number of two-variable multiplications in a formula that computes Φ, discarding the cost of other operations such as addition or multiplication by a constant. In particular when $\mathcal{A}$ is a finite dimensional algebra over $\mathbf{k}$, we define the bilinear complexity of $\mathcal{A}$ as $\mu(\mathcal{A}/\mathbf{k}) = \mu(m_{\mathcal{A}})$ where $m_{\mathcal{A}} : \mathcal{A} \times \mathcal{A} \to \mathcal{A}$ is the multiplication map in $\mathcal{A}$ seen as a $\mathbf{k}$-bilinear map.

Let $\mathbf{k}^{2 \times 2}$ be the algebra of 2×2 matrices over $\mathbf{k}$. We know thanks to Strassen's algorithm that

$$\mu(\mathbf{k}^{2 \times 2}/\mathbf{k}) \leq 7.$$

In fact, this is optimal, so we have exactly $\mu(\mathbf{k}^{2 \times 2}/\mathbf{k}) = 7$ [20, Thm. 3.1]. In general, it seems to be hard to find the bilinear complexity of a given algebra, for example the bilinear complexity of $\mathbf{k}^{3 \times 3}$ is not known. In the literature, work has been done both to algorithmically find the bilinear complexity of small algebras [5, 10] and to understand how the bilinear complexity asymptotically grows [2, 9]. Chudnovsky and Chudnovsky proved in 1988 that the bilinear complexity of an extension field $\mathbb{F}_{q^k}/\mathbb{F}_q$ is linear in the degree k of the extension, using an evaluation-interpolation method on curves. As the main contribution of this article, we investigate both questions for *trisymmetric* bilinear complexity, and solve a certain number of the open problems stated in [2, §5.2].

When a bilinear map admits certain invariance properties, it can be interesting, both for theoretical and for practical reasons, to find formulae for it that exhibit these same properties. For symmetric bilinear maps, and in particular for commutative algebras, this leads to the notion of symmetric bilinear complexity. A further refinement, the trisymmetric bilinear complexity of $\mathbb{F}_{q^k}$ over $\mathbb{F}_q$, was first introduced in [16], and rediscovered independently in [14, App. A].

In Sect. 2 we recall the definition of symmetric and trisymmetric formulae, and discuss further generalizations such as hypersymmetric formulae for higher multilinear maps, and Galois-invariant formulae. In Sect. 3 we describe algorithms to compute trisymmetric decompositions in small dimension. In all examples we were able to compute, the trisymmetric bilinear complexity is equal to the symmetric bilinear complexity. However we found an example where the Galois-invariant trisymmetric bilinear complexity is strictly larger. Finally, in Sect. 4, we prove that for all $q \geq 3$, the trisymmetric bilinear complexity of an extension of $\mathbb{F}_q$ is again linear in the degree, as well as similar results for higher multiplication maps.

2   Multiplication Formulae with Symmetries

Although we are mainly interested in bilinear multiplication formulae, the notions we will consider naturally involve higher multilinear maps.

Multilinear Complexity. Let $\Phi : V_1 \times \cdots \times V_t \to W$ be a t-multilinear map between finite dimensional vector spaces over **k**. A *multilinear algorithm*, or *multilinear decomposition*, or *multilinear formula* of length n for Φ is a collection of linear forms $(\varphi_i^{(j)})_{\substack{1 \leq i \leq n \\ 1 \leq j \leq t}}$, where $\varphi_i^{(j)}$ is in $V_j^\vee$, the dual vector space of V_j, and elements $(w_i)_{1 \leq i \leq n}$ in W, such that for all $v_1, \ldots, v_t$ we have

$$\Phi(v_1, \ldots, v_t) = \sum_{i=1}^{n} \varphi_i^{(1)}(v_1) \cdots \varphi_i^{(t)}(v_t) w_i.$$

The *multilinear complexity* $\mu(\Phi)$ is then defined as the smallest length n of such a decomposition. Equivalently, it is the rank of the tensor in $V_1^\vee \otimes \cdots \otimes V_t^\vee \otimes W$ corresponding to Φ.

Symmetric Multilinear Complexity. When $V_1 = \cdots = V_t = V$ and Φ is a *symmetric* multilinear map, it is natural to search for *symmetric multilinear decompositions*, *i.e.* formulae of the form

$$\Phi(v_1, \ldots, v_t) = \sum_{i=1}^{n} \varphi_i(v_1) \cdots \varphi_i(v_t) w_i$$

with $\varphi_i^{(1)} = \cdots = \varphi_i^{(t)} = \varphi_i \in V^\vee$ for all i. It is more space-efficient, since symmetric formulae admit a shorter description. From an algorithmic point of view, it should also be simpler to find symmetric formulae, because the search space is smaller. We define $\mu^{\mathrm{sym}}(\Phi)$, the *symmetric multilinear complexity* of Φ, as the minimal length n of such a symmetric decomposition, if it exists (otherwise we set $\mu^{\mathrm{sym}}(\Phi) = \infty$).

In the case $t = 2$, a symmetric bilinear map always admits a symmetric decomposition. However, when $t \geq 3$ and $\mathbf{k} = \mathbb{F}_q$ is a finite field, this can fail. When $t = 3$ and $q > 2$, it is shown in [16, Lemma 7] that a symmetric trilinear map Φ over $\mathbb{F}_q$ always admits a symmetric algorithm, while in the remaining case $t = 3$ and $q = 2$, as observed by Cascudo, a necessary condition is that Φ should satisfy $\Phi(x, x, y) = \Phi(x, y, y)$ for all entries x, y. These results were then combined and generalized into the following necessary and sufficient criterion:

Theorem 1 ([[14], Thm. A.7]). *Let* $\Phi : V^t \to W$ *be a t-multilinear map between finite dimensional vector spaces over* $\mathbb{F}_q$. *Then* Φ *admits a symmetric decomposition if and only if* Φ *is* Frobenius-symmetric, *i.e. if and only if it is symmetric and one of the following two conditions holds:*

- $t \leq q$
- $t \geq q + 1$ *and for all* $u, v, z_1, \ldots, z_{t-q-1}$ *in* V,

$$\Phi(\underbrace{u, \ldots, u}_{q \ times}, v, z_1, \ldots, z_{t-q-1}) = \Phi(u, \underbrace{v, \ldots, v}_{q \ times}, z_1, \ldots, z_{t-q-1}).$$

Observe that this criterion involves the *cardinality* of the field, not its characteristic.

Trisymmetric and Hypersymmetric Complexity. Now suppose furthermore that $V = W$, and that this space is equipped with a non-degenerate symmetric bilinear form, written as a scalar product

$$V \times V \to \mathbf{k}$$
$$(v, w) \mapsto \langle v, w \rangle.$$

This allows to identify V and $V^\vee$, *i.e.* any linear form $\varphi \in V^\vee$ is of the form $\varphi(x) = \langle a, x \rangle$ for a uniquely determined $a \in V$. As a consequence, a symmetric decomposition for $\Phi : V^t \to V$ can also be described as the data of elements $(a_i)_{1 \le i \le n}$ and $(b_i)_{1 \le i \le n}$ in V such that for all $v_1, \dots, v_t$ in V, we have $\Phi(v_1, \dots, v_t) = \sum_{i=1}^n \langle a_i, v_1 \rangle \cdots \langle a_i, v_t \rangle b_i$. In order to have an even more compact description, one could ask for b_i to be proportional to a_i, leading to the following:

Definition 1. *Let V be a finite dimensional $\mathbf{k}$-vector space equipped with a scalar product, and $\Phi : V^t \to V$ a symmetric t-multilinear map. Then a hypersymmetric formula for Φ is the data of elements $(a_i)_{1 \le i \le n}$ in V and scalars $(\lambda_i)_{1 \le i \le n}$ in $\mathbf{k}$ such that, for all $v_1, \dots, v_t \in V$,*

$$\Phi(v_1, \dots, v_t) = \sum_{i=1}^n \lambda_i \langle a_i, v_1 \rangle \cdots \langle a_i, v_t \rangle a_i.$$

The hypersymmetric complexity $\mu^{hyp}(\Phi)$ is then the minimal length n of such a hypersymmetric decomposition, if it exists. Obviously we always have $\mu^{sym}(\Phi) \le \mu^{hyp}(\Phi)$.

When $t = 2$, we will say trisymmetric *for hypersymmetric, and write $\mu^{tri}(\Phi)$ for $\mu^{hyp}(\Phi)$.*

As a further motivation, observe that to any t-multilinear map $\Phi : V^t \to V$ one can associate a $(t+1)$-multilinear *form* $\widetilde{\Phi} : V^{t+1} \to \mathbf{k}$, defined by

$$\widetilde{\Phi}(v_1, \dots, v_t, v_{t+1}) = \langle \Phi(v_1, \dots, v_t), v_{t+1} \rangle.$$

We then say that Φ is hypersymmetric (as a t-multilinear map) if $\widetilde{\Phi}$ is symmetric (as a $(t+1)$-multilinear form). It is easily seen that Φ hypersymmetric is a necessary condition for it to admit a hypersymmetric decomposition, and more precisely:

Lemma 1. *Elements $(a_i)_{1 \le i \le n}$ in V and scalars $(\lambda_i)_{1 \le i \le n}$ in $\mathbf{k}$ define a hypersymmetric formula for the t-multilinear map Φ,*

$$\Phi(v_1, \dots, v_t) = \sum_{i=1}^n \lambda_i \langle a_i, v_1 \rangle \cdots \langle a_i, v_t \rangle a_i,$$

if and only if they define a symmetric formula for the $(t+1)$-multilinear form $\widetilde{\Phi}$,

$$\widetilde{\Phi}(v_1, \ldots, v_s, v_{t+1}) = \sum_{i=1}^{n} \lambda_i \langle a_i, v_1 \rangle \cdots \langle a_i, v_t \rangle \langle a_i, v_{t+1} \rangle.$$

Thus, Φ admits a hypersymmetric formula if and only if $\widetilde{\Phi}$ is Frobenius-symmetric (in the sense of Theorem 1), and we have

$$\mu^{hyp}(\Phi) = \mu^{sym}\left(\widetilde{\Phi}\right).$$

In particular, if $q \geq t+1$, then any hypersymmetric t-multilinear map over $\mathbb{F}_q$ admits a hypersymmetric formula.

Proof. For the *only if* part in the first assertion, take scalar product with v_{t+1}. For the *if* part, use the fact that the scalar product is non-degenerate. The other assertions follow. □

Galois Invariance. Last we consider another type of symmetry. Let $\sigma : v \mapsto v^\sigma$ be a **k**-linear automorphism of V that respects the scalar product: $\langle v^\sigma, w^\sigma \rangle = \langle v, w \rangle$ for all v, w in V.

Lemma 2. *Let $\Phi : V^t \to V$ be a symmetric t-multilinear map that is compatible with σ, i.e.*

$$\Phi(v_1^\sigma, \ldots, v_t^\sigma) = \Phi(v_1, \ldots, v_t)^\sigma$$

for all $v_1, \ldots, v_t$ in V, and let $(a_i)_{1 \leq i \leq n}$ and $(b_i)_{1 \leq i \leq n}$ in V define a symmetric formula for Φ,

$$\Phi(v_1, \ldots, v_t) = \sum_{i=1}^{n} \langle a_i, v_1 \rangle \cdots \langle a_i, v_t \rangle b_i.$$

Then $(a_i^\sigma)_{1 \leq i \leq n}$ and $(b_i^\sigma)_{1 \leq i \leq n}$ also define a symmetric formula for Φ,

$$\Phi(v_1, \ldots, v_t) = \sum_{i=1}^{n} \langle a_i^\sigma, v_1 \rangle \cdots \langle a_i^\sigma, v_t \rangle b_i^\sigma.$$

Proof. Write $\Phi(v_1, \ldots, v_t) = \Phi(v_1^{\sigma^{-1}}, \ldots, v_t^{\sigma^{-1}})^\sigma$ and apply the formula. □

We then say that the symmetric formula given by $(a_i)_{1 \leq i \leq n}$ and $(b_i)_{1 \leq i \leq n}$ is σ-invariant if it is the same as the formula given by $(a_i^\sigma)_{1 \leq i \leq n}$ and $(b_i^\sigma)_{1 \leq i \leq n}$, i.e. if there is a permutation π of $\{1, \ldots, n\}$ such that $(a_i^\sigma, b_i^\sigma) = (a_{\pi(i)}, b_{\pi(i)})$ for all i. This applies also to hypersymmetric formulae, setting $b_i = \lambda_i a_i$.

If G is a group of **k**-linear automorphisms of V that respect the scalar product, and if $\Phi : V^t \to V$ is a symmetric t-multilinear map that is compatible with all elements in G, we then define $\mu^{\mathrm{sym},G}(\Phi)$ (resp. $\mu^{\mathrm{hyp},G}(\Phi)$), the *G-invariant* symmetric (resp. hypersymmetric) multilinear complexity of Φ, as the minimal length n of a symmetric (resp. hypersymmetric) multilinear formula for Φ that is G-invariant, i.e. σ-invariant for all σ in G.

Multiplication Formulae in Algebras. Let $\mathcal{A}$ be a finite dimensional commutative algebra over $\mathbf{k}$. We say a linear form $\tau : \mathcal{A} \to \mathbf{k}$ is trace-like if the symmetric bilinear form $\mathcal{A} \times \mathcal{A} \to \mathbf{k}$, $(x, y) \mapsto \tau(xy)$ is non-degenerate. If so, we set $\langle x, y \rangle = \tau(xy)$, which defines a scalar product on $\mathcal{A}$. In this work we will take $\mathbf{k} = \mathbb{F}_q$, and either:

- $\mathcal{A} = \mathbb{F}_{q^k}$ a finite field extension, and $\tau = \mathrm{Tr}_{\mathbb{F}_{q^k}/\mathbb{F}_q}$ the usual trace map; indeed it is well known that the trace bilinear form $\langle x, y \rangle = \mathrm{Tr}_{\mathbb{F}_{q^k}/\mathbb{F}_q}(xy)$ is non-degenerate
- $\mathcal{A} = \mathbb{F}_q[T]/(T^k)$ an algebra of truncated polynomials, and τ defined by $\tau(x) = x_{k-1}$ for $x = x_0 + x_1 T + \cdots + x_{k-1} T^{k-1}$ in $\mathcal{A}$; indeed, observe that for $x = x_0 + x_1 T + \cdots + x_{k-1} T^{k-1}$, $y = y_0 + y_1 T + \cdots + y_{k-1} T^{k-1}$, we then have $\langle x, y \rangle = \tau(xy) = x_0 y_{k-1} + x_1 y_{k-2} + \cdots + x_{k-1} y_0$, which is non-degenerate.

Let $\Phi : \mathcal{A} \times \mathcal{A} \to \mathcal{A}$ be the multiplication map, $\Phi(x, y) = xy$. It is easily seen that Φ is trisymmetric. Indeed $\widetilde{\Phi}$ is the trilinear form $x, y, z \mapsto \tau(xyz)$, which is symmetric. A symmetric bilinear multiplication formula for $\mathcal{A}$ is thus the data of $(a_i)_{1 \leq i \leq n}$ in $\mathcal{A}$ and $(\varphi_i)_{1 \leq i \leq n}$ in $\mathcal{A}^\vee$ such that

$$\forall x, y \in \mathcal{A}, \; xy = \sum_{i=1}^{n} \varphi_i(x)\varphi_i(y)a_i, \tag{1}$$

and a trisymmetric formula is the data of $(a_i)_{1 \leq i \leq n}$ in $\mathcal{A}$ and $(\lambda_i)_{1 \leq i \leq n}$ in $\mathbb{F}_q$ such that

$$\forall x, y \in \mathcal{A}, \; xy = \sum_{i=1}^{n} \lambda_i \langle a_i, x \rangle \langle a_i, y \rangle a_i. \tag{2}$$

We will write $\mu_q(k)$ (resp. $\hat{\mu}_q(k)$) for the bilinear complexity of multiplication in $\mathbb{F}_{q^k}$ (resp. in $\mathbb{F}_q[T]/(T^k)$) over $\mathbb{F}_q$, and we will write likewise $\mu_q^{\mathrm{sym}}(k)$, $\hat{\mu}_q^{\mathrm{sym}}(k)$, $\mu_q^{\mathrm{tri}}(k)$, $\hat{\mu}_q^{\mathrm{tri}}(k)$, $\mu_q^{\mathrm{sym},G}(k)$, $\hat{\mu}_q^{\mathrm{sym},G}(k)$, $\mu_q^{\mathrm{tri},G}(k)$, $\hat{\mu}_q^{\mathrm{tri},G}(k)$, etc. for the similar quantities with the corresponding symmetry conditions.

For $q \geq 3$ we have $\mu_q^{\mathrm{tri}}(k) < \infty$ and $\hat{\mu}_q^{\mathrm{tri}}(k) < \infty$ for all k, while for $q = 2$ we have $\mu_2^{\mathrm{tri}}(1) = \hat{\mu}_2^{\mathrm{tri}}(1) = 1$ and $\mu_2^{\mathrm{tri}}(2) = 3$, but $\mu_2^{\mathrm{tri}}(k) = \infty$ for $k \geq 3$ and $\hat{\mu}_2^{\mathrm{tri}}(k) = \infty$ for $k \geq 2$. This follows essentially from Theorem 1 and Lemma 1 (see also [14, Prop. A.14]).

Obviously we have $\mu_q(k) \leq \mu_q^{\mathrm{sym}}(k) \leq \mu_q^{\mathrm{tri}}(k)$ and $\hat{\mu}_q(k) \leq \hat{\mu}_q^{\mathrm{sym}}(k) \leq \hat{\mu}_q^{\mathrm{tri}}(k)$ for all q and k. But when all these quantities are finite, e.g. when $q \geq 3$, no example of strict inequality is known.

In the other direction, when $q \geq 4$ is not divisible by 3, [16, Thm. 2] gives $\mu_q^{\mathrm{tri}}(k) \leq 4\mu_q^{\mathrm{sym}}(k)$ and $\hat{\mu}_q^{\mathrm{tri}}(k) \leq 4\hat{\mu}_q^{\mathrm{sym}}(k)$. This allows to translate the many known upper bounds on symmetric complexity [2] into upper bounds on trisymmetric complexity. However the resulting upper bounds do not seem to be tight, so it would be desirable to have better estimates, and especially upper bounds that work also for q divisible by 3.

3   Finding Trisymmetric Decompositions

Algorithmic Search. Barbulescu *et al.* [5] and later Covanov [10] found clever ways of exhaustively searching for formulae for (symmetric) bilinear maps. Their method eliminates redundancy in the search but strongly relies on the fact that the vectors $a_i \in \mathcal{A}$ in the symmetric formulae (1) can be chosen independently of the linear forms $\varphi_i \in \mathcal{A}^\vee$, which is no longer the case when searching for trisymmetric decompositions. For this reason, we use another method that is once again a variant of an exhaustive search and thus still leads to an exponential complexity algorithm. Let Φ be the two-variable product in $\mathcal{A}$. Recall that we are looking for a trisymmetric decomposition:

$$\forall x, y \in \mathcal{A}, \ \Phi(x, y) = xy = \sum_{i=1}^{n} \lambda_i \langle x, a_i \rangle \langle y, a_i \rangle \, a_i,$$

with $a_i \in \mathcal{A}$ and $\lambda_i \in \mathbf{k}$ for all $1 \leq i \leq n$. Because we are allowed to use scalars $\lambda_i \in \mathbf{k}$, we can limit our search to "normalized" elements in $\mathcal{A}$, as follows. Choose a basis of $\mathcal{A}$, which gives an identification $\mathcal{A} \simeq \mathbf{k}^k$ as vector spaces. Then for all $1 \leq i \leq k$, let

$$\mathcal{E}_i = \left\{ x = (x_1, \ldots, x_k) \in \mathcal{A} \simeq \mathbf{k}^k \mid \forall j \leq i - 1, \ x_j = 0 \text{ and } x_i = 1 \right\}$$

and

$$\mathcal{E} = \bigcup_{i=1}^{k} \mathcal{E}_i.$$

We search for elements a_i in $\mathcal{E}$ instead of $\mathcal{A}$. We further use the vector space structure of $\mathcal{A}$ by searching for solutions on each coordinate. Let

$$xy = (\pi_1(x, y), \ldots, \pi_k(x, y)) \in \mathcal{A} \simeq \mathbf{k}^k,$$

where, for all $1 \leq i \leq k$, π_i is the bilinear form corresponding to the i-th coordinate of the product in $\mathbb{F}_{p^k}$. In other words,

$$\Phi = (\pi_1, \ldots, \pi_k).$$

We let $\mathcal{B}$ be the space of bilinear forms on $\mathcal{A}$ and we let f be the application mapping an element in $\mathcal{A}$ to its associated bilinear symmetric form:

$$f : \mathcal{A} \to \mathcal{B}$$
$$a \mapsto (x, y) \mapsto \langle x, a \rangle \langle y, a \rangle .$$

We then search for elements $a_1, \ldots, a_{n_1}$ in $\mathcal{E}_1$ and $\lambda_1, \ldots, \lambda_{n_1}$ in $\mathbf{k}$ such that

$$\pi_1 = \sum_{j=1}^{n_1} \lambda_j f(a_j), \tag{3}$$

and we obtain

$$\Phi - \sum_{j=1}^{n_1} \lambda_j f(a_j) a_j = (0, \pi'_2, \dots, \pi'_k),$$

where for $2 \leq i \leq k$, π'_i is some other bilinear form. We then continue the operation with π'_2 and elements $a_{n_1+1}, \dots, a_{n_2}$ in $\mathcal{E}_2$, then with π''_3 and elements in $\mathcal{E}_3$, and so on. In the end, we have n elements $a_1, \dots, a_n \in \mathcal{E}$ and $\lambda_1, \dots, \lambda_n \in \mathbf{k}$ such that

$$\Phi = \sum_{j=1}^{n} \lambda_j f(a_j) a_j.$$

Now, there is left to see how we compute the elements $a_1, \dots, a_{n_1} \in \mathcal{E}_1$ and $\lambda_1, \dots, \lambda_{n_1} \in \mathbf{k}$ in order to obtain (3). Let r_1 be the rank of π_1. We know that the number n_1 of elements in $\mathcal{E}_1$ such that we have (3) is at least r_1, but there also exist some trisymmetric decompositions where we need more than r_1 elements. To find these elements, we search through elements $a_1 \in \mathcal{E}_1$ such that there exists $\lambda_1 \in \mathbf{k}$ with

$$\text{rank}(\pi_1 - \lambda_1 f(a_1)) < \text{rank}(\pi_1),$$

then, for each such $a_1 \in \mathcal{E}_1$, we search through elements $a_2 \in \mathcal{E}_1$ such that there exists λ_2 with

$$\text{rank}(\pi_1 - \lambda_1 f(a_1) - \lambda_2 f(a_2)) < \text{rank}(\pi_1 - \lambda_1 f(a_1)),$$

and so on, eliminating a lot of unsuitable elements along the way. This method allows us to find decompositions of π_1 into a sum of exactly r_1 bilinear forms of rank 1. In order to find decompositions containing $r_1 + m_1$ bilinear forms, we repeat the same process, except that we allow the rank not to decrease m_1 times. Let m_j be the number of times we allow the rank not to decrease when dealing with the j-th coordinate in the algorithm. We let $\mathcal{M} = (m_1, \dots, m_k)$ and we call *margin* this k-tuple. This strategy was implemented in the Julia programming language [1] and a package searching for trisymmetric decompositions is available online[1], along with the source code.

This allowed us to compute $\mu_3^{\text{tri}}(3) = 6$, $\mu_p^{\text{tri}}(3) = 5$ for all primes $5 \leq p \leq 257$, $\mu_3^{\text{tri}}(4) = 9$, $\mu_5^{\text{tri}}(4) = 8$, and $\mu_p^{\text{tri}}(4) = 7$ for all primes $7 \leq p \leq 23$. Details about the computation can be found in Table 1, while examples of formulae obtained via our algorithm are given in Table 2 (actually the formulae in this table are *normalized* in the sense of [14, Def. A.16], *i.e.* they satisfy all $\lambda_i = 1$).

Galois Invariant Formulae. Let $\mathcal{A} = \mathbb{F}_{q^k}$ and G be the cyclic group generated by σ, the Frobenius automorphism over $\mathbb{F}_q$. In order to find G-invariant decompositions, we exhaustively search through orbits in $\mathbb{F}_{q^k}$, which is fast because the search space is smaller. This allows us to find Galois invariant trisymmetric formulae of length 11 for $\mathbb{F}_{3^5}$, and of length 10 for $\mathbb{F}_{5^5}$ and $\mathbb{F}_{7^5}$. Joint with the obvious inequalities $\mu_q(k) \leq \mu_q^{\text{sym}}(k) \leq \mu_q^{\text{tri}}(k) \leq \mu_q^{\text{tri},G}(k)$ and with known

---

[1] https://github.com/erou/TriSym.jl.

Table 1. Algorithmic results with various degrees, base fields and margins.

Field	Margin	Solutions	Length	Time (s)	Field	Margin	Solutions	Length	Time (s)
$\mathbb{F}_{3^2}$	$(0,0)$	1	3	$1.8 \cdot 10^{-4}$	$\mathbb{F}_{7^3}$	$(0,0,0)$	8	5	$7.0 \cdot 10^{-3}$
$\mathbb{F}_{3^3}$	$(0,0,0)$	1	6	$4.4 \cdot 10^{-4}$	$\mathbb{F}_{13^3}$	$(0,0,0)$	100	5	$2.9 \cdot 10^{-1}$
$\mathbb{F}_{3^4}$	$(0,0,0,0)$	2	9	$5.3 \cdot 10^{-3}$	$\mathbb{F}_{19^3}$	$(0,0,0)$	415	5	1.8
$\mathbb{F}_{3^4}$	$(2,1,0,0)$	18	9	$3.8 \cdot 10^{-1}$	$\mathbb{F}_{31^3}$	$(0,0,0)$	2031	5	29
$\mathbb{F}_{3^4}$	$(3,2,1,1)$	25	9	1.1	$\mathbb{F}_{47^3}$	$(0,0,0)$	7590	5	360

lower bounds from [2, Thm. 2.2] and [5], this gives $10 \leq \mu_3(5) \leq \mu_3^{\mathrm{sym}}(5) = \mu_3^{\mathrm{tri}}(5) = \mu_3^{\mathrm{tri},G}(5) = 11$, $\mu_5(5) = \mu_5^{\mathrm{sym}}(5) = \mu_5^{\mathrm{tri}}(5) = \mu_5^{\mathrm{tri},G}(5) = 10$, and $\mu_7(5) = \mu_7^{\mathrm{sym}}(5) = \mu_7^{\mathrm{tri}}(5) = \mu_7^{\mathrm{tri},G}(5) = 10$. Some examples of Galois invariant formulae can be found in Table 2.

For $q \geq 3$ we know no example where one of the inequalities in $\mu_q(k) \leq \mu_q^{\mathrm{sym}}(k) \leq \mu_q^{\mathrm{tri}}(k)$ is strict. However, it turns out that the inequality with $\mu_q^{\mathrm{tri},G}(k)$ can be strict. Indeed, let $q = 3$ and $k = 7$. In this setting our exhaustive search found no G-invariant decomposition of length up to 15. Since all orbits are of length 7, except the trivial orbit of length 1, the minimal length for a G-invariant decomposition is congruent to 0 or 1 modulo 7, so we deduce that it is at least 21. Furthermore, we know [2, table 2] that $\mu_3^{\mathrm{sym}}(7) \leq 19$, so we have

$$\mu_3(7) \leq \mu_3^{\mathrm{sym}}(7) \leq 19 < 21 \leq \mu_3^{\mathrm{tri},G}(7).$$

Table 2. Examples of trisymmetric multiplication formulae (the first three are Galois invariant).

Field	n	Field elements $a_1, \ldots, a_n$ such that $xy = \sum_{i=1}^{n} \langle a_i, x \rangle \langle a_i, y \rangle a_i$
$\mathbb{F}_{3^3} = \mathbb{F}_3[\alpha]/(\alpha^3 - \alpha + 1)$	6	$a_1 = \alpha,\ a_2 = a_1^\sigma,\ a_3 = a_2^\sigma,\ a_4 = 1 - \alpha^2,\ a_5 = a_4^\sigma,\ a_6 = a_5^\sigma$
$\mathbb{F}_{3^4} = \mathbb{F}_3[\alpha]/(\alpha^4 - \alpha^3 - 1)$	9	$a_1 = -1,\quad a_2 = -\alpha,\ a_3 = a_2^\sigma,\ a_4 = a_3^\sigma,\ a_5 = a_4^\sigma,$ $a_6 = \alpha^2 + \alpha + 1,\ a_7 = a_6^\sigma,\ a_8 = a_7^\sigma,\ a_9 = a_8^\sigma$
$\mathbb{F}_{3^5} = \mathbb{F}_3[\alpha]/(\alpha^5 - \alpha + 1)$	11	$a_1 = 1,\quad a_2 = \alpha - 1,\ a_3 = a_2^\sigma,\ a_4 = a_3^\sigma,\ a_5 = a_4^\sigma,\ a_6 = a_5^\sigma,$ $a_7 = 1 - \alpha - \alpha^2,\ a_8 = a_7^\sigma,\ a_9 = a_8^\sigma,\ a_{10} = a_9^\sigma,\ a_{11} = a_{10}^\sigma$
$\mathbb{F}_{5^3} = \mathbb{F}_5[\alpha]/(\alpha^3 + 3\alpha + 3)$	5	$a_1 = 3\alpha + 2,\ a_2 = -\alpha^2 - \alpha - 1,\ a_3 = 3\alpha^2 + 2\alpha + 2,\ a_4 = -\alpha,\ a_5 = 3\alpha^2 + 2\alpha$
$\mathbb{F}_{5^4} = \mathbb{F}_5[\alpha]/(\alpha^4 - \alpha^2 - \alpha + 2)$	8	$a_1 = -1,\ a_2 = 3\alpha^2 + 3\alpha + 3,\ a_3 = 3\alpha^3 - \alpha^2 + 2\alpha - 1,\ a_4 = 2\alpha^3 - \alpha^2 - \alpha + 1,$ $a_5 = \alpha,\ a_6 = -\alpha^2 + \alpha,\ a_7 = \alpha^3 + \alpha^2 + \alpha,\ a_8 = \alpha^3 + \alpha^2$

Universal Formulae. As mentioned in Sect. 2, for $q \geq 3$, we do not know any example of algebra $\mathcal{A} = \mathbb{F}_{q^k}$ or $\mathcal{A} = \mathbb{F}_q[T]/(T^k)$ where the bilinear complexity and the trisymmetric bilinear complexity are different. We can even prove that these quantities are the same in small dimension, by exhibiting trisymmetric *universal formulae*, *i.e.* trisymmetric decompositions that are true for (almost) any choice of $q \geq 3$. In order to obtain such formulae, it is useful to change our point of view on the problem. Assume we want to compute a trisymmetric decomposition of the product Φ in $\mathcal{A}$, a commutative algebra of degree k. After the choice of a basis of $\mathcal{A}$ and a basis of the space $\mathcal{B}$ of the bilinear forms on $\mathcal{A}$, we can represent

$$\Phi = (\pi_1, \ldots, \pi_k)$$

as a column vector B of length k^3. The first k^2 coordinates corresponding to π_1, the next k^2 coordinates corresponding to π_2 and so on up to π_k. Now, for each $a \in \mathcal{E}$, we note

$$\mathbf{f}(a) = a \otimes f(a),$$

where a is the column vector of length k corresponding to a in the basis of $\mathcal{A}$, $f(a)$ is the column vector of length k^2 corresponding to $f(a) \in \mathcal{B}$, and $\otimes$ is the Kronecker product. With these notations, finding a trisymmetric decomposition of the product in $\mathcal{A}$ is the same as finding elements $a_1 \ldots, a_n \in \mathcal{E}$ and $\lambda_1, \ldots, \lambda_n \in \mathbf{k}$ with

$$B = \sum_{j=1}^{n} \lambda_j \mathbf{f}(a).$$

Let A be the matrix which columns are the $\mathbf{f}(a)$ for all $a \in \mathcal{E}$, then the problem is to find a solution X of

$$AX = B$$

with the smallest possible number of nonzero entries in X.

We first consider the case $\mathcal{A} = \mathbb{F}_{q^2}$ over $\mathbf{k} = \mathbb{F}_q$, where the characteristic of $\mathbf{k}$ is not 2.

Proposition 1. *For any odd q we have*

$$\mu_q(2) = \mu_q^{tri}(2) = 3.$$

Proof. That $\mu_q(2) = 3$ follows e.g. from [2, Thm. 2.2]. In order to prove that $\mu_q^{tri}(2) = 3$, we find an *universal* trisymmetric formula of length 3. We know that we can find a non-square element ζ in $\mathbb{F}_q$, we can then define

$$\mathbb{F}_{q^2} \cong \mathbb{F}_q[T]/(T^2 - \zeta) = \mathbb{F}_q(\alpha),$$

where $\alpha = \bar{T}$ is the canonical generator of $\mathbb{F}_{q^2}$. Let $x = x_0 + x_1\alpha$ and $y = y_0 + y_1\alpha$ be two elements of $\mathbb{F}_{q^2}$, we have

$$xy = (x_0 + x_1\alpha)(y_0 + y_1\alpha) = x_0y_0 + \zeta x_1y_1 + (x_0y_1 + x_1y_0)\alpha.$$

We can lift the matrix B coming from the multiplication formula, that has coefficients in $\mathbb{F}_q$, to a matrix with coefficients in $\mathbb{Q}(\zeta)$, where ζ is an indeterminate. We can also lift the matrix A, because the map f (and therefore $\mathbf{f}$) has the same expression for all q not divisible by 2. Indeed, one can check that the map f is given by

$$f(x_0 + x_1\alpha) = \left(S \begin{bmatrix} x_0 \\ x_1 \end{bmatrix} \right) \left(S \begin{bmatrix} x_0 \\ x_1 \end{bmatrix} \right)^{\mathsf{T}} = 4 \begin{bmatrix} x_0^2 & \zeta x_0 x_1 \\ \zeta x_0 x_1 & \zeta^2 x_1^2 \end{bmatrix}.$$

where
$$S = [\langle \alpha^i, \alpha^j \rangle]_{0 \le i,j \le 1} = [\mathrm{Tr}(\alpha^{i+j})]_{0 \le i,j \le 1} = \begin{bmatrix} 2 & 0 \\ 0 & 2\zeta \end{bmatrix}.$$

We can then solve $AX = B$ over $\mathbb{Q}(\zeta)$ and finally check that
$$B = (1 - \zeta^{-1})4^{-1}\mathbf{f}(1) + (8\zeta)^{-1}\mathbf{f}(1 + \alpha) + (8\zeta)^{-1}\mathbf{f}(1 - \alpha),$$

so that the trisymmetric bilinear complexity of $\mathbb{F}_{q^2}/\mathbb{F}_q$ is 3.   □

Using the same strategy, we can also find universal formulae for another type of algebra $\mathcal{A} = \mathbb{F}_q[T]/(T^k)$, namely the truncated polynomials. In that context, we first observe that we have
$$\hat{\mu}_q^{tri}(k) \ge \hat{\mu}_q(k) \ge 2k - 1$$

for all q and k. Indeed this is a special case of [21, Thm. 4], which holds for any polynomial that is a power of an irreducible polynomial. Conversely we are able to find formulae for $2 \le k \le 4$ that match this lower bound.

Proposition 2. *For any odd q we have*
$$\hat{\mu}_q^{tri}(2) = 3.$$

Proof. Let $\mathcal{A} = \mathbb{F}_q[T]/(T^2) = \mathbb{F}_q[\alpha]$ with $\alpha = \bar{T}$, so $\alpha^2 = 0$. If $x = x_0 + x_1\alpha$ and $y = y_0 + y_1\alpha$ are two elements of $\mathcal{A}$, we have
$$xy = (x_0 + x_1\alpha)(y_0 + y_1\alpha) = x_0 y_0 + (x_0 y_1 + x_1 y_0)\alpha.$$

We can again construct the matrix B and A, and solve $AX = B$, this time simply over $\mathbb{Q}$. We obtain
$$B = -\mathbf{f}(1) + 2^{-1}\mathbf{f}(1 + \alpha) + 2^{-1}\mathbf{f}(1 - \alpha)$$

so that the trisymmetric bilinear complexity of $\mathcal{A} = \mathbb{F}_q[T]/(T^2)$ is at least 3, which concludes.   □

Proposition 3. *For any q not divisible by 2 nor 3 we have*
$$\hat{\mu}_q^{tri}(3) = 5 \qquad and \qquad \hat{\mu}_q^{tri}(4) = 7.$$

Proof. We use the same notations as before. For $\mathcal{A} = \mathbb{F}_q[T]/(T^3)$, we obtain
$$B = -\mathbf{f}(1 - \alpha - \alpha^2) + 3^{-1}\mathbf{f}(\alpha + 2\alpha^2) + 2^{-1}\mathbf{f}(1 - \alpha - 2\alpha^2) - 3^{-1}\mathbf{f}(\alpha - \alpha^2) + 2^{-1}\mathbf{f}(1 - \alpha).$$

Therefore the trisymmetric bilinear complexity of $\mathcal{A} = \mathbb{F}_q[T]/(T^3)$ is 5.
Finally, for $\mathcal{A} = \mathbb{F}_q[T]/(T^4)$, we obtain
$$\begin{aligned} B = {} & 2^{-1}\mathbf{f}(1 - \alpha^2 + \alpha^3) - \mathbf{f}(1 - \alpha^2) + 12^{-1}\mathbf{f}(\alpha + 2\alpha^2 + 2\alpha^3) - 12^{-1}\mathbf{f}(\alpha - 2\alpha^2 + 2\alpha^3) \\ & - 6^{-1}\mathbf{f}(\alpha + \alpha^2 - \alpha^3) + 6^{-1}\mathbf{f}(\alpha - \alpha^2 - \alpha^3) + 2^{-1}\mathbf{f}(1 - \alpha^2 - \alpha^3) \cdot (1 - \alpha^2 - \alpha^3). \end{aligned}$$

The trisymmetric bilinear complexity of $\mathcal{A} = \mathbb{F}_q[T]/(T^4)$ is then 7.   □

4 Asymptotic Bounds

In this section, we work with $\mathcal{A} = \mathbb{F}_{q^k}$ or $\mathbb{F}_q[T]/(T^k)$, seen as an algebra over $\mathbf{k} = \mathbb{F}_q$, and equipped with the trace-like linear form τ introduced at the end of Sect. 2. Our aim is to show that the trisymmetric bilinear complexities $\mu_q^{\mathrm{tri}}(k)$ and $\hat{\mu}_q^{\mathrm{tri}}(k)$ grow linearly as $k \to \infty$. Our proof will involve higher multilinear maps, and in turn, give results for them as well.

For any t we define the t-multilinear multiplication map in $\mathcal{A}$ over $\mathbf{k}$

$$m_t : \quad \begin{matrix} \mathcal{A}^t & \to & \mathcal{A} \\ (x_1, \ldots, x_t) & \mapsto & x_1 \cdots x_t \end{matrix}$$

and the t-multilinear trace form

$$\tau_t = \tau \circ m_t : \quad \begin{matrix} \mathcal{A}^t & \to & \mathcal{A} \\ (x_1, \ldots, x_t) & \mapsto & \tau(x_1 \cdots x_t). \end{matrix}$$

If needed, we will write $m_t{}^{\mathcal{A}/\mathbf{k}}$ or $\tau_t{}^{\mathcal{A}/\mathbf{k}}$ to keep $\mathcal{A}$ and $\mathbf{k}$ explicit.

The (symmetric) multilinear complexity of m_t has been considered in [7] in relation with the theory of testers.

Lemma 3. *The map m_t is hypersymmetric, and we have*

$$\mu^{hyp}(m_t) = \mu^{sym}(\tau_{t+1}) \leq \mu^{sym}(m_{t+1}).$$

Proof. Indeed we have $\tilde{m}_t = \tau_{t+1}$, and the equality on the left is a special case of Lemma 1. For the inequality on the right, take a symmetric formula for m_{t+1} and apply τ. $\qquad\qquad\square$

When studying the variation with the degree of the extension field $\mathbb{F}_{q^k}$ over $\mathbb{F}_q$, we will write $\mu_q^{\mathrm{sym}}(k, m_t)$ for $\mu^{\mathrm{sym}}\left(m_t{}^{\mathbb{F}_{q^k}/\mathbb{F}_q}\right)$, and we will also use the similar notations $\mu_q^{\mathrm{hyp}}(k, m_t)$, $\mu_q^{\mathrm{sym}}(k, \tau_t)$, etc. In particular for $t = 2$ we have

$$\mu_q^{\mathrm{tri}}(k) = \mu_q^{\mathrm{tri}}(k, m_2) = \mu_q^{\mathrm{sym}}(k, \tau_3).$$

When working in $\mathbb{F}_q[T]/(T^k)$ over $\mathbb{F}_q$, we will write likewise $\hat{\mu}_q^{\mathrm{sym}}(k, m_t)$, $\hat{\mu}_q^{\mathrm{hyp}}(k, m_t)$, etc.

Our aim is, for fixed q and t with $q \geq t + 1$, to show that $\mu_q^{\mathrm{hyp}}(k, m_t)$ and $\hat{\mu}_q^{\mathrm{hyp}}(k, m_t)$ grow linearly with $k \to \infty$. Thanks to Lemma 3, it suffices to show that $\mu_q^{\mathrm{sym}}(k, m_{t+1})$ and $\hat{\mu}_q^{\mathrm{sym}}(k, m_{t+1})$ grow linearly with $k \to \infty$. To ease notations we will set

$$M_{q,t}^{\mathrm{sym}} = \limsup_{k \to \infty} \frac{1}{k} \mu_q^{\mathrm{sym}}(k, m_t), \qquad M_{q,t}^{\mathrm{hyp}} = \limsup_{k \to \infty} \frac{1}{k} \mu_q^{\mathrm{hyp}}(k, m_t),$$

$$M_q^{\mathrm{tri}} = \limsup_{k \to \infty} \frac{1}{k} \mu_q^{\mathrm{tri}}(k) = M_{q,2}^{\mathrm{hyp}},$$

and likewise for $\hat{M}_{q,t}^{\mathrm{sym}}$, $\hat{M}_{q,t}^{\mathrm{hyp}}$, $\hat{M}_q^{\mathrm{tri}}$, etc.

Evaluation-Interpolation Method. We use the function field terminology and notations presented in [18]. Let $F/\mathbb{F}_q$ be an algebraic function field of one variable over $\mathbb{F}_q$ and let $\mathbb{P}_F$ be the set of places of F. Let $\mathcal{D}_F$ the set of divisors on F, and if $D \in \mathcal{D}_F$ is a divisor on F, we denote by $L(D)$ its Riemann-Roch space and $\ell(D) = \dim L(D)$.

Proposition 4. *Assume there exist a place* $Q \in \mathbb{P}_F$ *of* F *of degree* k, $P_1, \ldots, P_n \in \mathbb{P}_F$ *places of* F *of degree* 1, *and a divisor* $D \in \mathcal{D}_F$ *of* F *such that the places* Q *and* $P_1, \ldots, P_n$ *are not in the support of* D *and such that the following conditions hold.*

(i) *The evaluation map*
$$ev_{Q,D} : L(D) \to \mathbb{F}_{q^k}$$
$$f \mapsto f(Q)$$
  is surjective.
(ii) *The evaluation map*
$$ev_{\mathcal{P},tD} : L(tD) \to (\mathbb{F}_q)^n$$
$$h \mapsto (h(P_1), \ldots, h(P_n))$$
  is injective.

Then $m_t^{\mathbb{F}_{q^k}/\mathbb{F}_q}$ *admits a symmetric formula of length* n, *i.e. we have* $\mu_q^{sym}(k, m_t) \le n$.

Proof. Since the map $ev_{Q,D}$ is surjective, it admits a right inverse, *i.e.* a linear map $s : \mathbb{F}_{q^k} \to L(D)$ such that $ev_{Q,D} \circ s = \mathrm{Id}_{\mathbb{F}_{q^k}}$. For all $x \in \mathbb{F}_{q^k}$, we denote $s(x) \in L(D)$ by f_x, so the map $x \mapsto f_x$ is linear, and $f_x(Q) = x$. We also let
$$a : \mathbb{F}_{q^k} \to (\mathbb{F}_q)^n$$
$$x \mapsto (f_x(P_1), \ldots, f_x(P_n))$$

be the composite map $a = ev_{\mathcal{P},D} \circ s$. The situation is sumed up in the following drawing.

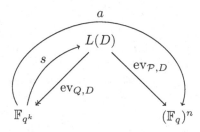

Observe that a is linear, so we can write
$$a(x) = (\varphi_1(x), \ldots, \varphi_n(x))$$

where $\varphi_i : \mathbb{F}_{q^k} \to \mathbb{F}_q$ is a linear form, namely $\varphi_i(x) = f_x(P_i)$.

Similarly, since the map $\mathrm{ev}_{\mathcal{P},tD}$ is injective, it admits a left inverse, *i.e.* a linear map $r : (\mathbb{F}_q)^n \to L(tD)$ such that $r \circ \mathrm{ev}_{\mathcal{P},tD} = \mathrm{Id}_{L(tD)}$. We also let $b : (\mathbb{F}_q)^n \to \mathbb{F}_{q^k}$ be the composite map $b = \mathrm{ev}_{Q,tD} \circ r$. The situation is sumed up in the following drawing.

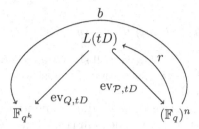

The map b is linear, so there are $b_1, \dots, b_n$ in $\mathbb{F}_{q^k}$ such that, for all $y = (y_1, \dots, y_n) \in (\mathbb{F}_q)^n$,

$$b(y) = \sum_{i=1}^{n} y_i b_i.$$

Now for $x, \dots, x_t \in \mathbb{F}_{q^k}$, let

$$p = (p_1, \dots, p_n) = ((\prod_{j=1}^{t} f_{x_j})(P_1), \dots, (\prod_{j=1}^{t} f_{x_j})(P_n))$$

in $(\mathbb{F}_q)^n$ be the coordinatewise product of the vectors $a(x_1)$, ..., $a(x_t)$. Then

$$h = r(p)$$

is an element of $L(tD)$ such that $h(P_i) = p_i = (\prod_{j=1}^{t} f_{x_j})(P_i)$ for all i. Since the map $\mathrm{ev}_{\mathcal{P},tD}$ is injective, this forces

$$h = \prod_{j=1}^{t} f_{x_j}.$$

Then, we have

$$b(p) = \mathrm{ev}_{Q,tD}(r(p)) = \mathrm{ev}_{Q,tD}(h) = h(Q) = \prod_{j=1}^{t} f_{x_j}(Q) = \prod_{j=1}^{t} x_j.$$

But we also have

$$b(p) = \sum_{i=1}^{n} p_i b_i = \sum_{i=1}^{n} (\prod_{j=1}^{t} f_{x_j}(P_i)) b_i = \sum_{i=1}^{n} (\prod_{j=1}^{t} \varphi_i(x_j)) b_i$$

and finally we get a symmetric formula for m_t:

$$\prod_{j=1}^{t} x_j = \sum_{i=1}^{n} (\prod_{j=1}^{t} \varphi_i(x_j)) b_i.$$

$\square$

Proposition 5. *Let $F/\mathbb{F}_q$ be an algebraic function field of genus g. Assume that F admits a place Q of degree k, and a set $\mathcal{S}$ of places of degree 1 of cardinality*

$$|\mathcal{S}| \geq (k + g - 1)t + 1.$$

Then we have

$$\mu_q^{sym}(k, m_t) \leq kt + (g - 1)(t - 1).$$

Proof. Set $n = kt + (g - 1)(t - 1)$. We will show that there are places $P_1, \ldots, P_n$ in $\mathcal{S}$, and a divisor D on F, such that Proposition 4 applies, which gives $\mu_q^{sym}(k, m_t) \leq n$ as desired.

Using e.g. [3, Lemma 2.1] we know F admits a non-special divisor R of degree $g - 1$. By the strong approximation theorem [18, Thm. 1.6.5] we can then find a divisor D linearly equivalent to $R + Q$ and of support disjoint from Q and $\mathcal{S}$.

Then $D - Q$ and D are non-special, with $\ell(D - Q) = 0$ and $\ell(D) = k$. We thus find

$$\mathrm{Ker}(\mathrm{ev}_{Q,D} : L(D) \to \mathbb{F}_{q^k}) = L(D - Q) = 0,$$

so $\mathrm{ev}_{Q,D}$ is injective, hence also surjective by equality of dimensions, *i.e.* the surjectivity condition (i) in Proposition 4 is satisfied.

Likewise, tD is non-special, with $\deg(tD) = (k + g - 1)t$ and $\ell(tD) = kt + (g - 1)(t - 1)$. Then the evaluation map

$$\begin{aligned}
\mathrm{ev}_{\mathcal{S},tD} : L(tD) &\to \quad (\mathbb{F}_q)^{|\mathcal{S}|} \\
h &\mapsto (h(P))_{P \in \mathcal{S}}
\end{aligned}$$

has kernel $L(tD - \sum_{P \in \mathcal{S}} P) = 0$, because $\deg(tD - \sum_{P \in \mathcal{S}} P) = (k+g-1)t - |\mathcal{S}| < 0$. So $\mathrm{ev}_{\mathcal{S},tD}$ is injective, with image of dimension $\dim \mathrm{Im}(\mathrm{ev}_{\mathcal{S},tD}) = \ell(tD) = n$. Then we can find a subset $\mathcal{P} = \{P_1, \ldots, P_n\} \subset \mathcal{S}$ of cardinality n, such that $\mathrm{ev}_{\mathcal{P},tD} : L(tD) \to (\mathbb{F}_q)^n$ is an isomorphism, and the injectivity condition (ii) in Proposition 4 is also satisfied. $\square$

Choice of the Curves for q a Large Enough Square

Proposition 6. *Let t be given, and assume q is a square, $q \geq (t + 2)^2$. Then we have*

$$M_{q,t}^{sym} \leq (1 + \epsilon_t(q))t$$

with $\epsilon_t(q) = \frac{t-1}{\sqrt{q}-t-1}$.

Proof. We know [17] that there exists a family of function fields $F_i/\mathbb{F}_q$ of genus $g_i \to \infty$ such that

(i) $\frac{g_{i+1}}{g_i} \to 1$
(ii) $N_i \sim (\sqrt{q} - 1)g_i$

where $N_i = \mathrm{Card}\,\{P \in \mathbb{P}_{F_i} \mid \deg P = 1\}$ is the number of places of degree 1 of F_i. We can also assume that the sequence g_i is increasing.

For any k let $i(k)$ be the smallest index such that

$$N_{i(k)} \geq (k + g_{i(k)} - 1)t + 1.$$

Such an $i(k)$ always exists since by (ii) we have $N_i \sim (\sqrt{q}-1)g_i$, with $\sqrt{q}-1 > t$.

By definition we thus have

$$N_{i(k)} \geq (k + g_{i(k)} - 1)t + 1 > (k + g_{i(k)-1} - 1)t + 1 > N_{i(k)-1}.$$

As $k \to \infty$ we have $i(k) \to \infty$, and by (i) we get $g_{i(k)} \sim g_{i(k)-1}$, so by (ii) we also get $N_{i(k)} \sim N_{i(k)-1}$. This then gives

$$N_{i(k)} \sim (k + g_{i(k)} - 1)t + 1$$
$$\sim (k + g_{i(k)})t$$

while by (ii),

$$N_{i(k)} \sim (\sqrt{q} - 1)g_{i(k)}.$$

From these two relations we deduce

$$g_{i(k)} \sim \frac{t}{\sqrt{q} - 1 - t}k.$$

For k large enough this implies in particular $2g_{i(k)} + 1 \leq q^{(k-1)/2}(\sqrt{q} - 1)$, so $F_{i(k)}$ admits a place of degree k by [18, Cor. 5.2.10].

From this we are allowed to apply Proposition 5 to $F_{i(k)}$, which gives

$$\mu_q^{\mathrm{sym}}(k, m_t) \leq kt + (g_{i(k)} - 1)(t - 1) \sim kt + g_{i(k)}(t - 1) \sim kt(1 + \epsilon_t(q))$$

as desired. $\square$

Corollary 1. *For q a square, $q \geq (t+3)^2$ we have*

$$M_{q,t}^{hyp} \leq (1 + \epsilon_{t+1}(q))(t + 1),$$

and in particular we have

$$M_q^{tri} \leq 3\left(1 + \frac{2}{\sqrt{q} - 4}\right)$$

for q a square, $q \geq 25$.

Conclusion for Arbitrary q

Lemma 4. *Let q be a prime power. Then for any integers t, d, k we have*

$$\mu_q^{sym}(k, m_t) \leq \mu_q^{sym}(dk, m_t) \leq \mu_q^{sym}(d, m_t)\mu_{q^d}^{sym}(k, m_t).$$

Proof. For the inequality on the left, there is nothing to prove if $\mu_q^{\mathrm{sym}}(dk, m_t) = \infty$. So let us assume $m_t^{\mathbb{F}_{q^{dk}}/\mathbb{F}_q}$ admits a symmetric multiplication formula of length $n = \mu_q^{\mathrm{sym}}(dk, m_t)$, *i.e.*

$$\forall x_1, \ldots, x_t \in \mathbb{F}_{q^{dk}}, \quad x_1 \cdots x_t = \sum_{i=1}^{n} \varphi_i(x_1) \cdots \varphi_i(x_t) a_i$$

for linear forms $\varphi_i : \mathbb{F}_{q^{dk}} \to \mathbb{F}_q$ and elements $a_i \in \mathbb{F}_{q^{dk}}$. Choose a linear projection

$$p : \mathbb{F}_{q^{dk}} \to \mathbb{F}_{q^k}$$

left inverse for the inclusion $\mathbb{F}_{q^k} \subseteq \mathbb{F}_{q^{dk}}$. Then we get

$$\forall x_1, \ldots, x_t \in \mathbb{F}_{q^k}, \quad x_1 \cdots x_t = p(x_1, \ldots, x_t) = \sum_{i=1}^{n} \varphi_i(x_1) \cdots \varphi_i(x_t) p(a_i)$$

which is a symmetric multiplication formula of length n for $m_t^{\mathbb{F}_{q^k}/\mathbb{F}_q}$.

Likewise, for the inequality on the right, there is nothing to prove if $\mu_q^{\mathrm{sym}}(d, m_t) = \infty$ or $\mu_{q^d}^{\mathrm{sym}}(k, m_t) = \infty$. So let us assume $m_t^{\mathbb{F}_{q^d}/\mathbb{F}_q}$ and $m_t^{\mathbb{F}_{q^{dk}}/\mathbb{F}_{q^d}}$ admit symmetric multiplication formulae of length $r = \mu_q^{\mathrm{sym}}(d, m_t)$ and $s = \mu_{q^d}^{\mathrm{sym}}(k, m_t)$ respectively, so

$$\forall y_1, \ldots, y_t \in \mathbb{F}_{q^d}, \quad y_1 \cdots y_t = \sum_{u=1}^{r} \psi_u(y_1) \cdots \psi_u(y_t) b_u$$

$$\forall z_1, \ldots, z_t \in \mathbb{F}_{q^{dk}}, \quad z_1 \cdots z_t = \sum_{v=1}^{s} \chi_v(z_1) \cdots \chi_v(z_t) c_v$$

for linear forms $\psi_u : \mathbb{F}_{q^d} \to \mathbb{F}_q$, $\chi_v : \mathbb{F}_{q^{dk}} \to \mathbb{F}_{q^d}$ and elements $b_u \in \mathbb{F}_{q^d}$, $c_v \in \mathbb{F}_{q^{dk}}$. Then setting $y_1 = \chi_v(z_1), \ldots, y_t = \chi_v(z_t)$ we find

$$\forall z_1, \ldots, z_t \in \mathbb{F}_{q^{dk}}, \quad z_1 \cdots z_t = \sum_{v=1}^{s} \sum_{u=1}^{r} (\psi_u \circ \chi_v)(z_1) \cdots (\psi_u \circ \chi_v)(z_t) \cdot (b_u c_v)$$

which is a symmetric multiplication formula of length rs for $m_t^{\mathbb{F}_{q^{dk}}/\mathbb{F}_q}$. $\square$

Theorem 2. *Let $t \geq 2$ be an integer and q a prime power. If $q < t$, then $\mu_q^{\mathrm{sym}}(k, m_t) = \infty$ for all $k \geq 2$.*

On the other hand, if $q \geq t$, then $\mu_q^{\mathrm{sym}}(k, m_t)$ grows at most linearly with k, i.e. we have

$$M_{q,t}^{\mathrm{sym}} \leq C_t(q)$$

for some real constant $C_t(q) < \infty$.

Proof. If $q < t$ and $k \geq 2$, then $\mu_q^{\mathrm{sym}}(k, m_t) = \infty$ follows from Theorem 1.

On the other hand, for $q \geq t$, we have $\mu_q^{\mathrm{sym}}(d, m_t) < \infty$ for any integer d. Choose d such that q^d is a square, $q^d \geq (t+2)^2$. Then Proposition 6 shows $\mu_{q^d}^{\mathrm{sym}}(k, m_t)$ grows linearly with k. The Theorem then follows thanks to Lemma 4, with $C_t(q) = \mu_q^{\mathrm{sym}}(d, m_t)(1 + \epsilon_t(q^d))t$. $\square$

Corollary 2. *For $q \geq t+1$ we have*

$$M_{q,t}^{hyp} \leq C_{t+1}(q)$$

and in particular for $q \geq 3$ we have

$$M_q^{tri} \leq C_3(q).$$

Further Remarks and Possible Improvements

1. When $q \geq 4$ is not divisible by 3, [16, Thm. 2] gives $\mu_q^{tri}(k) \leq 4\mu_q^{sym}(k)$. On the other hand, [9] shows that $\mu_q^{sym}(k)$ grows linearly with k (the result is stated for $\mu_q(k)$, but it is easily seen that the proof works for $\mu_q^{sym}(k)$). Taken together, these results show that $\mu_q^{tri}(k)$ grows linearly with k when $q \geq 4$ is not divisible by 3. One advantage of our method is that it works for all $q \geq 3$. Moreover it gives sharper bounds. For instance, when q is a square and large enough, joining [16, Thm. 2] with the best asymptotic upper bound known on $\mu_q^{sym}(k)$ [12, Thm. 6.4] gives $M_q^{tri} \leq 8\left(1 + \frac{1}{\sqrt{q}-2}\right)$, which is not as good as $M_q^{tri} \leq 3\left(1 + \frac{2}{\sqrt{q}-4}\right)$ from Corollary 1.

2. *Open question*: Lemma 3 reduces (upper) bounds on $\mu^{hyp}(m_t)$ to bounds on $\mu^{sym}(m_{t+1})$, and in particular it reduces bounds on M_q^{tri} to bounds on $M_{q,3}^{hyp}$, which does not seem optimal. Indeed we know no example where the inequality $\mu_q^{sym}(k) \leq \mu_q^{tri}(k)$ is strict. So, for instance for q square, $q \to \infty$, our method gives $M_q^{tri} \leq 3(1 + o(1))$, but one could ask whether it is possible to get a bound of the form $M_q^{tri} \leq 2(1 + o(1))$, as given by [12, Thm. 6.4] for M_q^{sym}.

3. *Open question*: The condition $|\mathcal{S}| \geq (k+g-1)t+1$ in Proposition 5 does not seem optimal since in the end we do evaluation-interpolation at only $kt+(g-1)(t-1)$ places. If one could relax this condition to $|\mathcal{S}| \geq kt+(g-1)(t-1)$, this would improve Proposition 6 to $M_{q,t}^{sym} \leq (1+\epsilon_t'(q))t$ for q square, $q \geq (t+1)^2$, with $\epsilon_t'(q) = \frac{t-1}{\sqrt{q}-t}$. For $t = 2$ this is done in [12,15] using techniques from [13]. However, as observed at the end of [13], a generalization to $t \geq 3$ would require new arguments.

4. Lemma 4, which generalizes [17, Lemma 1.2], is clearly not optimal. When deriving upper bounds on $\mu_q^{sym}(k, m_t)$ for non-square q, it might be better to use evaluation-interpolation at places of higher degree, as first introduced in [4], and further developped e.g. in [8,12]. To do this in an optimal way one needs function fields F_i defined over $\mathbb{F}_q$, of genus $g_i \to \infty$, with $\frac{g_{i+1}}{g_i} \to 1$ and $N_i^{(d)} \sim \frac{q^{d/2}-1}{d}g_i$ where $N_i^{(d)}$ is the number of places of degree d in F_i, for a convenient d. This improves the bound on $M_{q,t}^{sym}$ by a factor $\frac{1}{d}$. The existence of these function fields was first claimed in [8], but unfortunately with an incorrect proof. A corrected construction, based on Drinfeld modular curves, will be found in [6].

5. All our bounds for multiplication in extension fields also hold for truncated polynomials. For instance we have $\hat{M}_{q,t}^{\mathrm{sym}} \leq (1 + \epsilon_t(q))t$ for q square, $q \geq (t + 2)^2$, and $\hat{M}_{q,t}^{\mathrm{sym}} \leq C_t(q)$ for all $q \geq t$. This requires only minor changes in our constructions. In Proposition 4, instead of evaluation at a place Q of degree k, one uses evaluation at order k at an extra place P_0 of degree 1. Likewise in Proposition 5, one needs one more place of degree 1, but one does not need Q (then the proof of Proposition 6 is slightly simplified since one does not need to invoke [18, Cor. 5.2.10] anymore).

References

1. Julia: a high-level, high-performance dynamic language for technical computing. http://julialang.org
2. Ballet, S., Chaumine, J., Pieltant, J., Rambaud, M., Randriambololona, H., Rolland, R.: On the tensor rank of multiplication in finite extensions of finite fields and related issues in algebraic geometry. Russian Math. Surv. (to appear)
3. Ballet, S.: Curves with many points and multiplication complexity in any extension of $\mathbb{F}_q$. Finite Fields Appl. **5**, 364–377 (1999)
4. Ballet, S., Rolland, R.: Multiplication algorithm in a finite field and tensor rank of the multiplication. J. Algebra **272**, 173–185 (2004)
5. Barbulescu, R., Detrey, J., Estibals, N., Zimmermann, P.: Finding optimal formulae for bilinear maps. In: Özbudak, F., Rodríguez-Henríquez, F. (eds.) WAIFI 2012. LNCS, vol. 7369, pp. 168–186. Springer, Heidelberg (2012). https://doi.org/10.1007/978-3-642-31662-3_12
6. Bassa, A., Beelen, P., Rambaud, M., Randriam, H.: In preparation
7. Bshouty, N.H.: Multilinear complexity is equivalent to optimal tester size. Electron. Colloquium Comput. Complexity **20**, 11 (2013)
8. Cascudo, I., Cramer, R., Xing, C., Yang, A.: Asymptotic bound for multiplication complexity in the extensions of small finite fields. IEEE Trans. Inf. Theory **58**(7), 4930–4935 (2012)
9. Chudnovsky, D.V., Chudnovsky, G.V.: Algebraic complexities and algebraic curves over finite fields. J. Complexity **4**(4), 285–316 (1988)
10. Covanov, S.: Improved method for finding optimal formulas for bilinear maps in a finite field. Theoret. Comput. Sci. (2019)
11. Karatsuba, A.: Multiplication of multidigit numbers on automata. Soviet Physics Doklady **7**, 595–596 (1963)
12. Randriambololona, H.: Bilinear complexity of algebras and the Chudnovsky-Chudnovsky interpolation method. J. Complexity **28**(4), 489–517 (2012)
13. Randriambololona, H.: (2, 1)-separating systems beyond the probabilistic bound. Israel J. Math. **195**(1), 171–186 (2013)
14. Randriambololona, H.: On products and powers of linear codes under componentwise multiplication. In: Algorithmic Arithmetic, Geometry, and Coding Theory, Contemporary Mathematics, vol. 637, pp. 3–78. AMS (2015)
15. Randriam, H.: Gaps between prime numbers and tensor rank of multiplication in finite fields. Des. Codes Cryptogr. 627–645 (2018). https://doi.org/10.1007/s10623-018-0584-0
16. Seroussi, G., Lempel, A.: On symmetric algorithms for bilinear forms over finite fields. J. Algorithms **5**, 327–344 (1984)

17. Shparlinski, I.E., Tsfasman, M.A., Vladut, S.G.: Curves with many points and multiplication in finite fields. In: Stichtenoth, H., Tsfasman, M.A. (eds.) Coding Theory and Algebraic Geometry. LNM, vol. 1518, pp. 145–169. Springer, Heidelberg (1992). https://doi.org/10.1007/BFb0087999
18. Henning Stichtenoth. Algebraic Function Fields and Codes, vol. 254. Springer, Heidelberg (2009). https://doi.org/10.1007/978-3-540-76878-4
19. Strassen, V.: Gaussian elimination is not optimal. Numerische Mathematik **13**(4), 354–356 (1969)
20. Winograd, S.: On multiplication of 2×2 matrices. Linear Algebra Appl. **4**, 381–388 (1971)
21. Winograd, S.: Some bilinear forms whose multiplicative complexity depends on the field of constants. Math. Syst. Theory **10**, 169–180 (1977)

Coding Theory

A Construction of Self-dual Skew Cyclic and Negacyclic Codes of Length n over $\mathbb{F}_{p^n}$

Aicha Batoul[1] , Delphine Boucher[2]([✉]) , and Ranya Djihad Boulanouar[1]

[1] Faculty of Mathematics, University of Science and Technology Houari
Boumedienne (USTHB), 16111 Bab Ezzouar, Algiers, Algeria
[2] Univ Rennes, CNRS, IRMAR - UMR 6625, 35000 Rennes, France
delphine.boucher@univ-rennes1.fr

Abstract. The aim of this note is to give a construction and an enumeration of self-dual θ-cyclic and θ-negacyclic codes of length n over $\mathbb{F}_{p^n}$ where p is a prime number and θ is the Frobenius automorphism over $\mathbb{F}_{p^n}$. We use the notion of isodual codes to achieve this construction.

Keywords: Coding theory · Finite fields · Skew polynomial rings · Self-dual codes

1 Introduction

Isodual codes [17] have been recently studied on many aspects [1–3]. Meanwhile, in [5], a construction and an enumeration formula for self-dual θ-cyclic and θ-negacyclic codes of even length n over $\mathbb{F}_{p^2}$ were given in the case when p is a prime number and θ is the Frobenius automorphism over $\mathbb{F}_{p^2}$. The aim of this note is to give a construction and an enumeration formula for self-dual θ-cyclic and θ-negacyclic codes of length n over $\mathbb{F}_{p^n}$ when θ is the Frobenius automorphism over $\mathbb{F}_{p^n}$. To this end, we will use and develop the notion of (θ, ν)-isodual codes which form a subfamily of the family of isodual codes. Lastly we will consider the construction of some self-dual Gabidulin evaluation codes.

The text is organized as follows. In Sect. 2 we define the notion of (θ, ν)-isodual codes over $\mathbb{F}_q$ where θ is an automorphism of $\mathbb{F}_q$ and ν belongs to $\mathbb{F}_q^*$. We recall the definitions of (θ, a)-constacyclic, θ-cyclic and θ-negacyclic codes and some generalities on the dual of a (θ, a)-constacyclic code. Then we characterize (θ, ν)-isodual θ-cyclic and θ-negacyclic codes thanks to an equation satisfied by the skew check polynomials of the codes. In Sect. 3 we consider the special case when q is equal to p^n where p is a prime number and θ is the Frobenius automorphism over $\mathbb{F}_{p^n}$. After having given a necessary and sufficient condition for the existence of (θ, ν)-isodual θ-cyclic and θ-negacyclic codes, we give a construction and an enumeration formula for (θ, ν)-isodual and self-dual θ-cyclic and θ-negacyclic codes. In Sect. 4, we consider a subclass of self-dual θ-cyclic codes over $\mathbb{F}_{p^n}$ which are self-dual Gabidulin codes. We parametrize this family by a parameter which satisfies a polynomial system.

© Springer Nature Switzerland AG 2021
J. C. Bajard and A. Topuzoğlu (Eds.): WAIFI 2020, LNCS 12542, pp. 115–133, 2021.
https://doi.org/10.1007/978-3-030-68869-1_6

2   Some Generalities on Isodual Skew Codes

We first recall that a *linear code* C of length n and dimension k over $\mathbb{F}_q$ is a subspace of dimension k of $\mathbb{F}_q^n$. A *generator matrix* G of C is a $k \times n$ matrix with coefficients in $\mathbb{F}_q$ and rank k such that $C = \{m\,G \mid m \in \mathbb{F}_q^k\}$. Furthermore the dual of C is $C^\perp = \{x \in \mathbb{F}_q^n \mid \forall c \in C, < x, c >= 0\}$ where for $x = (x_0, \ldots, x_{n-1}), y = (y_0, \ldots, y_{n-1})$ in $\mathbb{F}_q^n$, $< x, y >:= \sum_{i=0}^{n-1} x_i y_i$ is the Euclidean scalar product of x and y. Isodual codes [17] have been recently studied on many aspects [1–3].

Definition 1 ([17] **page 199**). *A code C with generator matrix G is isodual if it is equivalent to its dual. That means that there exists a monomial matrix D such that $G \cdot D$ is a generator matrix of the dual $C^\perp$ of C.*

In what follows, we define a special class of isodual codes which are parameterized by an automorphism θ of $\mathbb{F}_q$ and an element ν of $\mathbb{F}_q^*$.

Definition 2. *Consider $n \in \mathbb{N}^*, \nu \in \mathbb{F}_q^*$ and $\theta \in Aut(\mathbb{F}_q)$. A linear code C of length n and generator matrix G is a (θ, ν)-isodual code if $G \cdot D$ is a generator matrix of $C^\perp$ where D is the $n \times n$ diagonal matrix with diagonal coefficients $\nu, \theta(\nu), \ldots, \theta^{n-1}(\nu)$.*

Remark 1. A code C is self-dual if and only if there exists ν fixed by θ such that C is (θ, ν)-isodual.

Recall that if θ is an automorphism of $\mathbb{F}_q$, the skew polynomial ring R is defined as $R = \mathbb{F}_q[X; \theta]$ under usual addition of polynomials and where multiplication is defined by the commutation law: $\forall a \in \mathbb{F}_q, X \cdot a = \theta(a)X$ ([16]). The ring R is noncommutative unless θ is the identity automorphism on $\mathbb{F}_q$. The ring R is right-Euclidean and left-Euclidean. For $f = \sum a_i X^i$ in R and α in $\mathbb{F}_q$, the *evaluation* $f(\alpha)$ of f at α is the remainder in the right division of f by $X - \alpha$. We have $f(\alpha) = \sum_i a_i N_i(\alpha)$ where $N_i(x) := x\theta(x) \cdots \theta^{i-1}(x)$ (see [14]). Recall also that if $q = p^n$ and θ is the Frobenius automorphism, then the center of R is $\mathbb{F}_p[X^n]$.

For a in $\mathbb{F}_q^*$ and θ in $Aut(\mathbb{F}_q)$, a (θ, a)-*constacyclic* code C of length n and dimension k is a left R-submodule $Rg/R(X^n - a) \subset R/R(X^n - a)$ where g is a monic skew polynomial of degree $n - k$ right-dividing $X^n - a$ in R ([7]). That means that a word $c = (c_0, \ldots, c_{n-1}) \in \mathbb{F}_q^n$ belongs to C if and only if the skew polynomial g right-divides the skew polynomial $c_0 + c_1 X + \cdots + c_{n-1} X^{n-1}$ in R. The skew polynomial g is called the *skew generator polynomial* of C. The monic skew polynomial h defined by

$$\Theta^n(h) \cdot g = X^n - a \tag{1}$$

is called *skew check polynomial* of C.

The (θ, a)-constacyclic code C is denoted $C = (g)_{n,\theta}^a$. If $a = 1$, the code is θ-*cyclic* and if $a = -1$, the code is θ-*negacyclic*.

A generator matrix of C is

$$
G = \begin{pmatrix}
g_0 & g_1 & \cdots & & \cdots & 1 & 0 \cdots 0 \\
0 & \theta(g_0) & \theta(g_1) & \cdots & & \cdots & 1 & \ddots & \vdots \\
\vdots & & \ddots & \ddots & & & & \ddots & 0 \\
0 & \cdots & 0 & \theta^{k-1}(g_0) & \theta^{k-1}(g_1) & & \cdots & \cdots & 1
\end{pmatrix}.
\tag{2}
$$

The *skew reciprocal polynomial* of $h = \Sigma_{i=0}^{k} h_i X^i \in R$ of degree k is $h^* = \Sigma_{i=0}^{k} \theta^i (h_{k-i}) X^i$. If $h_0 \neq 0$, the *left monic skew reciprocal polynomial of h* is $h^\natural = \frac{1}{\theta^k(h_0)} h^*$. The following technical lemma will be useful later. We will use the application $\Theta : R \mapsto R$ given by $\sum_{i=0}^{k} a_i X^i \mapsto \sum_{i=0}^{k} \theta(a_i) X^i$.

Lemma 1 (Lemma 1 of [8]). *Consider $\theta \in Aut(\mathbb{F}_q), R = \mathbb{F}_q[X; \theta], h$ and g in R. Then $(h \cdot g)^* = \Theta^{\deg(h)}(g^*) \cdot h^*$.*

Example 1. Consider $n = 4, \mathbb{F}_{2^4} = \mathbb{F}_2(a)$ with $a^4 + a + 1 = 0$ and $R = \mathbb{F}_{2^4}[X; \theta]$. We have

$$
X^4 + 1 = (X^2 + a^5 X + a^5) \cdot (X^2 + a^5 X + a^{10})
$$

therefore the skew polynomial $g_1 = X^2 + a^5 X + a^{10}$ generates a θ-cyclic code C_1 of length 4 and dimension 2 over $\mathbb{F}_{2^4}$. As Θ^4 is the identity over $\mathbb{F}_{2^4}$, the skew check polynomial of the code is $h_1 = X^2 + a^5 X + a^5$.
We have

$$
X^4 + 1 = (X^2 + aX + a^{14}) \cdot (X^2 + a^4 X + a)
$$

therefore the skew polynomial $g_2 = X^2 + a^4 X + a$ generates a θ-cyclic code C_2 of length 4 and dimension 2 over $\mathbb{F}_{2^4}$ with skew check polynomial $h_2 = X^2 + aX + a^{14}$.

The following proposition describes the dual of a (θ, a)-constacyclic code.

Proposition 1 (Theorem 1 and Lemma 2 of [8], Proposition 1 of [6]). *Consider $n \in \mathbb{N}^*, a \in \mathbb{F}_q^*, \theta \in Aut(\mathbb{F}_q)$ and C a (θ, a)-constacyclic code of length n with skew generator polynomial g and skew check polynomial h. Then the dual $C^\perp$ of C is a $(\theta, 1/a)$-constacyclic code with skew generator polynomial $h^\natural$.*

Proof. (proof of Proposition 1 of [6]) We consider the equality (1) in $R = \mathbb{F}_q[X; \theta]$ and we multiply both members of this equality by h on the right. We get $\Theta^n(h) \cdot g \cdot h = (X^n - a) \cdot h$ and we deduce from this equality that $\Theta^n(h) \cdot (X^n - g \cdot h) = a \cdot h$. As the skew polynomials $\Theta^n(h)$ and $a \cdot h$ have the same degrees, the skew polynomial $X^n - g \cdot h$ is a constant that we will denote λ and $\Theta^n(h) \cdot \lambda - a \cdot h = 0$. As the leading coefficient of $\Theta^n(h) \cdot \lambda - a \cdot h$ is equal to $\theta^k(\lambda) - a$, we get that $\lambda = \theta^{-k}(a)$.

Furthermore, as $\Theta^n(h) \cdot g = X^n - a$, according to Lemma 1, we have $-\frac{1}{a}\Theta^{k-n}(g^*) \cdot h^* = X^n - \frac{1}{a}$. Therefore $h^\natural$ right-divides $X^n - \frac{1}{a}$ and is the skew generator polynomial of a $(\theta, \frac{1}{a})$-constacyclic code of length n.

A quick computation gives that for all (i, j) in $\{0, \ldots, k-1\} \times \{0, \ldots, n-k-1\}$, the Euclidean scalar product of the words associated to $X^i \cdot g$ and $X^j \cdot h^*$ is equal to $\theta^i((g \cdot h)_{j-i+k})$. Therefore the scalar product is equal to 0 and the words of the code $(g)^a_{n,\theta}$ are orthogonal to the words of the code $(h^\natural)^{1/a}_{n,\theta}$.

In what follows, we characterize (θ, a)-constacyclic codes which are (θ, ν)-isodual.

Proposition 2. *Consider* $k \in \mathbb{N}^*, n = 2k, \nu \in \mathbb{F}_q^*, a \in \mathbb{F}_q, \theta \in Aut(\mathbb{F}_q), R = \mathbb{F}_q[X; \theta], h \in R$ *monic. The* (θ, a)-*constacyclic code of length* n, *dimension* k *and skew check polynomial* h *is* (θ, ν)-*isodual if and only if*

$$\Theta^n(h) \cdot \theta^k(\nu) \cdot h^\natural \cdot \frac{1}{\nu} = X^n - a. \tag{3}$$

In this case we have $a^2 = \theta^n(\nu)/\nu$.

Proof. Consider $C = (g)^a_{n,\theta}$ the (θ, a)-constacyclic code of length $n = 2k$, dimension k, skew check polynomial h and skew generator polynomial g. According to (1), we have $\Theta^n(h) \cdot g = X^n - a$. Therefore, the relation (3) is satisfied if and only if $h^\natural = \tilde{g}$ where $\tilde{g} = \theta^k(1/\nu) \cdot g \cdot \nu$.

Let us prove that C is (θ, ν)-isodual if and only if $h^\natural = \tilde{g}$. As g right-divides $X^n - a$, $\tilde{g}$ right-divides $X^n - \tilde{a}$ where $\tilde{a} = a\frac{\nu}{\theta^n(\nu)}$. Therefore we can consider the $(\theta, \tilde{a})$-constacyclic code of length n and skew generator polynomial $\tilde{g}$. Furthermore, according to Proposition 1, $C^\perp$ is a $(\theta, 1/a)$-constacyclic code of length n with skew generator polynomial $h^\natural$.

Let us prove that C is (θ, ν)-isodual if and only if $(h^\natural)^{1/a}_{n,\theta} = (\tilde{g})^{\tilde{a}}_{n,\theta}$.

Denote $g = \sum_{i=0}^{n-k} g_i X^i = \sum_{i=0}^{k} g_i X^i$ and $\tilde{g} = \sum_{i=0}^{k} \tilde{g}_i X^i$. We have $\tilde{g}_i = \theta^k(1/\nu)g_i\theta^i(\nu)$ for all i in $\{0, \ldots, k\}$. Therefore a generator matrix of $(\tilde{g})^{\tilde{a}}_{n,\theta}$ is

$$\tilde{G} = \begin{pmatrix} \tilde{g}_0 & \tilde{g}_1 & \cdots & & \cdots & 1 & 0 & \cdots 0 \\ 0 & \theta(\tilde{g}_0) & \theta(\tilde{g}_1) & & \cdots & & 1 & \ddots \vdots \\ \vdots & & \ddots & \ddots & & & & \ddots 0 \\ 0 & \cdots & 0 & \theta^{k-1}(\tilde{g}_0) & \theta^{k-1}(\tilde{g}_1) & \cdots & \cdots & 1 \end{pmatrix} = G \cdot D$$

where G is given by (2) and D is the diagonal matrix with diagonal elements $\nu, \theta(\nu), \ldots, \theta^{n-1}(\nu)$.

According to Definition 2, the code C is (θ, ν)-isodual if and only if a generator matrix of $C^\perp = (h^\natural)^{1/a}_{n,\theta}$ is $G \cdot D$. As $G \cdot D = \tilde{G}$ is a generator matrix of $(\tilde{g})^{\tilde{a}}_{n,\theta}$, we obtain that C is (θ, ν)-isodual if and only if $(h^\natural)^{1/a}_{n,\theta} = (\tilde{g})^{\tilde{a}}_{n,\theta}$.

Lastly as $\tilde{g}$ right-divides $X^n - a\frac{\nu}{\theta^n(\nu)}$ and $h^\natural$ right-divides $X^n - \frac{1}{a}$, we obtain $a^2 = \theta^n(\nu)/\nu$.

In the case when θ is the Frobenius automorphism over $\mathbb{F}_q$ and $q = p^n$ where p is prime and n is the length of the code we obtain the following corollary that will be useful in next section.

Corollary 1. *Consider* $k \in \mathbb{N}^*, n = 2k, p$ *a prime number,* $\nu \in \mathbb{F}_{p^n}^*, a \in \mathbb{F}_{p^n}, \theta : x \mapsto x^p \in Aut(\mathbb{F}_{p^n}), R = \mathbb{F}_{p^n}[X; \theta], h \in R$ *monic. The* (θ, a)-*constacyclic code of length* n *and skew check polynomial* h *is* (θ, ν)-*isodual if and only if*

$$h \cdot \theta^k(\nu) \cdot h^\natural \cdot \frac{1}{\nu} = h^\natural \cdot \frac{1}{\nu} \cdot h \cdot \theta^k(\nu) = X^n - a. \tag{4}$$

Furthermore, $a^2 = 1$.

Proof. As θ is the Frobenius automorphism over $\mathbb{F}_{p^n}$, the order of θ is equal to n. Therefore $\Theta^n(h) = h$ and $\theta^n(\nu) = \nu$. According to Proposition 2, the (θ, a)-constacyclic code of length n and skew check polynomial h is (θ, ν)-isodual if and only if $h \cdot \theta^k(\nu) \cdot h^\natural \cdot \frac{1}{\nu} = X^n - a$. In this case $a^2 = 1$. Therefore $X^n - a$ is central in R, and we have $h \cdot \theta^k(\nu) \cdot h^\natural \cdot \frac{1}{\nu} = X^n - a = h^\natural \cdot \frac{1}{\nu} \cdot h \cdot \theta^k(\nu)$.

Example 2. (Example 1 continued) The left monic skew reciprocal polynomial of $h_1 = X^2 + a^5 X + a^5$ is $h_1^\natural = X^2 + a^5 X + a^{10}$ and $X^4 + 1 = (X^2 + a^5 X + a^5) \cdot (X^2 + a^5 X + a^{10}) = (X^2 + a^5 X + a^{10}) \cdot (X^2 + a^5 X + a^5)$, therefore the θ-cyclic code C_1 with skew check polynomial h_1 is self-dual.

The left monic skew reciprocal polynomial of $h_2 = X^2 + aX + a^{14}$ is $h_2^\natural = X^2 + a^6 X + a^4$. Furthermore, $X^4 + 1 = (X^2 + aX + a^{14}) \cdot (X^2 + a^4 X + a) = (X^2 + aX + a^{14}) \cdot \frac{1}{a^4} \cdot (X^2 + a^6 X + a^4) \cdot \frac{1}{a^{14}}$, therefore the θ-cyclic code C_2 with skew check polynomial h_2 is (θ, a^{14})-isodual.

Lastly, we consider below a technical lemma which will be useful later and which deals with the factorization of skew polynomials right-dividing $X^n \pm 1$ in $\mathbb{F}_{p^n}[X; \theta]$ where θ is the Frobenius automorphism. These skew polynomials belong to a wide class of skew polynomials, called Wedderburn polynomials, which have been extensively studied (see Theorem 6.4 of [11] for the factorizations of these skew polynomials). Lemma 2 can be directly deduced from Theorem 6.4 of [11] as well as from Proposition 2.2.2. of [10]. We propose here a proof very specific to our special case.

Lemma 2. *Consider* $n \in \mathbb{N}^*$, p *a prime number,* $\theta : x \mapsto x^p \in Aut(\mathbb{F}_{p^n})$, $R = \mathbb{F}_{p^n}[X; \theta]$, f *in* R *of degree* d *and* $\epsilon \in \{-1, 1\}$ *such that* f *right-divides* $X^n - \epsilon$ *in* R. *Then* f *is the product of* d *linear factors right-dividing* $X^n - \epsilon$ *and*

$$\#\{(\alpha_1, \ldots, \alpha_d) \in \mathbb{F}_{p^n}^d \mid f = (X + \alpha_1) \cdots (X + \alpha_d)\} = \prod_{i=1}^{d} \frac{p^i - 1}{p - 1}.$$

Proof. Consider $y_1, \ldots, y_n$ in $\mathbb{F}_{p^n}$ linearly independent over $\mathbb{F}_p$. Consider ξ in $\mathbb{F}_{p^n}$ such that $X - \xi$ right-divides $X^n - \epsilon$, which means $N_n(\xi) = \epsilon$. Denote $\alpha_1 := \xi \frac{\theta(y_1)}{y_1}, \ldots, \alpha_n = \xi \frac{\theta(y_n)}{y_n}$. According to [14], the least common left multiple of $X - \alpha_1, \ldots, X - \alpha_n$ is $\text{lclm}_{1 \le i \le n}(X - \alpha_i) = X^n - \epsilon$. As f right-divides

$X^n - \epsilon$, according to Theorem 4 of [16], there exist $\beta_1, \ldots, \beta_d$ in $\mathbb{F}_{p^n}$ such that $f = \mathrm{lclm}_{1 \le i \le d}(X - \beta_i)$. Furthermore $N_n(\beta_i) = \epsilon = N_n(\xi)$. Therefore according to Theorem 27 of [13], there exists z_i in $\mathbb{F}_{p^n}^*$ such that $\beta_i = \xi \frac{\theta(z_i)}{z_i}$. According to [14], $z_1, \ldots, z_d$ are linearly independent over $\mathbb{F}_p$. Denote $f = \sum_{i=0}^d a_i X^i$, we have $\{\alpha \in \mathbb{F}_{p^n} \mid f(\alpha) = 0\} = \{\alpha = \xi \frac{\theta(y)}{y} = \xi y^{p-1} \in \mathbb{F}_{p^n} \mid L(y) = 0\}$ where $L(y) := \sum_{i=0}^d a_i N_i(\xi) \theta^i(y)$. As the equation $L(y) = 0$ has d solutions $(z_1, \ldots, z_d)$ in $\mathbb{F}_{p^n}$ linearly independent over $\mathbb{F}_p$, there are $p^d - 1$ nonzero y in $\mathbb{F}_{p^n}$ such that $L(y) = 0$. Therefore $\{\alpha \in \mathbb{F}_{p^n} \mid X - \alpha \text{ right-divides } f\}$ has $(p^d - 1)/(p - 1)$ elements. We conclude using an inductive argument.

Remark 2. It could be noted that the number $\displaystyle\prod_{i=1}^d \frac{p^i - 1}{p - 1}$ is the size of the general linear group $GL(d, p)$ modulo diagonal matrices, and corresponds to choosing a set of 1-dimensional vector spaces spanning a d-dimensional vector space.

Example 3. (Example 1 continued) The skew polynomial $h_1 = X^2 + a^5 X + a^5$ has 3 factorizations into the product of linear monic skew polynomials, namely $h_1 = (X + a^{14}) \cdot (X + a^6) = (X + a^{11}) \cdot (X + a^9) = (X + 1) \cdot (X + a^5)$. The skew polynomial $h_2 = X^2 + aX + a^{14}$ has also 3 factorizations into the product of linear monic skew polynomials, namely $h_2 = (X + a^7) \cdot (X + a^7) = (X + a^{11}) \cdot (X + a^3) = (X + a^{10}) \cdot (X + a^4)$.

3   Construction and Enumeration of (θ, ν)-isodual θ-cyclic and θ-negacyclic Codes of Length n over $\mathbb{F}_{p^n}$

The aim of this section is to construct and to enumerate self-dual θ-cyclic and θ-negacyclic codes of length n over $\mathbb{F}_{p^n}$ where θ is the Frobenius automorphism. Note that in this setting, θ-cyclic codes are called *Gabidulin p-cyclical codes* (page 6 of [12]).

To achieve this construction, we will consider θ-cyclic and θ-negacyclic codes which are (θ, ν)-isodual.

We introduce some notation. Consider $R = \mathbb{F}_{p^n}[X; \theta]$. We will denote, for $\epsilon \in \{-1, 1\}$ and $\nu \in \mathbb{F}_q^*$:

$$\mathcal{H}_{\nu, \epsilon} := \{h \in R \mid h \text{ monic}, h^\natural \cdot \frac{1}{\nu} \cdot h \cdot \theta^k(\nu) = X^n - \epsilon\}.$$

According to Corollary 1, the set $\mathcal{H}_{\nu, \epsilon}$ is the set of the skew check polynomials of (θ, ν)-isodual (θ, ϵ)-constacyclic codes. Following Remark 1, the set $\mathcal{H}_{1, \epsilon}$ is the set of the skew check polynomials of self-dual (θ, ϵ)-constacyclic codes.

3.1   Necessary and Sufficient Existence Condition for (θ, ν)-isodual θ-cyclic and θ-negacyclic Codes of Length n over $\mathbb{F}_{p^n}$

In [4] a necessary and sufficient condition for the existence of self-dual (θ, ϵ)-constacyclic codes over a finite field $\mathbb{F}_q$ was derived where θ is an automorphism of $\mathbb{F}_q$ and $\epsilon \in \{-1, 1\}$. In what follows we give a necessary and sufficient condition for the existence of (θ, ν)-isodual (θ, ϵ)-constacyclic codes of length n over $\mathbb{F}_{p^n}$ where p is a prime number and θ is the Frobenius automorphism.

Proposition 3. *Consider $k \in \mathbb{N}^*, n = 2k, p$ a prime number, $\theta : x \mapsto x^p \in \mathrm{Aut}(\mathbb{F}_{p^n}), \epsilon \in \{-1, 1\}$.*

(i) *If $p = 2$, then there exists a (θ, ν)-isodual θ-cyclic code of length n for all ν in $\mathbb{F}_{p^n}^*$.*

(ii) *If p is odd, then for $\nu \in \mathbb{F}_{p^n}^*$, there exists a (θ, ν)-isodual (θ, ϵ)-constacyclic code of length $2k$ if and only if*

$$\nu^{\frac{p^n-1}{2}} = -\epsilon(-1)^{k\frac{p-1}{2}}.$$

Proof. According to Corollary 1, there exists a (θ, ν)-isodual (θ, ϵ)-constacyclic code of length n if and only if the set $\mathcal{H}_{\nu, \epsilon}$ is nonempty.

– Assume that $p = 2$ (therefore $\epsilon = 1$) and $\nu \in \mathbb{F}_{2^n}^*$. Consider α in $\mathbb{F}_{2^n}$ such that $\alpha^2 = 1/\nu^{2^k-1} = \frac{\nu}{\theta^k(\nu)}$ and $h = X^k + \alpha$. We have

$$h^\natural \cdot \frac{1}{\nu} \cdot h \cdot \theta^k(\nu) = \left(X^k + \frac{1}{\theta^k(\alpha)}\right) \cdot \left(X^k + \frac{\theta^k(\nu)\alpha}{\nu}\right)$$
$$= X^{2k} + \theta^k\left(\frac{1}{\alpha} + \alpha\frac{\theta^k(\nu)}{\nu}\right)X^k + \frac{\theta^k(\nu)\alpha}{\theta^k(\alpha)\nu}.$$

As $\nu/\theta^k(\nu) = \alpha^2$, we obtain

$$h^\natural \cdot \frac{1}{\nu} \cdot h \cdot \theta^k(\nu) = X^{2k} + \frac{1}{\alpha\theta^k(\alpha)}.$$

As $\theta(\alpha) = \alpha^2 = \frac{\nu}{\theta^k(\nu)}$, we obtain $\theta^{k+1}(\alpha) = \frac{\theta^k(\nu)}{\nu}$, $\theta(\alpha)\theta^{k+1}(\alpha) = 1$ and $\alpha\theta^k(\alpha) = 1$. Therefore the skew polynomial h belongs to $\mathcal{H}_{\nu, \epsilon}$.

– Assume that p is odd and $\nu^{\frac{p^n-1}{2}} = -\epsilon(-1)^{k\frac{p-1}{2}}$. We have

$$\left(-\frac{\theta^k(\nu)}{\nu}\right)^{(p^n-1)/2} = (-1)^{(p^n-1)/2} \frac{\theta^k(\nu^{(p^n-1)/2})}{\nu^{(p^n-1)/2}}$$
$$= (-1)^{(p^n-1)/2}$$
$$= 1 \text{ (because } p \text{ is odd, therefore } p^2 \equiv 1 \pmod 4 \text{ and}$$
$$p^n \equiv 1 \pmod 4).$$

Therefore $-\frac{\nu}{\theta^k(\nu)}$ is a square. Consider α such that $-\frac{\nu}{\theta^k(\nu)} = \alpha^2$. Consider $h = X^k + \alpha$ in R.

$$h^\natural \cdot \frac{1}{\nu} \cdot h \cdot \theta^k(\nu) = \left(X^k + \frac{1}{\theta^k(\alpha)}\right) \cdot \left(X^k + \frac{\theta^k(\nu)\alpha}{\nu}\right)$$
$$= X^{2k} + \theta^k\left(\frac{1}{\alpha} + \alpha\frac{\theta^k(\nu)}{\nu}\right)X^k + \frac{\theta^k(\nu)\alpha}{\theta^k(\alpha)\nu}.$$

As $\nu/\theta^k(\nu) = -\alpha^2$, we obtain

$$h^\natural \cdot \frac{1}{\nu} \cdot h \cdot \theta^k(\nu) = X^{2k} + \frac{\theta^k(\nu)\alpha}{\theta^k(\alpha)\nu}.$$

Furthermore

$$\frac{\theta^k(\nu)\alpha}{\theta^k(\alpha)\nu} = \frac{\theta^k(\nu)}{\nu}\left(-\frac{\theta^k(\nu)}{\nu}\right)^{\frac{p^k-1}{2}} \quad \text{(because } -\theta^k(\nu)/\nu = 1/\alpha^2\text{)}$$

$$= (-1)^{\frac{p^k-1}{2}}\nu^{\frac{p^n-1}{2}}$$

$$= -\epsilon \quad \text{(because } \nu^{\frac{p^n-1}{2}} = -\epsilon(-1)^{k\frac{p-1}{2}}\text{)}.$$

Therefore

$$h^\natural \cdot \frac{1}{\nu} \cdot h \cdot \theta^k(\nu) = X^n - \epsilon.$$

– Assume that p is odd and that there exists a (θ, ϵ)-constacyclic code of length $2k$ which is (θ, ν)-isodual. Consider h its skew check polynomial and h_0 the constant term of h. Necessarily the degree of h is equal to k.
As the code is (θ, ϵ)-constacyclic of length $2k$, h right-divides $X^{2k} - \epsilon$. As the code is defined over $\mathbb{F}_{p^{2k}}$ with $\theta : x \mapsto x^p$, $X^{2k} - \epsilon$ is central of degree 1 in $\mathbb{F}_p[X^{2k}]$ and h is the product of k linear skew polynomials $X + \alpha_1, \ldots, X + \alpha_k$ right-dividing $X^{2k} - \epsilon$. Therefore h_0 is the product of the k constant terms $\alpha_1, \ldots, \alpha_k$. As $N_{2k}(-\alpha_i) = \epsilon$, we have:

$$N_{2k}(h_0) = \epsilon^k = N_k(h_0)N_k(\theta^k(h_0)).$$

As the code is (θ, ν)-isodual, we have $h^\natural \cdot \frac{1}{\nu} \cdot h \cdot \theta^k(\nu) = X^n - \epsilon$ and

$$h_0\theta^k(\nu) + \epsilon\theta^k(h_0)\nu = 0.$$

As $N_k(h_0)N_k(\theta^k(h_0)) = \epsilon^k$, we obtain $N_k(h_0)^2(-\epsilon)^k\frac{N_{2k}(\nu)}{N_k(\nu)^2} = \epsilon^k$ and

$$N_{2k}(\nu) = (-1)^k N_k(\nu/h_0)^2.$$

Furthermore, $\theta(N_k(\nu/h_0)) = N_k(\nu/h_0)\frac{\theta^k(\nu/h_0)}{\nu/h_0} = N_k(\nu/h_0)(-\epsilon)$. Therefore we have
$N_k(\nu/h_0)^{p-1} = -\epsilon$. To conclude, if p is odd,

$$\nu^{\frac{p^n-1}{2}} = N_{2k}(\nu)^{(p-1)/2} = (-1)^{k\frac{p-1}{2}}N_k(\nu/h_0)^{p-1} = -\epsilon(-1)^{k\frac{p-1}{2}}.$$

Remark 3. Consider $k \in \mathbb{N}^*, n = 2k, p$ an odd prime number, $\theta : x \mapsto x^p \in Aut(\mathbb{F}_{p^n}), \epsilon \in \{-1, 1\}$. According to Proposition 3 and Remark 1, there exists a self-dual (θ, ϵ)-constacyclic code of length $n = 2k$ over $\mathbb{F}_{p^n}$ if and only if $1 = -\epsilon(-1)^{k\frac{p-1}{2}}$. We therefore obtain the previous result of Proposition 5 of [4]: if p is odd, there exists a self-dual θ-cyclic code of dimension k if and only if $p \equiv 3 \pmod 4$ and k is odd; there exists a self-dual θ-negacyclic code of dimension k if and only if $p \equiv 1 \pmod 4$ or $p \equiv 3 \pmod 4$ and k even.

The existence result of Proposition 3 is not constructive and the aim of what follows is to design a construction of the set $\mathcal{H}_{\nu,\epsilon}$ based on the construction of the sets $\tilde{\mathcal{H}}_{\mu,\epsilon}$ defined for $\mu \in \mathbb{F}_{p^n}^*$ by

$$\tilde{\mathcal{H}}_{\mu,\epsilon} := \{h \in R \mid h \in \mathcal{H}_{h_0/\mu,\epsilon}\}.$$

Thanks to factorization properties of the elements of $\tilde{\mathcal{H}}_{\mu,\epsilon}$, we will give both a construction and an enumeration formula.

3.2   Construction and Enumeration Formula for the Set $\tilde{\mathcal{H}}_{\mu,\epsilon}$

The following technical lemma (Lemma 3) will be useful for the construction of $\tilde{\mathcal{H}}_{\mu,\epsilon}$ (Proposition 4).

Lemma 3. *Consider* $k \in \mathbb{N}^*, n = 2k, p$ *a prime number,* $\theta : x \mapsto x^p \in Aut(\mathbb{F}_{p^n}), R = \mathbb{F}_{p^n}[X;\theta], P \in R$ *and* $\ell \in \{0,\dots,k\}$ *such that* $\Theta^{k+\ell}(P^*) = P$ *and* $\deg(P) = 2k - 2\ell$. *If* $X + \alpha$ *right-divides* P *then there exists* $Q \in R$ *satisfying the two following properties:*

1. $P = \Theta^{k-\ell-1}((X + \alpha)^*) \cdot Q \cdot (X + \alpha);$
2. $\Theta^{k+\ell+1}(Q^*) = Q.$

Proof. Consider $f \in R$ such that $P = f \cdot (X + \alpha)$.

1. Let us prove that $P = \Theta^{k-1-\ell}((X + \alpha)^*) \cdot \Theta^{k+\ell}(f^*)$ and that $X + \alpha$ right-divides $\Theta^{k+\ell}(f^*)$.
   As $P = f \cdot (X + \alpha)$, according to Lemma 1, we have $P^* = \Theta^{2k-2\ell-1}((X + \alpha)^*) \cdot f^*$. Therefore $P = \Theta^{k+\ell}(P^*) = \Theta^{k-\ell-1}((X + \alpha)^*) \cdot \Theta^{k+\ell}(f^*)$.
   Denote $K = k - \ell$ and $f = \sum_{j=0}^{2K-1} a_j X^j$, then

$$f^* = \sum_{j=0}^{2K-1} \theta^{2K-1-j}(a_j)X^{2K-1-j}.$$

Consider $\beta = -\theta^K(\alpha)$, then $X + \alpha$ right-divides $\Theta^{k+\ell}(f^*)$ if and only if f^* cancels at β.
Let us prove that

$$\sum_{j=0}^{K-1} \theta^{2K-1-j}(a_j)N_{2K-1-j}(\beta) = -\sum_{j=K}^{2K-1} \theta^{2K-j-1}(a_j)N_{2K-j-1}(\beta).$$

As $P = f \cdot (X + \alpha) = \Theta^{K-1}((X + \alpha)^*) \cdot \Theta^{K+2\ell}(f^*)$, we obtain:
$\forall j \in \{1,\dots,K\}$,

$$a_{j-1} = \sum_{i=1}^{j} \frac{N_{i-1}(-\alpha)}{N_j(-\alpha)}\theta^K(\theta^{2\ell+i-1}(a_{2K-i}) + \alpha\theta^{2\ell+i-1}(a_{2K-i+1})).$$

Therefore, we get:

$$\sum_{j=0}^{K-1} \theta^{2K-1-j}(a_j) N_{2K-1-j}(\beta) =$$

$$= \sum_{j=1}^{K} \theta^{2K-j}(a_{j-1}) N_{2K-j}(\beta)$$

$$= -\sum_{j=1}^{K} N_{2K-j}(\beta)\theta^{2K-j}\left(\sum_{i=1}^{j}\frac{N_{i-1}(-\alpha)}{N_j(-\alpha)}\theta^K(\theta^{2\ell+i-1}(a_{2K-i}) + \alpha\theta^{2\ell+i-1}(a_{2K-i+1})))\right)$$

$$= -\sum_{j=1}^{K} N_{2K-j}(\beta)\sum_{i=1}^{j}\theta^{K-j}\left(\frac{N_{i-1}(\beta)}{N_j(\beta)}\right)\theta^{3K-j}\left(\theta^{2\ell+i-1}(a_{2K-i}) + \alpha\theta^{2\ell+i-1}(a_{2K-i+1})\right)$$

$$= -\sum_{i=1}^{K}\sum_{j=i}^{K} N_{2K-j}(\beta)\theta^{K-j}\left(\frac{N_{i-1}(\beta)}{N_j(\beta)}\right)\theta^{i+K-j-1}(a_{2K-i})$$

$$\quad - \sum_{i=1}^{K-1}\sum_{j=i+1}^{K} N_{2K-j}(\beta)\theta^{K-j}\left(\frac{N_i(\beta)}{N_j(\beta)}\right)\theta^{i+K-j}(a_{2K-i})\theta^{3K-j}(\alpha)$$

$$= -\sum_{i=1}^{K} N_K(\beta)\frac{N_{i-1}(\beta)}{N_K(\beta)}\theta^{i-1}(a_{2K-i})$$

$$\quad - \sum_{i=1}^{K-1}\sum_{j=i}^{K-1}\left(N_{2K-j}(\beta)\theta^{K-j}\left(\frac{N_{i-1}(\beta)}{N_j(\beta)}\right)\theta^{i+K-j-1}(a_{2K-i})\right.$$
$$\quad \left. + N_{2K-(j+1)}(\beta)\theta^{K-(j+1)}\left(\frac{N_i(\beta)}{N_{j+1}(\beta)}\right)\theta^{i+K-j-1}(a_{2K-i})\theta^{3K-j-1}(\alpha)\right)$$

$$= -\sum_{i=1}^{K} N_{i-1}(\beta)\theta^{i-1}(a_{2K-i})$$

$$\quad - \sum_{i=1}^{K-1}\sum_{j=i}^{K-1}\left(N_{2K-j-1}(\beta)\,\theta^{K-j-1}\left(\frac{N_i(\beta)}{N_{j+1}(\beta)}\right)\theta^{i+K-j-1}(a_{2K-i})\right.$$
$$\quad \left.(\theta^{2K-j-1}(\beta) + \theta^{3K-j-1}(\alpha))\right).$$

As $\beta = -\theta^K(\alpha)$, we obtain $\theta^{2K-j-1}(\beta) + \theta^{3K-j-1}(\alpha) = 0$. Therefore

$$\sum_{j=0}^{K-1}\theta^{2K-1-j}(a_j)N_{2K-1-j}(\beta) = -\sum_{i=1}^{K} N_{i-1}(\beta)\theta^{i-1}(a_{2K-i})$$

$$= -\sum_{j=K}^{2K-1}\theta^{2K-j-1}(a_j)N_{2K-j-1}(\beta).$$

We conclude that $f^*(\beta) = 0$ and that $X + \alpha$ right-divides $\Theta^{k+\ell}(f^*)$. We deduce the existence of a skew polynomial Q such that $\Theta^{k+\ell}(f^*) = Q \cdot (X+\alpha)$ and we obtain $P = \Theta^{k-1-\ell}((X+\alpha)^*) \cdot Q \cdot (X + \alpha)$.

2. Let us prove that $Q = \Theta^{k+1+\ell}(Q^*)$. According to Lemma 1, as $\Theta^{k+\ell}(f^*) = Q \cdot (X + \alpha)$, we obtain $f = \Theta^{k-1-\ell}((X + \alpha)^*) \cdot \Theta^{k+1+\ell}(Q^*)$. Therefore $P = \Theta^{k-1-\ell}((X + \alpha)^*) \cdot Q \cdot (X + \alpha) = \Theta^{k-1-\ell}((X + \alpha)^*) \cdot \Theta^{k+1+\ell}(Q^*) \cdot (X + \alpha)$ and $Q = \Theta^{k+1+\ell}(Q^*)$.

We are now giving a construction and an enumeration formula for the sets $\tilde{\mathcal{H}}_{\mu,\epsilon}$ (Proposition 4 and Algorithm 1).

Proposition 4. *Consider $k \in \mathbb{N}^*, n = 2k, p$ a prime number, $\theta : x \mapsto x^p \in Aut(\mathbb{F}_{p^n}), R = \mathbb{F}_{p^n}[X; \theta], \epsilon \in \{-1, 1\}, \mu \in \mathbb{F}_{p^n}^*$ and $P_k = -\mu\epsilon(X^{2k} - \epsilon) \in R$. The set $\tilde{\mathcal{H}}_{\mu,\epsilon}$ is nonempty if and only if $\theta^k(\mu) + \epsilon\mu = 0$. In this case we have*

$$\tilde{\mathcal{H}}_{\mu,\epsilon} = \{(X + \alpha_1) \cdots (X + \alpha_k) \mid X + \alpha_k \text{ right-divides } P_k$$
$$X + \alpha_{k-1} \text{ right-divides } P_{k-1}(\alpha_k)$$
$$\vdots$$
$$X + \alpha_1 \text{ right-divides } P_1(\alpha_2, \ldots, \alpha_k)\}$$

where for $i = k, k-1, \ldots, 2, P_{i-1}(\alpha_i, \ldots, \alpha_k)$ is the quotient in the left-division of Q_{i-1} by $\Theta^{i-1}((X + \alpha_i)^*)$ and Q_{i-1} the quotient in the right-division of $P_i(\alpha_{i+1}, \ldots, \alpha_k)$ by $X + \alpha_i$:

$$P_i(\alpha_{i+1}, \ldots, \alpha_k) = \underbrace{\Theta^{i-1}((X + \alpha_i)^*) \cdot P_{i-1}(\alpha_i, \ldots, \alpha_k)}_{Q_{i-1}} \cdot (X + \alpha_i). \qquad (5)$$

Furthermore $\tilde{\mathcal{H}}_{\mu,\epsilon}$ has $\prod_{i=1}^{k}(p^i + 1)$ elements.

Proof. Consider h in $\tilde{\mathcal{H}}_{\mu,\epsilon}$ and $\nu = h_0/\mu$ then $h^* \cdot \frac{1}{\nu} \cdot h = \theta^k(h_0) \cdot h^\natural \cdot \frac{1}{\nu} \cdot h = \theta^k(h_0) \cdot (X^n - \epsilon) \cdot \frac{1}{\theta^k(\nu)} = P_k$. Furthermore, $\Theta^k(P_k^*) = -(1 - \epsilon X^n) \cdot (\epsilon\theta^k(\mu)) = (X^n - 1)(-\epsilon\mu) = P_k$. As $X^n - \epsilon$ is central in R of degree one in $\mathbb{F}_p[X^n]$, the skew polynomial h is the product of k linear factors right-dividing $X^n - \epsilon$ and therefore P_k: $h = (X + \alpha_1) \cdots (X + \alpha_k)$. As the skew polynomial $X + \alpha_k$ right-divides P_k, according to Lemma 3 applied to $P = P_k$ and $\ell = 0$, there exists $P_{k-1}(\alpha_k) = P_{k-1} \in R$ such that

$$P_k = \Theta^{k-1}((X + \alpha_k)^*) \cdot P_{k-1} \cdot (X + \alpha_k)$$

and $\Theta^{k+1}(P_{k-1}^*) = P_{k-1}$. Consider $H = (X + \alpha_1) \cdots (X + \alpha_{k-1})$, according to Lemma 1, we have $h^* = \Theta^{k-1}((X + \alpha_k)^*) \cdot H^*$. We obtain

$$\Theta^{k-1}((X + \alpha_k)^*) \cdot P_{k-1} \cdot (X + \alpha_k) = \Theta^{k-1}((X + \alpha_k)^*) \cdot H^* \cdot \frac{1}{\nu} \cdot H \cdot (X + \alpha_k).$$

Therefore $P_{k-1} = H^* \cdot \frac{1}{\nu} \cdot H$ and $X + \alpha_{k-1}$ right-divides P_{k-1}. We conclude using an inductive argument and Lemma 3.

Conversely, consider $h = (X + \alpha_1) \cdots (X + \alpha_k)$ in R such that $X + \alpha_i$ right-divides $P_i(\alpha_{i+1}, \ldots, \alpha_k)$ defined by (5). According to Lemma 3, $\Theta^{k+1}(P_{k-1}^*) = $

$P_{k-1}, \ldots, \Theta^{2k-1}(P_1^*) = P_1$; furthermore, as $X + \alpha_1$ right-divides P_1, there exist P_0 such that $P_1 = (X + \alpha_1)^* \cdot P_0 \cdot (X + \alpha_1)$. According to Lemma 1, $h^* = \Theta^{k-1}((X + \alpha_k)^*) \cdots \Theta^0((X + \alpha_1)^*)$. Therefore, we obtain $h^* \cdot P_0 \cdot h = P_k$. In particular, the constant term of both polynomials is $P_0 h_0 = \mu$. Considering $\nu = h_0/\mu$, we obtain $h^* \cdot \frac{1}{\nu} \cdot h = -\epsilon\mu(X^n - \epsilon)$ and $h^\natural \cdot \frac{1}{\nu} \cdot h \cdot \theta^k(\nu) = -\frac{1}{\theta^k(h_0)}\epsilon\mu(X^n - \epsilon) \cdot \theta^k(\nu) = -\frac{1}{\theta^k(\mu)}\epsilon\mu(X^n - \epsilon) = X^n - \epsilon$.

Let us determine the cardinality of $\tilde{\mathcal{H}}_{\mu,\epsilon}$. According to Lemma 2, the number of factorizations (as a product of k linear monic factors) of any monic skew polynomial of degree k right-dividing $X^n - \epsilon$ in R is $\prod_{i=1}^{k} \frac{p^i-1}{p-1}$. Furthermore, as P_i has degree $2i$ and right-divides $X^n - \epsilon$, the number of $\alpha_i \in \mathbb{F}_q$ such that $X + \alpha_i$ right-divides P_i is $\frac{p^{2i}-1}{p-1}$. The number of elements of $\tilde{\mathcal{H}}_{\mu,\epsilon}$ is therefore

$$\frac{\prod_{i=1}^{k} \frac{p^{2i}-1}{p-1}}{\prod_{i=1}^{k} \frac{p^i-1}{p-1}} = \prod_{i=1}^{k}(p^i + 1).$$

Algorithm 1. Construction of the set $\tilde{\mathcal{H}}_{\mu,\epsilon}$

Require: : μ in $\mathbb{F}_{p^n}^*$ such that $\theta^k(\mu) + \epsilon\mu = 0$
Ensure: : $\tilde{\mathcal{H}}_{\mu,\epsilon}$
1: $P_k \leftarrow -\epsilon\mu(X^n - \epsilon)$
2: $S \leftarrow \emptyset$
3: Construct the sequences $(\alpha_i, P_{i-1})_{i=k,\ldots,1}$ such that

  - $X + \alpha_i$ right-divides P_i;
  - P_{i-1} is the quotient in the left-division of Q_{i-1} by $\Theta^{i-1}((X+\alpha_i)^*)$ where Q_{i-1} is the quotient in the right-division of P_i by $X + \alpha_i$.

4: **for** each sequence $(\alpha_1, \ldots, \alpha_k)$ **do**
5:     $S \leftarrow S \cup \{(X + \alpha_1) \cdots (X + \alpha_k)\}$
6: **end for**
7: **return** S

Example 4. Consider $\mathbb{F}_{2^4} = \mathbb{F}_2(a)$ with $a^4 + a + 1 = 0$ and $n = 4$. For $\mu \in \{1, a^5, a^{10}\}$, $\tilde{\mathcal{H}}_{\mu,1}$ has 15 elements:

$\tilde{\mathcal{H}}_{1,1} = \{X^2 + a^{14} X + a, X^2 + a X + a^{14}, X^2 + a^{11} X + a^4, X^2 + a^2 X + a^{13}, X^2 + a^{13} X + a^2, X^2 + a^4 X + a^{11}, X^2 + a^5 X + a^{10}, X^2 + 1, X^2 + a^8 X + a^7, X^2 + a^9, X^2 + a^7 X + a^8, X^2 + a^{12}, X^2 + a^3, X^2 + a^6, X^2 + a^{10} X + a^5\}$,

$\tilde{\mathcal{H}}_{a^5,1} = \{X^2 + a^5 X + a^5, X^2 + a^{14} X + a^{11}, X^2 + a^6 X + a^4, X^2 + a^8 X + a^2, X^2 + a^{11} X + a^{14}, X^2 + X + a^{10}, X^2 + a^2 X + a^8, X^2 + 1, X^2 + a^3 X + a^7, X^2 + a^9 X + a, X^2 + a^9, X^2 + a^{12} X + a^{13}, X^2 + a^{12}, X^2 + a^3, X^2 + a^6\}$,

$\tilde{\mathcal{H}}_{a^{10},1} = \{X^2 + a^{10} X + a^{10}, X^2 + a^{13} X + a^7, X^2 + a^{12} X + a^8, X^2 + a^9 X + a^{11}, X^2 + X + a^5, X^2 + a X + a^4, X^2 + 1, X^2 + a^6 X + a^{14}, X^2 + a^3 X + a^2, X^2 + a^9, X^2 + a^4 X + a, X^2 + a^{12}, X^2 + a^7 X + a^{13}, X^2 + a^3, X^2 + a^6\}$.

3.3 Enumeration of (θ, ν)-isodual and Self-dual θ-cyclic and θ-negacyclic Codes

From Proposition 4, we deduce the number of (θ, ϵ)-constacyclic codes of length $2k$ which are (θ, ν)-isodual. Let us first consider the particular case when $k = 1$:

Example 5. Consider $k = 1, n = 2$. If $p = 2$ and $\nu \in \mathbb{F}_{2^2}^*$, then $\mathcal{H}_{\nu,\epsilon} = \{X + \alpha\}$ where $\alpha^2 = \nu$. If p is odd and $\nu^{\frac{p^2-1}{2}} = -\epsilon(-1)^{\frac{p-1}{2}}$, then $\mathcal{H}_{\nu,\epsilon} = \{X + \alpha \mid \alpha^2 = -\nu^{p-1}\}$. Namely, consider $h = X + \alpha \in R$, then

$$h^{\natural} \cdot \nu \cdot h \cdot \tfrac{1}{\theta(\nu)} = (X + \tfrac{1}{\theta(\alpha)}) \cdot (X + \tfrac{\nu\alpha}{\theta(\nu)})$$
$$= X^2 + \frac{\nu + \theta(\nu)\theta(\alpha^2)}{\theta(\alpha)\nu} X + \frac{\nu\alpha}{\theta(\nu\alpha)}.$$

Therefore h belongs to $\mathcal{H}_{\nu,\epsilon}$ if and only if $\alpha^2 = -\theta(\nu)/\nu$.

Proposition 5. *Consider $k \in \mathbb{N}^*, n = 2k, p$ a prime number, $\theta : x \mapsto x^p \in Aut(\mathbb{F}_{p^n}), \epsilon \in \{-1, 1\}$ and $\nu \in \mathbb{F}_{p^n}^*$. If $\nu^{\frac{p^n-1}{2}} = -\epsilon(-1)^{k\frac{p-1}{2}}$ then the number of (θ, ν)-isodual (θ, ϵ)-constacyclic codes of length n over $\mathbb{F}_{p^n}$ is*

$$N \prod_{i=1}^{k-1} (p^i + 1)$$

where $N = 1$ if $p = 2$ and $N = 2$ if p is odd.

Proof. We first prove that for ν in $\mathbb{F}_{p^n}^*$, if $\mathcal{H}_{\nu,\epsilon}$ is nonempty, then its cardinality does not depend on ν. Consider ν, ν' in $\mathbb{F}_{p^n}^*$ such that

$$\nu^{\frac{p^n-1}{N}} = (\nu')^{\frac{p^n-1}{N}} = -\epsilon(-1)^{k\frac{p-1}{N}}.$$

Then according to Proposition 3, $\mathcal{H}_{\nu,\epsilon}$ and $\mathcal{H}_{\nu',\epsilon}$ are nonempty. Consider $\xi = \frac{\nu}{\nu'}$ and a a square root of $\theta^k(\xi)$. As $\xi^{\frac{p^n-1}{N}} = 1$, a is well defined. The application

$$f : \begin{cases} \mathcal{H}_{\nu,\epsilon} \to \mathcal{H}_{\nu',\epsilon} \\ h \mapsto \frac{1}{\theta^k(a)} \cdot h \cdot a \end{cases}$$

is also well defined: namely, consider for h in $\mathcal{H}_{\nu,\epsilon}$, $H = \frac{1}{\theta^k(a)} \cdot h \cdot a$, then $H^* = (h \cdot a)^* \cdot \frac{1}{\theta^k(a)} = \theta^k(a) \cdot h^* \cdot \frac{1}{\theta^k(a)}$. Therefore $H^{\natural} = \theta^k(\theta^k(a)/(h_0 a))\theta^k(a)\theta^k(h_0) \cdot h^{\natural} \cdot \frac{1}{\theta^k(a)} = a \cdot h^{\natural} \cdot \frac{1}{\theta^k(a)}$ and $H^{\natural} \cdot \frac{1}{\nu'} \cdot H \cdot \theta^k(\nu') =$

$$a \cdot h^{\natural} \cdot \frac{1}{\theta^k(a)} \cdot \frac{\xi}{\nu} \frac{1}{\theta^k(a)} \cdot h \cdot a\theta^k(\nu') = a \cdot h^{\natural} \cdot \frac{1}{\nu} \cdot h \cdot a \frac{\theta^k(\nu)}{\theta^k(\xi)} = a(X^n - \epsilon)\frac{a}{\theta^k(\xi)} = X^n - \epsilon.$$

Therefore H belongs to $\mathcal{H}_{\nu',\epsilon}$.

Consider H in $\mathcal{H}_{\nu',\epsilon}$ then $h = \theta^k(a)H \cdot \frac{1}{a}$ is the unique pre-image of H in $\mathcal{H}_{\nu,\epsilon}$. Therefore f is bijective and all nonempty sets $\mathcal{H}_{\nu,\epsilon}$ have the same number of elements, M: for all ν in $\mathbb{F}_{p^n}$ such that $\nu^{\frac{p^n-1}{N}} = -\epsilon(-1)^{k\frac{p-1}{N}}$, the number of (θ,ν)-isodual (θ,ϵ)-constacyclic codes of length $2k$ is M. Furthermore, we have

$$\forall h \in R, (\exists \mu : h \in \tilde{\mathcal{H}}_{\mu,\epsilon} \Leftrightarrow \exists \mu : h \in \mathcal{H}_{h_0\mu,\epsilon} \Leftrightarrow \exists \nu : h \in \mathcal{H}_{\nu,\epsilon}). \tag{6}$$

Therefore

$$\cup_\mu \tilde{\mathcal{H}}_{\mu,\epsilon} = \cup_\nu \mathcal{H}_{\nu,\epsilon}.$$

Now consider the union of the intersections $\tilde{\mathcal{H}}_{\mu,\epsilon} \cap \tilde{\mathcal{H}}_{\mu',\epsilon}$ for $\mu \neq \mu'$ and $\tilde{\mathcal{H}}_{\mu,\epsilon}, \tilde{\mathcal{H}}_{\mu',\epsilon}$ nonempty. We have, according to (6):

$$\bigcup_{\mu \neq \mu'} (\tilde{\mathcal{H}}_{\mu,\epsilon} \cap \tilde{\mathcal{H}}_{\mu',\epsilon}) = \bigcup_{\nu \neq \nu'} (\mathcal{H}_{\nu,\epsilon} \cap \mathcal{H}_{\nu',\epsilon}).$$

Similarly, we get

$$\bigcup_{\mu \neq \mu' \neq \mu''} (\tilde{\mathcal{H}}_{\mu,\epsilon} \cap \tilde{\mathcal{H}}_{\mu',\epsilon} \cap \tilde{\mathcal{H}}_{\mu'',\epsilon}) = \bigcup_{\nu \neq \nu' \neq \nu''} (\mathcal{H}_{\nu,\epsilon} \cap \mathcal{H}_{\nu',\epsilon} \cap \mathcal{H}_{\nu'',\epsilon}) \dots$$

$$\vdots$$

where the involved sets $\mathcal{H}_{\nu,\epsilon}$ and $\tilde{\mathcal{H}}_{\mu,\epsilon}$ are nonempty. Furthermore, $\#(\cup_\mu \tilde{\mathcal{H}}_{\mu,\epsilon}) = \sum_\mu \#\tilde{\mathcal{H}}_{\mu,\epsilon} - \sum_{\mu \neq \mu'} \#(\tilde{\mathcal{H}}_{\mu,\epsilon} \cap \tilde{\mathcal{H}}_{\mu',\epsilon}) + \sum_{\mu \neq \mu' \neq \mu''} \#(\tilde{\mathcal{H}}_{\mu,\epsilon} \cap \tilde{\mathcal{H}}_{\mu',\epsilon} \cap \tilde{\mathcal{H}}_{\mu'',\epsilon}) - \cdots$ and $\#(\cup_\nu \mathcal{H}_{\nu,\epsilon}) = \sum_\nu \#\mathcal{H}_{\nu,\epsilon} - \sum_{\nu \neq \nu'} \#(\mathcal{H}_{\nu,\epsilon} \cap \mathcal{H}_{\nu',\epsilon}) + \sum_{\nu \neq \nu' \neq \nu''} \#(\mathcal{H}_{\nu,\epsilon} \cap \mathcal{H}_{\nu',\epsilon} \cap \mathcal{H}_{\nu'',\epsilon}) - \cdots$. Therefore

$$\sum_\mu \#\tilde{\mathcal{H}}_{\mu,\epsilon} = \sum_\nu \#\mathcal{H}_{\nu,\epsilon}.$$

Lastly, according to Proposition 4, the $p^k - 1$ nonempty sets $\tilde{\mathcal{H}}_{\mu,\epsilon}$ all have $\prod_{i=1}^k (1 + p^i)$ elements. As the $\frac{p^{2k}-1}{N}$ nonempty sets $\mathcal{H}_{\nu,\epsilon}$ all have M elements, we obtain

$$\prod_{i=1}^k (1 + p^i)(p^k - 1) = M \frac{p^{2k} - 1}{N}.$$

Example 6. Consider the (θ,ν)-isodual θ-cyclic codes of length 4 over $\mathbb{F}_{2^4} = \mathbb{F}_2(a)$ where $a^4 + a + 1 = 0$. For $\mu \in \{1, a^5, a^{10}\}$, $\tilde{\mathcal{H}}_{\mu,1}$ has 15 elements (Example 4) and $\bigcup \tilde{\mathcal{H}}_{\mu,1} = \bigcup \mathcal{H}_{\nu,\epsilon}$ has 35 elements. For $\nu \in \mathbb{F}_{2^4}^*$, $\mathcal{H}_{\nu,\epsilon}$ has 3 elements: there are 3 (θ,ν)-isodual θ-cyclic codes:

$\mathcal{H}_{1,1} = \{X^2 + 1, X^2 + a^{10} X + a^{10}, X^2 + a^5 X + a^5\}$, $\mathcal{H}_{a^{14},1} = \{X^2 + a X + a^{14}, X^2 + a^9, X^2 + a^6 X + a^4\}$, $\mathcal{H}_{a^{13},1} = \{X^2 + a^2 X + a^{13}, X^2 + a^{12} X + a^8, X^2 + a^3\}$,

$\mathcal{H}_{a^{12},1} = \{X^2 + a^8\ X + a^2, X^2 + a^{13}\ X + a^7, X^2 + a^{12}\}, \mathcal{H}_{a^{11},1} = \{X^2 + a^9\ X + a, X^2 + a^6, X^2 + a^4\ X + a^{11}\}, \mathcal{H}_{a^{10},1} = \{X^2 + 1, X^2 + X + a^5, X^2 + a^5\ X + a^{10}\}, \mathcal{H}_{a^9,1} = \{X^2 + a\ X + a^4, X^2 + a^{11}\ X + a^{14}, X^2 + a^9\}, \mathcal{H}_{a^8,1} = \{X^2 + a^7\ X + a^8, X^2 + a^{12}\ X + a^{13}, X^2 + a^3\}, \mathcal{H}_{a^7,1} = \{X^2 + a^8\ X + a^7, X^2 + a^{12}, X^2 + a^3\ X + a^2\}, \mathcal{H}_{a^6,1} = \{X^2 + a^{14}\ X + a^{11}, X^2 + a^4\ X + a, X^2 + a^6\}, \mathcal{H}_{a^5,1} = \{X^2 + 1, X^2 + X + a^{10}, X^2 + a^{10}\ X + a^5\}, \mathcal{H}_{a^4,1} = \{X^2 + a^{11}\ X + a^4, X^2 + a^9, X^2 + a^6\ X + a^{14}\}, \mathcal{H}_{a^3,1} = \{X^2 + a^2\ X + a^8, X^2 + a^7\ X + a^{13}, X^2 + a^3\}, \mathcal{H}_{a^2,1} = \{X^2 + a^3\ X + a^7, X^2 + a^{13}\ X + a^2, X^2 + a^{12}\}, \mathcal{H}_{a,1} = \{X^2 + a^{14}\ X + a, X^2 + a^9\ X + a^{11}, X^2 + a^6\}.$

We check that

$$\bigcup_{\mu \neq \mu'} (\tilde{\mathcal{H}}_{\mu,1} \cap \tilde{\mathcal{H}}_{\mu',1}) = \bigcup_{\nu \neq \nu'} (\mathcal{H}_{\nu,\epsilon} \cap \mathcal{H}_{\nu',\epsilon}) = \{X^2 + 1, X^2 + a^9, X^2 + a^3, X^2 + a^{12}, X^2 + a^6\},$$

$$\tilde{\mathcal{H}}_{1,1} \cap \tilde{\mathcal{H}}_{a^5,1} \cap \tilde{\mathcal{H}}_{a^{10},1} = \bigcup_{\nu \neq \nu' \neq \nu''} (\mathcal{H}_{\nu,\epsilon} \cap \mathcal{H}_{\nu',\epsilon} \cap \mathcal{H}_{\nu'',\epsilon})$$

$$= \{X^2 + 1, X^2 + a^9, X^2 + a^3, X^2 + a^{12}, X^2 + a^6\},$$

$$\emptyset = \bigcup_{\nu \neq \nu' \neq \nu'' \neq \nu'''} (\mathcal{H}_{\nu,\epsilon} \cap \mathcal{H}_{\nu',\epsilon} \cap \mathcal{H}_{\nu'',\epsilon} \cap \mathcal{H}_{\nu''',\epsilon}).$$

In [5], a formula for the number of self-dual (θ, ϵ)-constacyclic codes of length n is given over $\mathbb{F}_{p^2} \subset \mathbb{F}_{p^n}$ when $\epsilon^2 = 1$. In what follows, we deduce from Proposition 5 the number of self-dual (θ, ϵ)-constacyclic codes of length n over $\mathbb{F}_{p^n}$.

Proposition 6. *Consider $k \in \mathbb{N}^*, n = 2k, \epsilon \in \{-1, 1\}, p$ a prime number and $\theta : x \mapsto x^p \in \mathrm{Aut}(\mathbb{F}_{p^n})$. If $p = 2$, there are $\prod_{i=1}^{k-1}(p^i + 1)$ self-dual θ-cyclic codes of length n over $\mathbb{F}_{p^n}$. If $p \neq 2$, and if $(-1)^k \frac{p-1}{2} \epsilon = -1$, there are $2 \prod_{i=1}^{k-1}(p^i + 1)$ self-dual (θ, ϵ)-constacyclic codes of length n over $\mathbb{F}_{p^n}$.*

Proof. Self-dual (θ, ϵ)-constacyclic codes are $(\theta, 1)$-isodual (θ, ϵ)-constacyclic codes. The result follows from Proposition 5.

Example 7. There are 3 self-dual θ-cyclic codes of length 4 over $\mathbb{F}_{2^4}$, given in Example 6 by $\mathcal{H}_{1,1}$. There are 15 self-dual θ-cyclic codes of length 6 over $\mathbb{F}_{2^6} = \mathbb{F}_2(a)$ where $a^6 + a^4 + a^3 + a + 1 = 0$. Their skew check polynomials are:
$X^3 + a^{52}X^2 + a^{23}X + a^{54}, X^3 + X^2 + a^9\ X + a^{36}, X^3 + a^{36}\ X^2 + a^{36}\ X + a^{45}, X^3 + 1,$
$X^3 + a^{19}\ X^2 + a^{29}\ X + a^{27}, X^3 + a^{41}\ X^2 + a^{46}\ X + a^{45}, X^3 + a^{18}\ X^2 + a^{18}\ X + a^{54},$
$X^3 + a^{21}\ X^2 + a^{42}\ X + 1, X^3 + a^{26}\ X^2 + a^{43}\ X + a^{27}, X^3 + a^9\ X^2 + a^9\ X + a^{27},$
$X^3 + a^{42}\ X^2 + a^{21}\ X + 1, X^3 + a^{13}\ X^2 + a^{53}\ X + a^{45}, X^3 + a^{38}\ X^2 + a^{58}\ X + a^{54},$
$X^3 + X^2 + a^{18}\ X + a^9$ and $X^3 + X^2 + a^{36}\ X + a^{18}.$

Algorithm 2. (θ, ν)-isodual (θ, ϵ)-constacyclic codes of length n over $\mathbb{F}_{p^n}$

Require: : $k \in \mathbb{N}^*$, $n = 2k$, p, prime number, $\epsilon \in \{-1, 1\}$, $\nu \in \mathbb{F}_{p^n}^*$

Ensure: : $\mathcal{H}_{\nu, \epsilon}$: the set of the skew check polynomials h of (θ, ν)-isodual (θ, ϵ)-constacyclic codes of length n and dimension k over $\mathbb{F}_{p^n}$.

1: $S \leftarrow \emptyset$

2: **for** μ in $\mathbb{F}_{p^n}^*$ such that $\theta^k(\mu) + \epsilon\mu = 0$ **do**

3:    $P_k \leftarrow -\epsilon\mu(X^n - \epsilon)$

4:    Construct the sequences (α_i, P_{i-1}) for $i = k, \ldots, 1$ such that

     – $X + \alpha_i$ right-divides P_i;

     – P_{i-1} is the quotient in the left-division of Q_{i-1} by $\Theta^{i-1}((X + \alpha_i)^*)$ where Q_{i-1} is the quotient in the right-division of P_i by $X + \alpha_i$;

     – $P_0 = \mu / \prod_{i=1}^{k} \alpha_i = \nu$.

5:    **for** each $(\alpha_1, \ldots, \alpha_k)$ **do**

6:       $S \leftarrow S \cup \{(X + \alpha_1) \cdots (X + \alpha_k)\}$

7:    **end for**

8: **end for**

9: **return** S

---

4   A Construction of Self-dual θ-cyclic Codes over $\mathbb{F}_{p^n}$ Which are Gabidulin Evaluation Codes

We consider here a special subfamily of self-dual θ-cyclic codes of length n over $\mathbb{F}_{p^n}$ which are in fact Gabidulin evaluation codes and therefore Maximum Rank Distance (MRD) codes (see [12] for the theory of MRD codes). In previous sections we have provided a construction of self-dual codes based on the factorization of the skew check polynomials into the product of linear skew polynomials. Here we change the point of view by writting the skew generator polynomials as least common left multiples (lclm) of special linear skew polynomials, namely skew polynomials of the form $X - \theta^i(\alpha)$.

Definition 3. [12] *Consider* $k \leq n \in \mathbb{N}^*, p$ *a prime number,* $\theta : x \mapsto x^p$ $\in Aut(\mathbb{F}_{p^n}), y_1, \ldots, y_n \in \mathbb{F}_{p^n}$ *linearly independent over* $\mathbb{F}_p$. *The* Gabidulin evaluation code *of length* n, *dimension* k *and support* $(y_1, \ldots, y_n)$ *is the code with generator matrix*

$$G = \begin{pmatrix} y_1 & y_1 & \cdots & y_n \\ \theta(y_1) & \theta(y_2) & \cdots & \theta(y_n) \\ \vdots & & & \\ \theta^{k-1}(y_1) & \theta^{k-1}(y_2) & \ldots & \theta^{k-1}(y_n) \end{pmatrix}. \tag{7}$$

The following Proposition 7 can be found in [9] (Proposition 3, part 1).

Proposition 7 (Proposition 3 of [9]). *Consider* $k \leq n \in \mathbb{N}^*, p$ *a prime number,* $\theta : x \mapsto x^p \in Aut(\mathbb{F}_{p^n}), R = \mathbb{F}_{p^n}[X; \theta], \alpha \in \mathbb{F}_{p^n}, \epsilon = N_n(\alpha)$. *Assume that* $\epsilon^2 = 1$ *and that* $1, \alpha, N_2(\alpha), \ldots, N_{n-1}(\alpha)$ *are linearly independent over* $\mathbb{F}_p$. *The*

θ-cyclic code of length n and skew generator polynomial $g = \text{lclm}_{0 \leq i \leq k-1}(X - \theta^i(\alpha))$ is the dual of the Gabidulin evaluation code of length n, dimension k and support $(1, \alpha, N_2(\alpha), \ldots, N_{n-1}(\alpha))$.

Proof. Consider $c = (c_0, \ldots, c_{n-1}) \in \mathbb{F}_{p^n}^n$. The word c belongs to the θ-cyclic code generated by g if and only if for all i in $\{0, \ldots, k-1\}$, $X - \theta^i(\alpha)$ right-divides g. As the remainder in the right division of c by $X - \theta^i(\alpha)$ is $\sum_{j=0}^{n-1} c_j N_j(\theta^i(\alpha)) = \sum_{j=0}^{n-1} c_j \theta^i(N_j(\alpha))$, we obtain that a check matrix for the code is the matrix G given by (7) where $y_1 = 1, y_2 = \alpha, \ldots, y_n = N_{n-1}(\alpha)$.

According to Theorem 4.10 of [15], if a Gabidulin evaluation code of length $2k$ and dimension k is self-dual then $p \equiv 3 \pmod 4$ and k is odd. In what follows we construct a family of self-dual Gabidulin evaluation codes parameterized by an element α of $\mathbb{F}_{p^n}$.

Proposition 8. *Consider* $k \in \mathbb{N}^*, n = 2k, p$ *a prime number,* $\theta : x \mapsto x^p$ $\in Aut(\mathbb{F}_{p^n}), R = \mathbb{F}_{p^n}[X; \theta], \alpha \in \mathbb{F}_{p^n}$ *such that* $N_n(\alpha) = 1$ *and* $1, \alpha, N_2(\alpha), \ldots,$ $N_{n-1}(\alpha)$ *are linearly independent over* $\mathbb{F}_p$. *The* θ-*cyclic code of length* n *and skew generator polynomial* $g = \text{lclm}_{0 \leq i \leq k-1}(X - \theta^i(\alpha))$ *is self-dual if and only if* $\sum_{i=0}^{n-1} N_i(\alpha)^{1+p^\ell} = 0, \forall \ell \in \{0, \ldots, k-1\}$.

Proof. The code is self-dual if and only if the lines of G are pairwise orthogonal i.e. $G G^T = 0$ where G is the matrix defined by (7) with $y_1 = 1, y_2 = \alpha, \ldots, y_n = N_{n-1}(\alpha)$.

Example 8. For $p = 3$ and $k = 3$, according to Proposition 5, there are 80 self-dual θ-cyclic codes of length 6 over $\mathbb{F}_{3^3}$. The polynomial system

$$\begin{cases} \sum_{i=0}^{6} N_i(\alpha)^2 = 0 \\ \sum_{i=0}^{6} N_i(\alpha)^4 = 0 \\ \sum_{i=0}^{6} N_i(\alpha)^{10} = 0 \end{cases}$$

has 18 solutions α: $a^{580}, a^{406}, a^{378}, a^{436}, a^{124}, a^8, a^{648}, a^{126}, a^{388}, a^{216}, a^{42}, a^{72},$ $a^{14}, a^{284}, a^{24}, a^{372}, a^{488}, a^{490}$ and we get 18 self-dual θ-cyclic codes generated by the skew polynomials $g = \text{lclm}(X - \alpha, X - \theta(\alpha), X - \theta^2(\alpha))$. These codes are self-dual Gabidulin evaluation codes (and therefore self-dual MRD codes). For example, take $\alpha = a^8$, then $g = \text{lclm}(X - a, X - \theta(a), X - \theta^2(a)) = X^3 + a^{185} X^2 + a^{383} X + a^{322}$ generates a self-dual θ-cyclic code of length 6 which is the Gabidulin evaluation code of dimension 3 and support $(1, a, \ldots, N_5(a)) = (1, a^8, a^{32}, a^{104}, a^{320}, a^{240})$.

An open question is to determine the number of self-dual θ-cyclic codes generated by $g = \text{lclm}_{0 \leq i \leq k-1}(X - \theta^i(\alpha))$ over $\mathbb{F}_{p^n}$ with $n = 2k$. More generally, it could be interesting to construct and count self-dual θ-cyclic codes which are MRD.

5 Conclusion

This note was devoted to the construction of self-dual θ-constacyclic codes of length n over $\mathbb{F}_{p^m}$ when m is equal to n and θ is the Frobenius automorphism over $\mathbb{F}_{p^n}$. This work completes previous works on self-dual θ-constacyclic codes over $\mathbb{F}_{p^m}$ when $m = 1$ (then θ is the identity and codes are classical constacyclic codes) and when $m = 2$. As a further work, it could be interesting to study the cases when $2 < m < n$ and to have a more general classification only based on the order of the automorphism θ in $Aut(\mathbb{F}_{p^n})$.

Acknowledgments. The authors thank the referees for their fruitful remarks. The second author is supported by the French government "Investissements d'Avenir" program ANR-11-LABX-0020-01.

References

1. Alahmadi, A., Alsulami, S., Hijazi, R., Solé, P.: Isodual cyclic codes over finite fields of odd characteristic. Discrete Math. **339**(1), 344–353 (2016)
2. Batoul, A., Guenda, K., Gulliver, T.A.: Repeated-root isodual cyclic codes over finite fields. In: El Hajji, S., Nitaj, A., Carlet, C., Souidi, E.M. (eds.) C2SI 2015. LNCS, vol. 9084, pp. 119–132. Springer, Cham (2015). https://doi.org/10.1007/978-3-319-18681-8_10
3. Batoul, A., Guenda, K., Kaya, A., Yildiz, B.: Cyclic isodual and formally self-dual codes over $\mathbb{F}_q + \nu \mathbb{F}_q$. Eur. J. Pure Appl. Math. **8**(1), 64–80 (2015)
4. Boucher, D.: A note on the existence of self-dual skew codes over finite fields. In: El Hajji, S., Nitaj, A., Carlet, C., Souidi, E.M. (eds.) C2SI 2015. LNCS, vol. 9084, pp. 228–239. Springer, Cham (2015). https://doi.org/10.1007/978-3-319-18681-8_18
5. Boucher, D.: Construction and number of self-dual skew codes over $\mathbb{F}_{p^2}$. Adv. Math. Commun. **10**(4), 765–795 (2016)
6. Boucher, D.: Autour de codes définis à l'aide de polynômes tordus. Habilitation à diriger des recherches de l'Université Rennes 1, 2 juin 2020
7. Boucher, D., Ulmer, F.: Codes as modules over skew polynomial rings. In: Parker, M.G. (ed.) IMACC 2009. LNCS, vol. 5921, pp. 38–55. Springer, Heidelberg (2009). https://doi.org/10.1007/978-3-642-10868-6_3
8. Boucher, D., Ulmer, F.: A note on the dual codes of module skew codes. In: Chen, L. (ed.) IMACC 2011. LNCS, vol. 7089, pp. 230–243. Springer, Heidelberg (2011). https://doi.org/10.1007/978-3-642-25516-8_14
9. Boucher, D., Ulmer, F.: Linear codes using skew polynomials with automorphisms and derivations. Des. Codes Cryptogr. **70**(3), 405–431 (2012). https://doi.org/10.1007/s10623-012-9704-4
10. Caruso, X., Le Borgne, J.: A new faster algorithm for factoring skew polynomials over finite fields. J. Symbolic Comput. **79**(part 2), 411–443 (2017)
11. Delenclos, J., Leroy, A.: Noncommutative symmetric functions and W-polynomials. J. Algebra Appl. **6**(5), 815–837 (2007)
12. Gabidulin, È.M.: Theory of codes with maximum rank distance. Problemy Peredachi Informatsii **21**(1), 3–16 (1985)
13. Jacobson, N.: The Theory of Rings. American Mathematical Society Mathematical Surveys, vol. II. American Mathematical Society, New York (1943)

14. Lam, T.Y., Leroy, A.: Vandermonde and Wronskian matrices over division rings. Bull. Soc. Math. Belg. Sér. A **40**(2), 281–286 (1988). Deuxième Contact Franco-Belge en Algèbre (Faulx-les-Tombes, 1987)
15. Nebe, G., Willems, W.: On self-dual MRD codes. Adv. Math. Commun. **10**(3), 633–642 (2016)
16. Ore, O.: Theory of Non-commutative Polynomials. Ann. Math. **34**(3), 480–508 (1933)
17. Rains, E.M., Sloane, N.J.A.: Self-dual codes. In: Handbook of Coding Theory, vol. I, II, pp. 177–294. North-Holland, Amsterdam (1998)

Decoding up to 4 Errors
in Hyperbolic-Like Abelian Codes
by the Sakata Algorithm

José Joaquín Bernal(ID) and Juan Jacobo Simón$^{(\boxtimes)}$(ID)

Departamento de Matemáticas, Universidad de Murcia, 30100 Murcia, Spain
{josejoaquinbb,jsimon}@um.es
http://www.um.es

Abstract. We deal with two problems related with the use of the
Sakata's algorithm in a specific class of bivariate codes (see [2,8,9]).
The first one is to improve the general framework of locator decoding
in order to apply it on such abelian codes. The second one is to find
sufficient conditions to guarantee that the minimal set of polynomials
given by the algorithm is exactly a Groebner basis of the locator ideal.

Keywords: Abelian codes · Decoding · Berlakamp-Massey-Sakata
Algorithm

1 Introduction

The Sakata algorithm (or Berlekamp-Massey-Sakata algorithm, BMSa, for short)
is one of the best known procedures to find Groebner basis (see [4,11]) for the so
called ideal of linear recurrence relations on a doubly periodic array [7,8]. It is
specially used for decoding algebraic geometric codes [2,10]. Another use of the
BMSa, which is also well-known but less studied or understood, may be found in
the context of the locator decoding method for general bivariate abelian codes
[2,9].

This paper deals with two questions related with the use of the Sakata's
algorithm in bivariate codes. Firstly, we give an improvement of the framework
for applying the locator decoding algorithm in a class of abelian codes that we
call Hyperbolic-like abelian codes. Secondly, and in our opinion the main goal
of this paper, we give a sufficient condition to guarantee that the minimal set of
polynomials (given by the algorithm) is exactly the Groebner basis of the ideal
of linear recurrence relations, and so, to get the defining set for the locator ideal
(see [8, Section 6]). This condition is given in terms of the defining sets of these
codes. This paper is a portion of a study in progress of the BMSa in a more
general frame.

This work was partially supported by MINECO, project MTM2016-77445-P, and Fun-
dación Séneca of Murcia, project 19880/GERM/15.

J. C. Bajard and A. Topuzoğlu (Eds.): WAIFI 2020, LNCS 12542, pp. 134–146, 2021.
https://doi.org/10.1007/978-3-030-68869-1_7

2   Bivariate Codes

Let $\mathbb{F}$ be a finite field with q elements, with q a power of a prime number, let r_i be positive integers, for $i \in \{1, 2\}$, and $r = r_1 \cdot r_2$. We denote by $\mathbb{Z}_{r_i}$ the ring of integers modulo r_i. We always write its elements as canonical representatives. When necessary we write $\bar{a} \in \mathbb{Z}_k$ for any $a \in \mathbb{Z}$.

A **bivariate code**, or 2-dimensional abelian code, of length r (see [6]) is an ideal in the algebra $\mathbb{F}(r_1, r_2) = \mathbb{F}[X_1, X_2]/\langle X_1^{r_1} - 1, X_2^{r_2} - 1 \rangle$. Throughout this work, we assume that this algebra is semisimple; that is, $\gcd(r_i, q) = 1$, for $i \in \{1, 2\}$. The codewords are identified with polynomials. The weight of a codeword c is denoted by $\omega(c)$. We denote by I the set $\mathbb{Z}_{r_1} \times \mathbb{Z}_{r_2}$ and we write the elements $f \in \mathbb{F}(r_1, r_2)$ as $f = \sum a_m \mathbf{X}^m$, where $m = (m_1, m_2) \in I$ and $\mathbf{X}^m = X_1^{m_1} \cdot X_2^{m_2}$. Given a polynomial $f \in \mathbb{F}[X_1, X_2]$, we denote by $\bar{f}$ its image under the canonical projection onto $\mathbb{F}(r_1, r_2)$, when necessary.

For each $i \in \{1, 2\}$, we denote by R_{r_i} (resp. $\mathcal{R}_{r_i}$) the set of r_i-th roots of unity (resp. r_i-th primitive roots of unity) and define $R = R_{r_1} \times R_{r_2}$ ($\mathcal{R} = \mathcal{R}_{r_1} \times \mathcal{R}_{r_2}$). Throughout this paper, we fix $\mathbb{L}|\mathbb{F}$ as a extension field containing R_{r_i}.

For $f = f(X_1, X_2) \in \mathbb{F}[X_1, X_2]$ and $\bar{\alpha} = (\alpha_1, \alpha_2) \in R$, we write $f(\bar{\alpha}) = f(\alpha_1, \alpha_2)$. For $m = (m_1, m_2) \in I$, we write $\bar{\alpha}^m = (\alpha_1^{m_1}, \alpha_2^{m_2})$.

It is a known fact that, in the semi simple case, every abelian code C in $\mathbb{F}(r_1, r_2)$ is totally determined by its **root set** or **set of zeros**, namely

$$Z(C) = \{\bar{\alpha} \in R \mid f(\bar{\alpha}) = 0, \quad \text{for all } f \in C\}.$$

For a fixed $\bar{\alpha} \in \mathcal{R}$, the code C is determined by its **defining set**, with respect to $\bar{\alpha}$, which is defined as

$$\mathcal{D}_{\bar{\alpha}}(C) = \{m \in I \mid \bar{\alpha}^m \in Z(C)\}.$$

It is easy to see that the notions of set of zeros and defining set may be considered for any set of either polynomials or ideals in $\mathbb{F}(r_1, r_2)$ (or $\mathbb{L}(r_1, r_2)$); moreover, it is known that for any $G \subset \mathbb{F}(r_1, r_2)$ (or $\mathbb{L}(r_1, r_2)$) and $\bar{\alpha} \in \mathcal{R}$ we have $\mathcal{D}_{\bar{\alpha}}(G) = \mathcal{D}_{\bar{\alpha}}(\langle G \rangle)$. In [2,4], the defining set is considered for ideals P in $\mathbb{L}[\mathbf{X}]$. From the definition, we have $\mathcal{D}_{\bar{\alpha}}(P) = \mathcal{D}_{\bar{\alpha}}(\overline{P})$, where $\overline{P} \in \mathbb{L}[\mathbf{X}]$ is the canonical projection of P onto $\mathbb{L}(r_1, r_2)$.

We also recall the extension of the concept of q-cyclotomic coset of an integer to two components.

Given an element $(a_1, a_2) \in I$, we define its q-*orbit* modulo (r_1, r_2) as

$$Q(a_1, a_2) = \{(a_1 \cdot q^i, a_2 \cdot q^i) \mid i \in \mathbb{N}\} \subseteq I = \mathbb{Z}_{r_1} \times \mathbb{Z}_{r_2}. \tag{1}$$

It is easy to see that for every abelian code $C \subseteq \mathbb{F}(r_1, r_2)$, $\mathcal{D}_{\bar{\alpha}}(C)$ is closed under multiplication by q in I, and then $\mathcal{D}_{\bar{\alpha}}(C)$ is necessarily a disjoint union of q-orbits modulo (r_1, r_2). Conversely, every union of q-orbits modulo (r_1, r_2) defines an abelian code in $\mathbb{F}(r_1, r_2)$. For the sake of simplicity we only write q-orbit, and the tuple of integers will be clear from the context.

3   Apparent Distance and Multilevel Bound

In [2, p. 1614] Blahut considered hyperbolic codes with multilevel bound δ, that is, bivariate codes in $\mathbb{F}(r_1, r_2)$ whose defining set with respect to $\bar{\alpha}$ is

$$\mathcal{D}_{\bar{\alpha}}(C) = \{(i,j) \in I \mid (i+1)(j+1) \le \delta\}.$$

For this family of codes it has been proved that δ is a lower bound for the minimum distance. From this idea we are considering bivariate codes in $\mathbb{F}(r_1, r_2)$ whose defining sets contain some sets very similar to those used for the definition of hyperbolic codes. In detail, we take $\delta \in \mathbb{N}$ and define

$$\mathcal{B}_\delta = \{(i,j) \in I \mid (i+1)(j+1) \le \delta\} \setminus \{(\delta-1, 0), (0, \delta-1)\} \qquad (2)$$

Now, for any $m \in I$ we set $m + \mathcal{B}_\delta = \{m + a \mid a \in \mathcal{B}_\delta\} \subset I$. We shall see that every abelian code having a subset $(m + \mathcal{B}_\delta) \subset \mathcal{D}_{\bar{\alpha}}(C)$ verifies that $\delta \le d(C)$. To do this, we shall use the Algorithm 1 in [1, p. 662] to compute a bound for the minimum strong apparent distance (msd) of the matrix afforded by $\mathcal{D}_{\bar{\alpha}}(C)$ (see [1,3]); that is, $M = (a_n)_{n \in I}$ where $a_n = 1$ if $n \notin \mathcal{D}_{\bar{\alpha}}(C)$ and $a_n = 0$ otherwise. In fact we prove the following lemma.

Lemma 1. Let $m = (m_1, m_2) \in I$ and $0 \ne M = (a_n)_{n \in I}$ such that $a_n = 0$ for all $n \in m + \mathcal{B}_\delta$, with $\delta \in \mathbb{N}$. Then the strong apparent distance of M, denoted by $sd^*(M)$ [1, Definition 10], satisifies $sd^*(M) \ge \delta$.

Proof. We shall follow the notation in [1, Remark 11]. Let R_i the i-th row of M. Then $m + \{(0,0), \ldots, (0, \delta-2)\} \subset R_{m_1}$ modulo r_1, for some $m_1 \in \mathbb{Z}_{r_1}$. If $R_{m_1} \ne 0$ then $\epsilon_M(X_1) \ge sd^*(R_{m_1}) \ge \delta$ and hence $sd^*(M) \ge \delta$. So, suppose that $0 = R_{m_1} = \ldots = R_{m_1+k}$ and $R_{m_1+k+1} \ne 0$, for $0 \le k < \delta - 2$. Then $\epsilon_M(X_1) \ge sd^*\left(R_{m_1+k+1}\right) \ge \left(\lfloor \frac{\delta}{k+2} \rfloor + 1\right)$ and $\omega_M(X_1) \ge k+1$. Then $sd^*(M) \ge \left(\lfloor \frac{\delta}{k+2} \rfloor + 1\right)(k+2) = \lfloor \frac{\delta}{k+2} \rfloor(k+2) + (k+2) > \delta$. If $0 = R_{m_1} = \ldots = R_{m_1+\delta-2}$ then $\omega_M(X_1) \ge \delta - 1$ and we are done.

Corollary 1. Let $m = (m_1, m_2) \in I$ and $\delta \in \mathbb{N}$. Let $0 \ne M = (a_n)_{n \in I}$ such that $a_n = 0$ for all $n \in m + \mathcal{B}_\delta$. Then $msd(M) \ge \delta$.

Consequently if C is a bivariate code in $\mathbb{F}(r_1, r_2)$ such that $m + \mathcal{B}_\delta \subset \mathcal{D}_{\bar{\alpha}}(C)$ then $d(C) \ge \delta$.

Proof. Comes directly from Definition 15, Theorem 16, Algorithm 1 and Theorem 18 in [1], and the lemma above.

4   The Berlekamp-Massey-Sakata Algorithm

As it is commented in [2,7,8,11], the BMSa is an iterative procedure (w.r.t. a total ordering) in order to construct a Groebner basis for the ideal of polynomials satisfying some linear recurring relations for a doubly periodic array. Let us recall some terminology and some facts about it. We shall introduce some minor modifications in order to improve its application.

We denote by $\mathbb{N}$ the set of natural numbers (including 0) and we define $\Sigma_0 = \mathbb{N} \times \mathbb{N}$. We consider the partial ordering in Σ_0 given by $(n_1, n_2) \preceq (m_1, m_2) \Longleftrightarrow$ $n_1 \leq m_1$ and $n_2 \leq m_2$. On the other hand, we will use a (total) monomial ordering [4, Definition 2.2.1], denoted by "$\leq_T$", as in [8, Section 2]. This ordering will be either the lexicographic order (with $X_1 > X_2$) [4, Definition 2.2.3] or the (reverse) graded order (with $X_2 > X_1$) [4, Definition 2.2.6]. Of course, any result in this paper may be obtained under the alternative lexicographic or graded orders. The meaning of "$\leq_T$" will be specified as required.

Definition 1. *For $s, k \in \Sigma_0$, we define*

1. *$\Sigma_s = \{m \in \Sigma_0 \mid s \preceq m\}$,*
2. *$\Sigma_s^k = \{m \in \Sigma_0 \mid s \preceq m \text{ and } m <_T k\}$ and*
3. *$\Delta_s = \{n \in \Sigma_0 \mid n \preceq s\}$.*

Given $m, n \in \Sigma_0$, we define $m + n$, $m - n$ (provided that $n \preceq m$) and $n \cdot m$, coordinatewise, as it is usual. An infinite array or matrix is defined as $U = (u_n)_{n \in \Sigma_0}$; where the u_n will always belong to the extension field $\mathbb{L}$. In practice, we work with finite arrays defined as infinite doubly periodic (see [8, p. 324]) and consider subarrays, as follows.

Definition 2. *Let $U = (u_n)_{n \in \Sigma_0}$ be an infinite array.*

1. *We say that U is a doubly periodic array of period $r_1 \times r_2$ if the following property is satisfied: for $n = (n_1, n_2)$ and $m = (m_1, m_2)$ we have that $n_i \equiv m_i$ mod r_i for $i = 1, 2$ implies that $u_n = u_m$.*
2. *If U is a doubly periodic array of period $r_1 \times r_2$, a finite subarray $u^l \subset U$, with $l \in \Sigma_0$ is the array $u^l = \left(u_m \mid m \in \Sigma_0^l \cap \Delta_{(r_1-1, r_2-1)} \right)$*

Note that, in the case of period $r_1 \times r_2$ we may identify $I = \mathbb{Z}_{r_1} \times \mathbb{Z}_{r_2} = \Delta_{(r_1-1, r_2-1)}$; so that, $u^l = (u_m \mid m \in I)$ for $l >_T (r_1, r_2)$.

As it is well known, every monomial ordering is a well order, so that any $n \in \Sigma_0$ has a succesor. For the graded order we have

$$n + 1 = \begin{cases} (n_1 - 1, n_2 + 1) & \text{if } n_1 > 0 \\ (n_2 + 1, 0) & \text{if } n_1 = 0 \end{cases}.$$

In the case of the lexicographic order, we have to introduce, besides the unique succesor with respect to the monomial ordering, another succesor that we will only use for the recursion steps over $n \in \Delta_{(r_1-1, r_2-1)}$. We also denote it by $n + 1$ as follows:

$$n + 1 = \begin{cases} (n_1, n_2 + 1) & \text{if } n_2 < r_2 - 1 \\ (n_1 + 1, 0) & \text{if } n_2 = r_2 - 1 \end{cases}.$$

So, during the implementation of the BMSa (that is, results related with it), the succesor of $n \in \Delta_{(r_1-1,r_2-1)}$ will be donted by $n+1$, independently of the monomial ordering considered.

Now we recall some definitions that may be found in [8, pp. 322–323]. For any $f \in \mathbb{L}[\mathbf{X}]$ or $f \in \mathbb{L}(r_1, r_2)$, we denote the leading power product exponent of f, with respect to "$\leq_T$" by $LP(f)$. Of course $LP(f) \in \Sigma_0$. For $F \subset \mathbb{L}[\mathbf{X}]$, we denote $LP(F) = \{LP(f) \mid f \in F\}$.

Definition 3. *Let U be a doubly periodic array, $f \in \mathbb{L}[\mathbf{X}]$, $n \in \Sigma_0$ and $LP(f) = s$. We write $f = \sum_{m \in supp(f)} f_m \mathbf{X}^m$ and define*

$$f[U]_n = \begin{cases} \sum_{m \in supp(f)} f_m u_{m+n-s} & \text{if } n \in \Sigma_s \\ 0 & \text{otherwise} \end{cases}.$$

*The equality $f[U]_n = 0$ will be called a **linear recurring relation** and in this case, we will say that the polynomial f **is valid for** U **at** n.*

Definition 4. *Let U be a doubly periodic array and $f \in \mathbb{L}[\mathbf{X}]$ with $LP(f) = s$.*

1. *We say that f generates U and write $f[U] = 0$, if $f[U]_n = 0$ at any $n \in \Sigma_0$.*
2. *For any $u = u^k \subset U$, we say that f generates u if $f[U]_n = 0$ at every $n \in \Sigma_s^k$ and we write $f[u] = f[u^k] = 0$. In case $\Sigma_s^k = \emptyset$ we define $f[u] = 0$.*
3. *For any $u = u^k \subset U$, we say that f generates u, up to $l <_T k$, if $f[u^l] = 0$.*
4. *Let $u = u^k \subset U$.*
   (a) We write the set of generating polynomials for u as
   $$\mathbf{\Lambda}(u) = \{f \in \mathbb{L}[\mathbf{X}] \mid \mathbf{f}[\mathbf{u}] = \mathbf{0}\}.$$
   (b) We write the set of generating polynomials for U as
   $$\mathbf{\Lambda}(U) = \{f \in \mathbb{L}[\mathbf{X}] \mid \mathbf{f}[\mathbf{U}] = \mathbf{0}\},$$
   which was originally called $VALPOL(U)$ [8, p. 323].

Remark 1. By results in [2,8,9] we have the following facts:

1. $\mathbf{\Lambda}(U)$ is an ideal of $\mathbb{L}[\mathbf{X}]$.
2. Setting $\overline{\mathbf{\Lambda}(U)} = \{\bar{g} \mid g \in \mathbf{\Lambda}(U)\}$, and viewing the elements of $\mathbb{L}(r_1, r_2)$ as polynomials, we have that the ideal $\overline{\mathbf{\Lambda}(U)} = \mathbb{L}(r_1, r_2) \cap \mathbf{\Lambda}(U)$.

Let $0 < d \in \mathbb{N}$ and consider the sequence $s^{(1)}, \ldots, s^{(d)}$ in Σ_0 satisfying

$$s_1^{(1)} > \ldots > s_1^{(d)} = 0 \quad \text{and} \quad 0 = s_2^{(1)} < \ldots < s_2^{(d)}. \tag{3}$$

Now we set

$$\begin{aligned} \Delta_i &= \left\{ m \in \Sigma_0 \mid m \preceq \left(s_1^{(i)} - 1, s_2^{(i+1)} - 1 \right) \right\}_{1 \leq i \leq d-1} \\ &= \Delta_{(s_1^{(i)}-1, s_2^{(i+1)}-1)} \end{aligned} \tag{4}$$

and define $\Delta = \bigcup_{i=1}^{d-1} \Delta_i$, which is called a Δ-**set** or delta-set, and the elements $s^{(1)}, \ldots, s^{(d)}$ are called its **defining points**.

We denote by $\mathfrak{F}$ the collection of sets $F = \{f^{(1)}, \ldots, f^{(d)}\} \subset \mathbb{F}[\mathbf{X}]$ where $\{LP(f^{(i)}) = s^{(i)} \mid i = 1, \ldots, d\}$ satisfy the condition (3). We shall say that the elements $F \in \mathfrak{F}$ are of type Δ and we denote by $\Delta(F)$ the Δ-sets determined by them.

Definition 5. *Let U be doubly periodic and $u = u^k \subset U$. We say that the set $F = \{f^{(1)}, \ldots, f^{(d)}\}$ is a minimal set of polynomials for u if:*

1. $F \subset \mathbf{\Lambda}(u)$.
2. $F \in \mathfrak{F}$; *that is $\Delta(F)$ exists.*
3. *If $g \in \mathbb{F}[\mathbf{X}]$ verifies $LP(g) \in \Delta(F)$ then $g \notin \mathbf{\Lambda}(u)$ (i.e. $g[u] \neq 0$).*

We denote by $\mathfrak{F}(u)$ the collection of the minimal sets of u.

For any minimal set of polynomials $F = \{f^{(1)}, \ldots, f^{(d)}\}$ one may see that [8, p. 327] the sets Δ_i in (4) for $i \in \{1, \ldots, d-1\}$ always are nonempty and they are determined by corresponding polynomials that we call $g^{(i)}$. Using this fact, in each iteration, one may construct a set $G = \{g^{(i)} \mid i = 1, \ldots, d-1\}$.

4.1   The Algorithm

From [2, 8, 9], we have the following facts:

Remark 2. Let U be a doubly periodic array.

1. For any $l \in \Sigma_0$, $u^l \subset U$ and $F, F' \in \mathfrak{F}(u^l)$ we have that $\Delta(F) = \Delta(F')$, so that we may write $\Delta(u^l)$.
2. $\Delta(u^l) \subseteq \Delta(U)$ for all $l \in \Sigma_0$ and if $k <_T l \in \Sigma_0$ then $\Delta(u^k) \subseteq \Delta(u^l)$.
3. For any $l \in \Sigma_0$, the set $\Delta(u^l)$ always exists.
4. The set $\Delta(U)$ is exactly the footprint (see [2, p. 1615]) of $\mathbf{\Lambda}(U)$, and it is completely determined by any of its Groebner basis.
5. For any $F \in \mathfrak{F}(u^l)$ we have $F \subset \mathbf{\Lambda}(U)$ implies $\langle F \rangle = \mathbf{\Lambda}(U)$. In fact, F is a Groebner basis for $\mathbf{\Lambda}(U)$ by Definition 5 (3) and [4, Definition 2.5].
6. For any $F \in \mathfrak{F}(u^l)$, we always may construct a "normalized set" $F' \in \mathfrak{F}(u^l)$; that is, satisfying the following property: for any $f \in F'$ and for all $m \in \mathrm{supp}(f) \setminus \{LP(f)\}$ we have $m \preceq LP(f')$, for all $f' \in F'$; that is, $m \in \Delta(u^l)$ [8, Section 6].
7. As we have commented, for any $\overline{\alpha} \in \mathcal{R}$, the equality $\mathcal{D}_{\overline{\alpha}}\left(\overline{\mathbf{\Lambda}(U)}\right) = \mathcal{D}_{\overline{\alpha}}\left(\mathbf{\Lambda}(U)\right)$ holds. Then by [4, Proposition 5.3.1] or [2, p. 1617, Theorem] we have that $\left|\mathcal{D}_{\overline{\alpha}}\left(\overline{\mathbf{\Lambda}(U)}\right)\right| = |\Delta(U)|$ (see also [9, p. 1202]).
8. If F is a reduced Groebner basis for $\mathbf{\Lambda}(U)$ then $LP(F) \subset I$ and, for any $\overline{\alpha} \in \mathcal{R}$, $D_{\overline{\alpha}}(F) = D_{\overline{\alpha}}\left(\overline{\mathbf{\Lambda}(U)}\right)$

Each iteration in the BMSa gives us a minimal set of polynomials for $u = u^{l+1}$ from such a set u^l and the Δ-set $\Delta(u^l)$. The construction of $\Delta(u^{l+1})$ is based on the following remark.

Remark 3. Suppose that $f \in F \in \mathfrak{F}(u^l)$, and $f[u]_l \neq 0$. Then, by the Agreement Theorem and Sakata-Massey Theorem in [2], and Lemma 5 and Lemma 6 in [8] it must happen one of the two following options:

1. $l - LP(f) \in \Delta(u^l)$ and then $LP(f)$ will be a defining point of $\Delta(u^{l+1})$.
2. $l - LP(f) \notin \Delta(u^l)$ and then, $\Delta(u^{l+1})$ will have at least one point more, $l - LP(f)$ itself; in fact, $\Delta_{l-LP(f)} \subset \Delta(u^{l+1})$

Before giving a brief description of the Sakata's algorithm we show some previous basic procedures used in it.

For a minimal set of polynomials $F = \{f^{(1)}, \ldots, f^{(d)}\}$ of $\Lambda(u^l)$, with $LP(f^{(i)}) = \left(s_1^{(i)}, s_2^{(i)}\right)$, for $i = 1, \ldots, d$, we set $F_{\Lambda} = F \cap \Lambda(u^{l+1})$ and $F_N = F \setminus F_{\Lambda}$. We also consider $G = \{g^{(1)}, \ldots, g^{(d-1)}\}$, mentioned in the paragraph below Definition 5.

Theorem 1 (Berlekamp procedure. Lemmas 5, 6 in [8]). *Let $f^{(a)} \in F$ and $g^{(b)} \in G$ such that $f^{(a)} \in \Lambda(u^l)$, $g^{(b)} \in \Lambda(u^k)$, for some $k <_T l \in I$, with $f^{(a)}[u]_l = w_a \neq 0$ and $g^{(b)}[u]_k = v_b \neq 0$.*
We define

$$r_1 = \max\{s_1^{(a)}, l_1 - s_1^{(b)} + 1\},$$
$$r_2 = \max\{s_2^{(a)}, l_2 - s_2^{(b+1)} + 1\} \text{ and}$$
$$\mathbf{e} = \left(r_1 - l_1 + s_1^{(b)} - 1, r_2 - l_2 + s_2^{(b+1)} - 1\right).$$

Then, setting $r = (r_1, r_2)$, we have that

$$h_{f^{(a)}, g^{(b)}} = \mathbf{X}^{r-s^{(a)}} f^{(a)} - \frac{w_a}{v_b} \mathbf{X}^{\mathbf{e}} g^{(b)} \in \Lambda(u^{l+1}).$$

We note that $s_1^{(b)}$ and $s_2^{(b+1)}$ refers to elements of F and not G. Now, we establish two procedures to be used in the algorithm.

Procedure 1. [8, Theorem 1]. *If $f^{(i)} \in F_N$ and $l \in s^{(i)} + \Delta(u^l)$.*

1. *Find $1 \leq j \leq d - 1$ such that $l_1 < s_1^{(i)} + s_1^{(j)}$ and $l_2 < s_2^{(i)} + s_2^{(j+1)}$.*
2. *In the set F we replace $f^{(i)}$ by $h_{f^{(i)}, g^{(j)}}$ obtained by the Berlekamp procedure. The point $s^{(i)}$ will be a defining point of $\Delta(u^{l+1})$ as well.*

Procedure 2. [8, Theorem 2]. *If $f^{(i)} \in F_N$ and $l \notin s^{(i)} + \Delta(u^l)$ then one consider all the following defining points and constructions $h_{f^{(a)}, g^{(b)}}$ to replace $f^{(i)}$ (and, possibly, some elements of G) with the suitable new polynomials in order to get a new $F \in \mathfrak{F}(u^{l+1})$.*

1. $S = \left(l_1 - s_1^{(i)} + 1, l_2 - s_2^{(i+1)} + 1\right)$; *with $f^{(i+1)} \in F_N$ and $1 \leq i < d$. Then find $k \in \{1, \ldots, d\}$ such that $s^{(k)} \prec S$ and set $h_{f^{(k)}, g^{(i)}}$.*

2. $S = \left(l_1 - s_1^{(k)} + 1, s_2^{(i)}\right)$; for some $k < d$, with $f^{(k)} \in F_N$ and $s^{(i)} \prec S$. Then set $h_{f^{(i)}, g^{(k)}}$.

3. $S = \left(l_1 + 1, s_2^{(i)}\right)$ with $i < d$. Then set $h = X_1^{l_1 - s_1^{(i)} + 1} \cdot f^{(i)}$.

4. $S = \left(s_1^{(i)}, l_2 - s_2^{(j)} + 1\right)$ for $j > 2$ with $f^{(j)} \in F_N$ and $s^{(i)} \prec S$. Then set $h_{f^{(i)}, g^{(j-1)}}$.

5. $S = \left(s_1^{(i)}, l_2 + 1\right)$. Then set $h = X_2^{l_2 - s_2^{(i)} + 1} \cdot f^{(i)}$.

Now, we can show a brief scheme of the Sakata's algorithm. See [8, p. 331] for a detailed description.

Algorithm 1 (Sakata). *We start from a finite doubly periodic array, $u \subset U$.*

☐ *Initialize* $|l| = 0$; *that is* $l = (0,0)$, $F = \{1\}$, $G = \emptyset$ *and* $\Delta = \emptyset$.
☐ *For* $l \geq (0,0)$,

1. *For each* $f^{(i)} \in F$ *for which* $f^{(i)} \in F_N$ *we do*
   - *If* $l \in s^{(i)} + \Delta(u^l)$ *then replace* $f^{(i)}$ *by Procedure 1.*
   - *Otherwise, replace* $f^{(i)}$ *by one or more polynomials by Procedure 2.*
2. *Then form the new* F, G *and* $\Delta(u^{l+1})$.
3. *Set* $l := l + 1$.

Let $l \in \Sigma_0$, $F \in \mathfrak{F}(u^l)$ and consider the ideal $\langle F \rangle$ in $\mathbb{L}[\mathbf{X}]$. We suppose WLOG that the elements in F are written in their normal form. Then, on the one hand, it may happen that F is not a Groebner basis for $\langle F \rangle$; on the other hand, even if F is a Groebner basis for $\langle F \rangle$, it may happen that F is not a Groebner basis for $\Lambda(U)$. In [5,8] sufficient conditions on $l \in \Sigma_0$ and F are given to ensure that F is a Groebner basis for $\Lambda(U)$; however, in general, such conditions are not satisfied neither for hyperbolic codes nor hyperbolic-like codes. As we comment in Introduction, we will study this problem in the next section.

5   A New Framework for Locator Decoding

Locator decoding in (bivariate) abelian codes was introduced in [9] (see also [2]). Let us recall, and extend slightly, the basic ideas.

Let C be a bivariate code over $\mathbb{F}(r_1, r_2)$ with defining set $\mathcal{D}_{\bar{\alpha}}(C)$, with respect to some fixed $\bar{\alpha} \in \mathcal{R}$. Suppose a word $c \in C$ was sended and the polynomial $c + e$ in $\mathbb{F}(r_1, r_2)$ has been received. So that, the polynomial e represents the error that we want to find out. To do this, we define the locator ideal in $\mathbb{L}(r_1, r_2)$, which is defined originally in $\mathbb{L}[\mathbf{X}]$ (see [2,9]).

Definition 6. *In the setting above, the locator ideal for e is*

$$L(e) = \{f \in \mathbb{L}(r_1, r_2) \mid f(\bar{\alpha}^n) = 0,\ \forall n \in \mathrm{supp}(e)\}.$$

Having in mind that $\mathbb{L}|\mathbb{F}$ is a splitting field for U, it is easy to see that $\mathcal{D}_{\bar{\alpha}}(L(e)) = \text{supp}(e)$. Our objective is to find the defining set of $L(e)$ and hence $\text{supp}(e)$. The final step (that we will not comment) will be to solve a system of equations to get the coefficients of e (in case $q > 2$). To do this, we shall connect $L(e)$ to the linear recurring relations as follows. Based on the so called syndromes of the received polynomial, we are going to determine a suitable doubly periodic array $U = (u_n)_{n \in \Sigma_0}$ such that the equality $L(e) = \overline{\Lambda(U)}$ holds (see Remark 1). We begin dealing with syndromes. As it is usual in locator decoding, we first consider (theoretically) the syndrome values of $e \in \mathbb{F}(r_1, r_2)$: let $m \in \mathbb{Z}_{r_1} \times \mathbb{Z}_{r_2}$ and define $U - (u_n)_{n \in \Sigma_0}$, such that $u_n = e(\bar{\alpha}^{m+n})$. Clearly, U is an infinite doubly periodic array.

Definition 7. *Let $e \in \mathbb{F}(r_1, r_2)$, $m \in \mathbb{Z}_{r_1} \times \mathbb{Z}_{r_2} = I$ and define $U = (u_n)_{n \in \Sigma_0}$, such that $u_n = e(\bar{\alpha}^{m+n})$. We call U the syndrome table afforded by e and m.*

In practice, we do not know all values of U. Let us return, for a moment, to the error correcting context. By the notion of defining set, one has, for each $m + n \in \mathcal{D}_{\bar{\alpha}}(C)$, that $(c + e)(\bar{\alpha}^{n+m}) = e(\bar{\alpha}^{n+m})$; so, the syndrome values of the error polynomial e are known for all elements in $\mathcal{D}_{\bar{\alpha}}(C)$.

Now we state the mentioned equality of ideals. The proof of the following theorem is (*mutatis mutandi*) similar to that of [9, p. 1202].

Theorem 2. *Let U be the syndrome table afforded by e and m. For any $f \in \mathbb{L}(r_1, r_2)$ the following conditions are equivalent:*

1. $f \in L(e)$.
2. $\sum_{s \in \text{supp}(e)} e_s \bar{\alpha}^{s \cdot n} f(\bar{\alpha}^s) = 0$, for all $n \in \Sigma_m$.
3. $f \in \overline{\Lambda(U)}$.

Consequently, $L(e) = \overline{\Lambda(U)}$.

Theorem 2, together with Remark 2, say that if F is a Groebner basis of $\Lambda(U)$, then
$$\mathcal{D}_{\bar{\alpha}}(L(e)) = \mathcal{D}_{\bar{\alpha}}(\overline{\Lambda(U)}) = \mathcal{D}_{\bar{\alpha}}(F)$$
according to the notation of Sect. 2.

The ideal $\Lambda(U)$ drives us to the framework used in the BMSa in the specific case of U, the syndrome table afforded by e.

5.1 Sufficient Conditions to Obtain a True Groebner Basis for the Ideal $\Lambda(U)$

Now, we present our sufficient condition to obtain a Groebner basis for $\Lambda(U)$ by the BMSa, under the assumption $\omega(e) \le 4$. It is essential to note that $\omega(e) \le t$ implies $|\Delta(U)| \le t$ (see Remark 2).

Lemma 2. *Let U be the syndrome table afforded by e and m, with $\omega(e) \leq t \leq 4$. Suppose that, following the BMSa we have constructed, for $l = (l_1, l_2)$, with $u = u^l$, the sets $\Delta(u) = \Delta$ and $F \in \mathfrak{F}(u)$. We also suppose that there is $f \in F$ such that $f[u]_l \neq 0$ and that $l \notin LP(F) + \Delta$; that is, the delta-set will increase (see Remark 3). Then*

$$(l_1 + 1)(l_2 + 1) \leq 2t + 1.$$

Proof. We shall prove the result for $t = 4$. The other cases are similar and simpler than this. Suppose that $F = \{f^{(1)}, \ldots, f^{(d)}\}$ with $LP(f^i) = s^{(i)}$ for $i = 1, \ldots, d \geq 2$. Setting $f = f^{(i)}$ we have, by hypothesis, $f^{(i)}[u]_l \neq 0$ and $l \notin s^{(i)} + \Delta$.

First note that $|\Delta| \leq 3$ because the size will be increased. So let us list all possible delta-sets: $\Delta_{11} = \{(0,0)\}$, $\Delta_{21} = \{(0,0),(0,1)\}$, $\Delta_{22} = \{(0,0),(1,0)\}$, $\Delta_{31} = \{(0,0),(0,1),(0,2)\}$, $\Delta_{32} = \{(0,0),(1,0),(0,1)\}$ and $\Delta_{33} = \{(0,0),(1,0),(2,0)\}$.

We also note that, by definition of delta-set, if $l \notin LP(F) + \Delta$ then $\Sigma_l \cap (LP(F) + \Delta) = \emptyset$.

Case a: $l_1 > 6$. By paragraph above, we only have to consider $l_1 = 7$. As $s_1^{(1)} \leq 3$, we have that $l_1 - s_1^{(i)} \geq 7 - 3 = 4$, thus, at least $(3,0),(4,0)$ increase $\Delta(u^{l+1})$, which is impossible. So we should have $l_1 \leq 6$.

Case b: $l_1 = 6$ and $l_2 \geq 1$. Again, we only have to consider $l = (6,2)$. Then, the points $(3,0)$ and $(3,1)$ will be added. If $|\Delta| = 2$ then we have to add, in addition, $(2,0)$ and $(2,1)$, and for $\Delta = \Delta_{11}$ we have to add besides the points below, $(1,0)$ and $(1,1)$. In all cases we get $|\Delta(u^{l+1})| > 4$, which is impossible.

Case c: $l_1 = 5$ and $l_2 \geq 1$, so we set $l = (5,1)$. If $s_1^{(1)} = 3$ and $i = 1$ then $(2,1) \in \Delta(u^{l+1})$ which implies that $(0,1),(1,1) \in \Delta(u^{l+1})$ too. In case $i = 2$, then at least $(3,0)$ and $(4,0)$ will be added. If $s_1^{(1)} = 2$ then $l_1 - s_1^{(i)} \geq 3$ so that, for $i = 1$ we have that $(2,0)$, $(2,1)$, $(3,0)$, $(3,1) \in \Delta(u^{l+1})$; for $i = 2$ then $l_1 - s_1^{(i)} \geq 4$, so $(2,0)$, $(3,0)$, $(4,0) \in \Delta(u^{l+1})$. The case $s_1^{(1)} = 1$ is trivial and then in all cases we get $|\Delta(u^{l+1})| > 4$, which is impossible.

Case d: $l_1 = 4$ and $l_2 \geq 1$, so that, set $l = (4,1)$. If $s_1^{(i)} = 3$ then we have to add at least $(0,1)$ and $(1,1)$, if $s_2^{(i)} = 2$ then we must have $i = 2$ and we should add at least $(1,1)$ and $(2,1)$, for Δ_{32} and $(0,1)$ in addition, for Δ_{21}. For $s_1^{(i)} = 1$ then $(1, l_2 - s_2^{(d)})$, $(2, l_2 - s_2^{(d)})$ and $(3, l_2 - s_2^{(d)})$ should be added. All of them are impossible.

Case e: $l_1 = 3$ and $l_2 \geq 2$; so that $l = (3,2)$. If $i = d$ then we add at least $(2, l_2 - s_2^{(d)})$ and $(3, l_2 - s_2^{(d)})$ for those $|\Delta| = 3$ and, in addition, $(0, l_2 - s_2^{(d)})$ and $(1, l_2 - s_2^{(d)})$ for those $|\Delta| \leq 2$. For Δ_{32} and $i = 2$, we have $l - s^{(2)} = (l_1 - s_1^{(2)}, 1)$ so we add at least $(1,1)$ and $(2,1)$. Finally, the case $i = 1$ is obvious and then in all cases we get $|\Delta(u^{l+1})| > 4$, which is impossible.

Case f: $l_1 = 2$ and $l_2 \geq 3$. Take $l = (2,3)$ and repeat **Case e** changing l_2 by l_1; $s_2^{(d)}$ by $s_1^{(1)}$ and so.

Case g: $l_1 = 1$ and $l_2 \geq 4$. Take $l = (1, 4)$ and repeat **Case d** with the adecuate changes, as above.

The last case, $l_1 = 0$ is immediate by Procedure 2.

The proof of the next lemma is a direct computation similar to that we have done above.

Lemma 3. *Let U be the syndrome table afforded by e and m, with $\omega(e) \leq t \leq 4$. Suppose that, following the BMSa we have constructed, for $l = (l_1, l_2)$, with $u = u^l$ the sets $\Delta(u) = \Delta$ and $F \in \mathfrak{F}(u)$. If $l = (l_1, l_2)$ is such that $(l_1 + 1)(l_2 + 1) > 2t + 1$ then $l \notin LP(F) + \Delta$ and hence $\Sigma_l \cap (\Delta + LP(F)) = \emptyset$.*

Thus, if $n \in \Sigma_{(i,j)}$, with $(i,j) \in I$ satisfying $(i + 1)(j + 1) > 2t + 1$ then $f[u]_n = 0$, for any $f \in F$.

Let us summarize the results above in the following theorem.

Theorem 3. *Let U be the syndrome table afforded by e and m, with $\omega(e) \leq t \leq 4$. Suppose that, following the BMSa we have constructed, for $l = (l_1, l_2)$, and $u = u^l$, the sets $\Delta(u) = \Delta$ and $F \in \mathfrak{F}(u)$. For any $f \in F$, we have that:*

1. *If $f \in F$ is such that $f[u]_l \neq 0$ and $l \notin LP(f) + \Delta$ then $(l_1 + 1)(l_2 + 1) \leq 2t + 1$.*
2. *If $l_k > 2t - 1$, for $k \in \{1, 2\}$ then $f[u]_l = 0$. If $l = (0, l_2)$ or $l = (l_1, 0)$ with $l_k \leq 2t - 1$, for $k \in \{1, 2\}$ then it may happen that $f[u]_l \neq 0$.*
3. *If $l_1, l_2 \neq 0$ and $(l_1 + 1)(l_2 + 1) > 2t + 1$ then $f[u]_l = 0$ for any $f \in F$.*

Let C be a bivariate code with error-correction capability $t = \lfloor \frac{d(C)-1}{2} \rfloor$ and let $g = c + e$ the received polynomial. Let U be the syndrome table afforded by e and $m \in I$, and assume that $\omega(e) \leq t \leq 4$. Suppose that $\overline{m + \mathcal{B}_{2t+1}} \subset \mathcal{D}_{\bar{\alpha}}(C)$ (see (2)). Then, for all $n \in \mathcal{B}_{2t+1}$, the values $u_n = e(\bar{\alpha}^{m+n}) = g(\bar{\alpha}^{m+n})$ are known.

In practice we only know and work with a bound of the error-correction capability; that is, $t^* = \lfloor \frac{sd^*(C)-1}{2} \rfloor$. We suppose that $\omega(e) \leq t^* \leq 4$ and we consider $\mathcal{B}_{2t^*+1}$.

Theorem 4. *In the setting described in paragraph above, if $u_{(0,j)} \neq 0$, for some $j < t$ (respectively if $u_{(i,j)} \neq 0$ with $i + j = 1$) we may find a Groebner basis for the locator ideal $L(e)$ following the BMSa with the lexicographic order (respectively the graded order).*

Proof. We begin by considering the lexicographic order.

We recall that at initializing the BMSa we take $F = \{1\}$, so that $1[u]_{(0,j)} = u_{(0,j)}$ and the first two defining points are $(1, 0)$ and $(0, j + 1)$, which indicate us the necessity $u_{(0,j)} \neq 0$ for some $j < t$.

Now, to do all steps for the pair of the form $(0, *)$ we have to compute at most $l = (0, j)$ for $j = 0, \ldots 2t - 1$. Now suppose we have compute $\Delta(u^l)$ for all $l = (l_1, l_2)$ with $l_1, l_2 \neq 0$ and $(l_1 + 1)(l_2 + 1) \leq 2t + 1$, which is equivalent for $t \leq 4$ to the values $l_2 = 0, \ldots, t - l_1$. Then any step considered after that, say again l, must verify $f[u]_l = 0$, by Theorem 3(3).

Clearly, the last point for which our Δ may be increased is $(2t-1,0)$. After that, Theorem 3 guarantees us that Δ cannot increase their size. However, any step of the form $l=(l_1,0)$ with $l_1 \leq 2t-1$ may satisfy $l \in LP(F)+\Delta$ and so F may be changed. So we have to consider them.

For any step of the form $l \geq_T (2t,0)$ it happens that $l \notin LP(F)+\Delta$ and clearly $f[u]_l = 0$, for all $f \in F \in \mathfrak{F}(u^l)$ because $|\Delta(u^{l+1})| \leq t$.

Now we deal with graded order. Suppose we compute $\Delta(u^l)$ for all $\{l = (l_1,l_2) \mid l_1+l_2 \leq t\}$, and F is the minimal set of polynomials obtained in the last iteration, with $\Delta(F) = \Delta$. Consider a point $l = (l_1,l_2)$ such that $l_1 + l_2 \geq t+1$. Then one may check that $(l_1+1)(l_2+1) > 2t+1$, for $t \leq 4$; so, if one has that $l_1,l_2 \neq 0$ then Theorem 3 says that $f[u]_l = 0$ for all $f \in F$. Finally, it may happen that $f[u]_l \neq 0$ for $l = (a,0),(0,a)$ with $a \in \{t+1,\ldots,2t-1\}$ (the cases $l = (j,0)$, with $j \geq 2t$ has been already seen). We will continue forming minimal sets of polynomials until consider all of them.

Therefore in any of the monomial orders considered, the polynomials of F are valid in I; so that $F \subset \Lambda(U)$ and then $\langle F \rangle = \Lambda(U)$. By Remark 2 and Theorem 2 we are done.

Example 1. Consider the code C, in $\mathbb{F}_2(5,15)$ with primitive root a, and $\mathcal{D}_{(\alpha,\beta)}(C) = Q(0,13) \cup Q(1,13) \cup Q(2,13) \cup Q(3,13) \cup Q(4,13) \cup Q(0,0) \cup Q(0,1)$. One may check that the strong apparent distance $sd^*(C) = 6$, so that $t = 2$ is a lower bound for the error correction capability of C. For the error polynomial $e = X_1^2 X_2^2 + X_2$ and $m = (0,13)$ we have the first value $u_{(0,0)} = e(\alpha^0,\beta^{13}) = a^4$ and the last one $u_{(4,0)} = e(\alpha^4,\beta^{13}) = a^2$. So that we arrange

$$(u_n \mid n \in \mathcal{S}(2)) = \begin{pmatrix} a^4 \ a^2 \ 0 \ a^5 \\ a^{14} \ a^9 \\ a^3 \\ a^2 \end{pmatrix}.$$

Next table summarizes all computation with respect to the lexicographic order.

l	$F \subset \Lambda(u^{l+1})$	G	$\Delta(u^{l+1})$
Initializing	$\{1\}$	$\emptyset$	$\emptyset$
$(0,0) \rightarrow$	$\{X_1, X_2\}$	$\{1\}$	$\{(0,0)\}$
$(0,1) \rightarrow$	$\{X_1, X_2 + a^{13}\}$	$\{1\}$	$\{(0,0)\}$
$(0,2) \rightarrow$	$\{X_1, X_2^2 + a^{13}X_2 + a^{11}\}$	$\{X_2 + a^{13}\}$	$\{(0,0),(0,1)\}$
$(0,3) \rightarrow$	$\{X_1, X_2^2 + a^5 X_2 + a^3\}$	$\{X_2 + a^{13}\}$	$\{(0,0),(0,1)\}$
$(1,0) \rightarrow$	$\{X_1 + a^6 X_2 + a^2,$ $X_2^2 + a^5 X_2 + a^3\}$	$\{X_2 + a^{13}\}$	$\{(0,0),(0,1)\}$
$(1,1) \rightarrow$	$\{X_1 + a^8 X_2 + a^7,$ $X_2^2 + a^5 X_2 + a^3\}$	$\{X_2 + a^{13}\}$	$\{(0,0),(0,1)\}$
$(2,0),(3,0) \rightarrow$	Same	Same	Same

The reader may check that $\mathcal{D}_{\bar{\alpha}}(\Lambda(U)) = \mathcal{D}_{\bar{\alpha}}(\langle F \rangle) = \{(2,2),\ (0,1)\}$.

References

1. Bernal, J.J., Bueno-Carreño, D.H., Simón, J.J.: Apparent distance and a notion of BCH multivariate codes. IEEE Trans. Inf. Theory **62**(2), 655–668 (2016)
2. Blahut, R.E.: Decoding of cyclic codes and codes on curves. In: Huffman, W.C., Pless, V. (eds.) Handbook of Coding Theory, vol. II, pp. 1569–1633 (1998)
3. Camion, P.: Abelian codes. MCR Tech. Sum. Rep. 1059, University of Wisconsin, Madison (1970)
4. Cox, D.A., Little, J., O'Shea, D.: Ideals, Varieties, and Algorithms. Springer, Heidelberg (1998). https://doi.org/10.1007/978-3-662-41154-4
5. Hackl, M.: Multivariate polynomial codes. Johannes-Kepler-University (2000)
6. Imai, H.: A theory of two-dimensional cyclic codes. Inf. Control **34**(1), 1–21 (1977)
7. Rubio, I.M., Sweedler, M., Heegard, C.: Finding a Gröbner basis for the ideal of recurrence relations on m-dimensional periodic arrays. In: 12th International Conference on Finite Fields and Their Applications, Contemporary Developments in Finite Fields and Applications, pp. 296–320. World Scientific (2016)
8. Sakata, S.: Finding a minimal set of linear recurring relations capable of generating a given finite two-dimensional array. J. Symb. Comput. **5**, 321–337 (1988)
9. Sakata, S.: Decoding binary cyclic 2-D codes by the 2-D Berlekamp-Massey algorithm. IEEE Trans. Inf. Theory **37**(4), 1200–1203 (1991)
10. Sakata, S.: The BMS algorithm and decoding of AG codes. In: Sala, M., Sakata, S., Mora, T., Traverso, C., Perret, L. (eds.) Gröbner Basis, Coding, and Cryptography, pp. 165–185. Springer, Heidelberg (2010). https://doi.org/10.1007/978-3-540-93806-4_10
11. Sakata, S.: The BMS Algorithm. In: Sala, M., Sakata, S., Mora, T., Traverso, C., Perret, L. (eds.) Gröbner Basis, Coding, and Cryptography, pp. 143–163. Springer, Heidelberg (2010). https://doi.org/10.1007/978-3-540-93806-4_9

Dihedral Codes with Prescribed Minimum Distance

Martino Borello[1(⊠)] and Abdelillah Jamous[2]

[1] Université Paris 8, Laboratoire de Géométrie, Analyse et Applications, LAGA,
Université Sorbonne Paris Nord, CNRS, UMR 7539, 93430 Villetaneuse, France
`borello@math.univ-paris13.fr`
[2] Faculty of Mathematics, University of Sciences and Technology Houari
Boumediene, Algiers, Algeria

Abstract. Dihedral codes, particular cases of quasi-cyclic codes, have
a nice algebraic structure which allows to store them efficiently. In this
paper, we investigate it and prove some lower bounds on their dimension
and minimum distance, in analogy with the theory of BCH codes. This
allows us to construct dihedral codes with prescribed minimum distance.
In the binary case, we present some examples of optimal dihedral codes
obtained by this construction.

Keywords: Group algebras · Dihedral codes · BCH bound

1 Introduction

Block codes were invented in the 1940s to correct errors in the communication
through noisy channels (see [18] for more details), and they are used nowadays
in different areas of information security. Originally, they were thought of just
as subsets of (code)words of n letters chosen in an alphabet K, which are far
enough apart from each other with respect to the Hamming distance. However,
they usually need to have more algebraic structure to be stored efficiently. By
considering *linear codes* of length n over a finite field K, that is subspaces of the
vector space K^n, we have a compact description given, for example, by the *parity
check matrix*, which is a matrix H such that $c \in \mathcal{C}$ if and only if $cH = 0$. Such a
description reduces exponentially the size of the data to be stored with respect
to general block codes. However, this reduction reveals to be insufficient in the
context of code-based cryptography ([20,24] and many others), where the public
key is related to the parity check matrix of a code of large length and dimension.
The size of the public key constitutes one of the main practical disadvantages in
the use of code-based cryptography and many efforts have been made to reduce
it by preserving the security of the system. One option may be to use codes
with symmetries, like cyclic or quasi-cyclic codes (see for example [3]). However,
since decoding of general quasi-cyclic codes is difficult, the algebraic structure
that one needs to add may also turn out to be a weakness of the system (see for
example [14]).

© Springer Nature Switzerland AG 2021
J. C. Bajard and A. Topuzoğlu (Eds.): WAIFI 2020, LNCS 12542, pp. 147–159, 2021.
https://doi.org/10.1007/978-3-030-68869-1_8

A natural generalisation of cyclic codes is given by the family of group codes: a linear code $\mathcal{C}$ is called a *G-code* (or a group code) if $\mathcal{C}$ is a right (or left) ideal in the group algebra $KG = \{a = \sum_{g \in G} a_g g \mid a_g \in K\}$ where G is a finite group. Reed Muller codes over prime fields $\mathbb{F}_p$ are group codes for an elementary abelian p-group G [4,10], and there are many other remarkable optimal codes which have been detected as group codes [5,12,15,21]. If G is cyclic, then all right (or left) ideals of KG afford only one check equation (and then only a small amount of data has to be stored). In the case G is a general finite group there are only particular right (or left) ideals which satisfy this property, called *checkable* codes [19]. In [6] it is proved that such codes are the duals of principal ideals, and group algebras KG for which all right (or left) ideals are checkable (or equivalently principal), called *code-checkable group algebras*, are characterised: KG is a code-checkable group algebra if and only if G is p-nilpotent with a cyclic Sylow p-subgroup, where p is the characteristic of K. This is a consequence of an early result by Passman ([25, Theorem 4.1]). Checkable codes are asymptotically good [2,7] and many optimal codes are checkable [6, Remark 2.9].

In the next table, the results of some simulations by MAGMA of random binary dihedral codes are presented. The first line is the cardinality $2n$ of the group, the second line is the percentage of optimal checkable codes among 1000 tested of dimension $3 \leq k \leq 2n - 3$, the third line is the percentage of optimal principal codes among 1000 tested of dimension $3 \leq k \leq 2n - 3$.

10	14	18	22	26	30	34	38	42
51.4	5.4	4.4	48.4	42.2	0.3	1.3	2.871	0.7
23.2	5.6	70.2	47.5	20.5	54.1	27.2	1.105	69.3

All these data seem to suggest that the family of principal and checkable codes is worth further investigation. In particular, it is desirable to prove some bounds on the dimension and minimum distance for principal or checkable codes and to introduce families of principal or checkable codes with prescribed minimum distance (in analogy with BCH codes).

To our knowledge, there are very few results concerning the parameters of group codes, both for general and particular groups. In [13], an algorithm for computing the dimension of general group codes is given. In a very recent paper [11], several relations and bounds for the dimension of principal ideals in group algebras are determined by analysing minimal polynomials of regular representations. The concatenated structure of dihedral codes is investigated in [9]. However, we are not aware of results which allow to construct group codes with a prescribed minimum distance or explicit lower bounds on both dimension and minimum distance, even in the easiest case of dihedral codes. This paper wants to be a first contribution in this direction. In Sect. 2 we will recall some results of the theory of quasi-cyclic codes. In Sect. 3 we will recall the definition of dihedral codes, present some results about their algebraic structure, make some remarks about the dual codes, prove a BCH bound for principal dihedral codes, propose a definition of principal BCH-dihedral codes, consider the particular case of binary dihedral codes and give some construction of optimal codes. Finally, in Sect. 4 we will present some open problems. In particular, an efficient decoding algorithm would be a necessary prerequisite for applications in cryptography.

2   Quasi-cyclic Codes

We recall in this section some definitions and known results about quasi-cyclic codes. As we will see in the next section, dihedral codes, as all group codes, form a subfamily of quasi-cyclic codes.

Let q be a power of a prime and $\mathbb{F}_q$ the finite field with q elements. Let $n \in \mathbb{N}$. The symmetric group S_n acts on the vector space $\mathbb{F}_q^n$ as follows:

$$v^\sigma := \left(v_{\sigma^{-1}(1)}, v_{\sigma^{-1}(2)}, \dots, v_{\sigma^{-1}(n)}\right)$$

for $v := (v_1, v_2, \dots, v_n) \in \mathbb{F}_q^n$ and $\sigma \in S_n$. For a linear code $\mathcal{C} \subseteq \mathbb{F}_q^n$, the set of permutations such that $\mathcal{C}^\sigma := \{c^\sigma \mid c \in \mathcal{C}\}$ is equal to $\mathcal{C}$ is a group which is called the *permutation automorphism group* of $\mathcal{C}$ and which is denoted by $\mathrm{PAut}(\mathcal{C})$.

In this context, a remarkable transformation is the so-called *shift map*, that is

$$T_n : \mathbb{F}_q^n \to \mathbb{F}_q^n \qquad c \mapsto c^{(1 \ \cdots \ n)} = (c_n, c_1, \dots, c_{n-1}).$$

Linear codes which are invariant under the shift or its power are the so-called quasi-cyclic codes, which are defined as follows.

Definition 1. *Let $\mathcal{C} \subseteq \mathbb{F}_q^n$ be a linear code. Suppose that $n = \ell m$, for some positive integers ℓ and m. The code $\mathcal{C}$ is* quasi-cyclic of index ℓ *if $T_n^\ell(\mathcal{C}) = \mathcal{C}$, that is if*

$$(1 \ \dots \ n)^\ell = \prod_{j=1}^{\ell} (j \ \ell+j \ 2\ell+j \ \dots \ (m-1)\ell+j) \in \mathrm{PAut}(\mathcal{C}).$$

If $\ell = 1$, the code $\mathcal{C}$ is called cyclic.

Let $R := \mathbb{F}_q[x]/(x^m - 1)$. We may relabel the coordinates and consider the bijective $\mathbb{F}_q$-linear map

$$\varphi : \mathbb{F}_q^n = (\mathbb{F}_q^\ell)^m \to R^\ell \tag{1}$$

$$(c_{11}, \dots, c_{1\ell}, \dots, c_{m1}, \dots, c_{m\ell}) \mapsto (c_{11} + \dots + c_{m1}x^{m-1}, \dots, c_{1\ell} + \dots + c_{m\ell}x^{m-1}).$$

The image of a quasi-cyclic code in R^ℓ is an R-submodule. Actually, the multiplication by x corresponds to the ℓ-th power of the shift.

Remark 1. There is a one-to-one correspondence between the R-submodules of R^ℓ and left ideals of $\mathrm{Mat}_\ell(R)$ (which is isomorphic, as a ring, to $\mathrm{Mat}_\ell(\mathbb{F}_q)[x]/(x^m - 1)$). This is a particular case of the Morita equivalence for modules [23]. The explicit one-to-one map is given as follows: to any R-submodule N of R^ℓ we associate the left ideal $\mathcal{I}_N$ of $\mathrm{Mat}_\ell(R)$ composed by matrices whose rows are elements in N. As already observed in [1], since R is a commutative principal ideal ring, every R-submodule N of R^ℓ has at most ℓ generators, so that the left ideal $\mathcal{I}_N$ is principal (it suffices to consider the matrix whose rows are the generators and eventually some zeros). So there exists a generator of $\mathcal{I}_N$ which can be seen a polynomial in $\mathrm{Mat}_\ell(\mathbb{F}_q)[x]/(x^m - 1)$.

Let ℓ be a positive integer, and $\nu \in \mathbb{F}_{q^\ell}$ be a primitive element of $\mathbb{F}_{q^\ell}/\mathbb{F}_q$. Recall that $\{1, \nu, \ldots, \nu^{\ell-1}\}$ is an $\mathbb{F}_q$-base of the vector space $\mathbb{F}_{q^\ell}$. The *folding* is the $\mathbb{F}_q$-linear map

$$\phi : \mathbb{F}_q^\ell \to \mathbb{F}_{q^\ell} = \mathbb{F}_q[\nu]$$
$$(a_1, \ldots, a_\ell) \mapsto a_1 + a_2\nu + \cdots + a_\ell\nu^{\ell-1}.$$

Definition 2. *Let $\mathcal{C} \subseteq \mathbb{F}_q^n = (\mathbb{F}_q^\ell)^m$ be a linear code. The* folded code *of $\mathcal{C}$ is $\mathcal{C}' = \phi^m(\mathcal{C}) \subseteq (\mathbb{F}_{q^\ell})^m$ (where ϕ^m is the tensor power of ϕ). In this case, $\mathcal{C}$ is the* unfolded code *of $\mathcal{C}'$.*

Remark 2. Note that the definition of the folding depends on the choice of ν. However the properties that we will consider in this paper do not depend on it and this is the reason why the definite article "the" makes sense in this context. Note that the folded code $\mathcal{C}'$ of a linear code $\mathcal{C}$ is an $\mathbb{F}_q$-linear code. Moreover, $\mathcal{C}$ is quasi-cyclic if and only if $\mathcal{C}'$ is invariant under the shift T_m.

In the next section we will often use the above equivalence and the following definition.

Definition 3. *An $\mathbb{F}_q$-linear code $\mathcal{C} \subseteq (\mathbb{F}_{q^\ell})^m$ which is invariant under the shift T_m is called an $\mathbb{F}_q$-linear cyclic code.*

Barbier *et al.* define in [1] the analogue of BCH codes in the quasi-cyclic case. They call them *quasi-BCH* codes. In [16], the algebraic structure of $\mathbb{F}_q$-linear cyclic codes over $\mathbb{F}_{q^\ell}$ is studied. In next section we will explore the same concepts in the context of dihedral codes.

3   Dihedral Codes

Let $m \geq 3$ be an integer and

$$D_{2m} := \langle \alpha, \beta \mid \alpha^m = 1, \beta^2 = 1, \beta\alpha = \alpha^{m-1}\beta \rangle,$$

be the *dihedral group* of order $2m$. The *group algebra* $\mathbb{F}_q D_{2m}$ is the set

$$\mathbb{F}_q D_{2m} := \left\{ \sum_{\gamma \in D_{2m}} a_\gamma \gamma \ \middle|\ a_\gamma \in \mathbb{F}_q \right\},$$

which is a vector space over $\mathbb{F}_q$ with canonical basis $\{\gamma\}_{\gamma \in D_{2m}}$. The operations of sum and multiplication by scalars are defined in the following natural way: for any $a_\gamma, b_\gamma \in \mathbb{F}_q$ and $c \in \mathbb{F}_q$

$$\sum_{\gamma \in D_{2m}} a_\gamma \gamma + \sum_{\gamma \in D_{2m}} b_\gamma \gamma = \sum_{\gamma \in D_{2m}} (a_\gamma + b_\gamma)\gamma,$$

$$c \cdot \left(\sum_{\gamma \in D_{2m}} a_\gamma \gamma\right) = \sum_{\gamma \in D_{2m}} ca_\gamma \gamma.$$

Moreover, $\mathbb{F}_q D_{2m}$ is an algebra with the product

$$\left(\sum_{\gamma \in D_{2m}} a_\gamma \gamma\right) \bullet \left(\sum_{\gamma \in D_{2m}} b_\gamma \gamma\right) = \sum_{\gamma \in D_{2m}} \left(\sum_{\mu\nu = \gamma} a_\mu b_\nu\right) \gamma.$$

Definition 4. *A* dihedral code, *or a* D_{2m}-code, *is a left ideal of* $\mathbb{F}_q D_{2m}$.

As observed in [8], a linear code of length $2m$ can be seen as a D_{2m}-code if and only if its permutation automorphism group contains a subgroup isomorphic to D_{2m} all of whose nontrivial elements act fixed point free on the coordinates $\{1, \ldots, 2m\}$. In particular, if we consider the ordering

$$D_{2m} = \{\underbrace{1}_{b_1}, \underbrace{\beta}_{b_2}, \underbrace{\alpha}_{b_3}, \underbrace{\alpha\beta}_{b_4}, \underbrace{\alpha^2}_{b_5}, \underbrace{\alpha^2\beta}_{b_6}, \ldots, \underbrace{\alpha^{m-1}}_{b_{2m-1}}, \underbrace{\alpha^{m-1}\beta}_{b_{2m}}\}, \tag{2}$$

and the $\mathbb{F}_q$-linear isomorphism between $\mathbb{F}_q^{2m}$ and $\mathbb{F}_q D_{2m}$ given by $e_i \mapsto b_i$ (where $\{e_i\}$ is the canonical basis of $\mathbb{F}_q^{2m}$), a linear code $\mathcal{C} \subseteq \mathbb{F}_q^{2m}$ is a D_{2m}-code if and only if

$$\alpha' := (1\ 3\ 5\ \ldots\ 2m-1)(2\ 4\ 6\ \ldots\ 2m)$$

and

$$\beta' := (1\ 2)(3\ 2m)(4\ 2m-1)(5\ 2m-2) \cdots (m+1\ m+2)$$

are in $\mathrm{PAut}(\mathcal{C})$. These elements correspond to the permutation representation of the left multiplication by α and by β respectively in $\mathbb{F}_q D_{2m}$. In particular, since $\alpha' = (1\ \ldots\ 2m)^2$, a dihedral code is a quasi-cyclic code of index 2.

From now on, we will always consider the ordering (2) fixed and we will identify $\mathbb{F}_q^{2m}$ and $\mathbb{F}_q D_{2m}$.

3.1 Algebraic Structure

Let $\mathcal{C}$ be a D_{2m}-code over $\mathbb{F}_q$. As we observed above, since $\mathcal{C}$ is a quasi-cyclic codes of index 2, $\mathcal{C}$ is a free left module of rank 2 over $R := \mathbb{F}_q[x]/(x^m - 1)$, which is a commutative principal ideal ring. As we have already seen in Remark 1, this means that $\mathcal{C}$ has at most two generators as a module over R. These are also two generators of $\mathcal{C}$ viewed as an ideal in $\mathbb{F}_q D_{2m}$. We have one generator of $\mathcal{C}$ as an ideal in $\mathrm{Mat}_2(\mathbb{F}_q)[x]/(x^m - 1)$, given by the polynomial with coefficients in the ring of matrices with first row given by the first generator and second row given by the second one. However, it may happen that $\mathcal{C}$ is not principal as an ideal in $\mathbb{F}_q D_{2m}$.

Remark 3. As observed in [6], an early result by Passman ([25, Theorem 4.1]) gives us that all D_{2m}-codes over a field $\mathbb{F}_q$ of characteristic p are principal if and only if D_{2m} is p-nilpotent with a cyclic Sylow p-subgroup (we recall that a group G is p-nilpotent if it admits a normal subgroup N of order coprime with p and such that G/N is a p-group). This is the case if and only if p does not divide m. So

- if $(m, q) = 1$, all D_{2m}-codes over $\mathbb{F}_q$ are principal;
- otherwise, a D_{2m}-code over $\mathbb{F}_q$ is either principal or the sum of two principal ideals.

We will study then the algebraic structure of principal left ideals in $\mathbb{F}_q D_{2m}$, that is principal dihedral codes. Via the map φ defined as in (1), we can consider $\varphi(\mathcal{C})$ inside R^2. The automorphism α' corresponds to the multiplication by x in R^2, whereas the automorphism β' acts on R^2 as follows: for $(a(x), b(x)) \in R^2$,

$$(a(x), b(x))^{\beta'} = (b(x^{m-1}), a(x^{m-1})).$$

So, $\mathcal{C}$ is a D_{2m}-code if and only if $\varphi(\mathcal{C})$ is an R-submodule of R^2 invariant under the action of β', that is such that $(b(x^{m-1}), a(x^{m-1})) \in \varphi(\mathcal{C})$ for all $(a(x), b(x)) \in \varphi(\mathcal{C})$.

If $\mathcal{C}$ is principal, then $\varphi(\mathcal{C})$ is an R-submodule of R^2 generated, as a module, by

$$(a(x), b(x)) \quad \text{and} \quad (b(x^{m-1}), a(x^{m-1})).$$

Remark 4. We have already mentioned the Morita correspondence between R-sub-modules and left ideals in $\mathrm{Mat}_2(R) \cong \mathrm{Mat}_2(\mathbb{F}_q)[x]/(x^m - 1)$. In this case, the left ideal $I_{\mathcal{C}} \subseteq \mathrm{Mat}_2(\mathbb{F}_q)[x]/(x^m - 1)$ associated to $\mathcal{C}$ is the principal ideal

$$I_{\mathcal{C}} = \left\langle \begin{pmatrix} a_0 & b_0 \\ b_0 & a_0 \end{pmatrix} + \begin{pmatrix} a_1 & b_1 \\ b_{m-1} & a_{m-1} \end{pmatrix} x + \cdots + \begin{pmatrix} a_{m-1} & b_{m-1} \\ b_1 & a_1 \end{pmatrix} x^{m-1} \right\rangle,$$

where $a(x) := a_0 + a_1 x + \ldots + a_{m-1} x^{m-1}$ and $b(x) := b_0 + b_1 x + \ldots + b_{m-1} x^{m-1}$.

Considering the folding $(\mathbb{F}_q)^2 \to \mathbb{F}_{q^2} = \mathbb{F}_q[\nu]$ (where ν is a primitive element of $\mathbb{F}_{q^2}/\mathbb{F}_q$), we can see the two polynomials $a(x), b(x)$ as a unique polynomial over $\mathbb{F}_{q^2}$, that is

$$p(x) := (a_0 + b_0 \nu) + (a_1 + b_1 \nu)x + \ldots + (a_{m-1} + b_{m-1} \nu)x^{m-1},$$

so that a principal dihedral code can be seen as the sum (as vector spaces) of the two $\mathbb{F}_q$-linear cyclic codes over $\mathbb{F}_{q^2}$, that is the one generated by $p(x)$ and the one generated by $\bar{p}(x^{m-1})$, where

$$\bar{p}(x) := (b_0 + a_0 \nu) + (b_1 + a_1 \nu)x + \ldots + (b_{m-1} + a_{m-1} \nu)x^{m-1}.$$

For $\tau := a + b\nu \in \mathbb{F}_{q^2}$, let $\overline{\tau} := b + a\nu$. The $\mathbb{F}_q$-linear map $\tau \mapsto \overline{\tau}$ can be expressed by the following linearised polynomial:

$$\tau \mapsto L(\tau) := \left(\frac{1 - \nu^2}{\nu^q - \nu}\right)\tau^q + \left(\frac{\nu^{q+1} - 1}{\nu^q - \nu}\right)\tau.$$

so that, if

$$p(x) := \tau_0 + \tau_1 x + \ldots + \tau_{m-1} x^{m-1},$$

we have

$$\overline{p}(x) = \overline{\tau_0} + \overline{\tau_1} x + \ldots + \overline{\tau_{m-1}} x^{m-1} =$$
$$\left(\frac{1 - \nu^2}{\nu^q - \nu}\right) p(x^{1/q})^q + \left(\frac{\nu^{q+1} - 1}{\nu^q - \nu}\right) p(x).$$

Note that $\overline{p}(x)$ is taken modulo $x^m - 1$ and fractional exponents are only apparent, since we consider the q-th power of $p(x^{1/q})$.

Definition 5. *For a polynomial $r(x) \in \mathbb{F}_{q^2}[x]/(x^m - 1)$, we denote by $\langle r(x) \rangle_{\mathbb{F}_q}$ the unfolded $\mathbb{F}_q$-linear cyclic code generated by $r(x)$, i.e. the unfolded of*

$$\{t(x)r(x) \in \mathbb{F}_{q^2}[x]/(x^m - 1) \mid t(x) \in \mathbb{F}_q[x]\}.$$

We can summarise all the discussion in the following.

Theorem 1. *Let $\mathbb{F}_{q^2} = \mathbb{F}_q[\nu]$ and $\mathcal{C}$ be a principal D_{2m}-code over $\mathbb{F}_q$. There exists $p(x) \in \mathbb{F}_{q^2}[x]/(x^m - 1)$ such that*

$$\mathcal{C} = \langle p(x) \rangle_{\mathbb{F}_q} + \langle \overline{p}(x^{m-1}) \rangle_{\mathbb{F}_q},$$

where

$$\overline{p}(x^{m-1}) = \left(\frac{1 - \nu^2}{\nu^q - \nu}\right) p(x^{(m-1)/q})^q + \left(\frac{\nu^{q+1} - 1}{\nu^q - \nu}\right) p(x^{m-1}) \in \mathbb{F}_{q^2}[x]/(x^m - 1).$$

In particular, as we have already observed in Remark 3, all D_{2m}-codes over $\mathbb{F}_q$ are principal if $(m, q) = 1$ and they are a sum (as vector spaces) of at most two principal D_{2m}-codes otherwise.

Definition 6. *We call the polynomial $p(x)$ a generator of the principal dihedral code.*

Corollary 1. *Let $\mathcal{C}$ be a principal D_{2m}-code over $\mathbb{F}_q$ generated by $p(x)$. Then*

$$\dim_{\mathbb{F}_q} \mathcal{C} \geq \max\{m - \deg p(x), m - \deg \overline{p}(x^{m-1})\}.$$

Proof. This follows from the fact that the vectors in $\mathbb{F}_q^{2m}$ corresponding to the polynomials

$$\{p(x), xp(x), \ldots, x^{m - \deg p(x) - 1} p(x)\}$$

are linearly independent, and the same holds for the ones corresponding to

$$\{\overline{p}(x^{m-1}), x\overline{p}(x^{m-1}), \ldots, x^{m - \deg \overline{p}(x^{m-1}) - 1} \overline{p}(x^{m-1})\}.$$

Remark 5. For calculations, it may be interesting to have integer exponents. In the case $(m, q) = 1$, we can take m' to be the inverse of m modulo q, so that $m'm - 1$ is divisible by q. Let $r := (m'm - 1)/q$. Then

$$\overline{p}(x^{m-1}) = \left(\frac{1 - \nu^2}{\nu^q - \nu}\right) p(x^r)^q + \left(\frac{\nu^{q+1} - 1}{\nu^q - \nu}\right) p(x^{m-1}).$$

3.2   Dual Code

In analogy with the theory of cyclic and quasi-cyclic codes, it it interesting to investigate the dual codes of dihedral codes, which are still dihedral.

Proposition 1. *The dual code $C^{\perp}$ of a dihedral code C is a dihedral code.*

Proof. This follows trivially from the fact that $\mathrm{PAut}(C^{\perp}) = \mathrm{PAut}(C)$.

The dual of a principal dihedral code is not necessarily principal. But if $(m, q) = 1$, as we mentioned already, all dihedral codes are principal. So it makes sense to investigate the relation between the generator of a code and a generator of its dual.

Let $p(x)$ and $q(x)$ be two polynomial in $\mathbb{F}_{q^2}[x]/(x^m - 1)$ and let v and w the two vectors in $\mathbb{F}_q^{2m}$ corresponding to $p(x)$ and $q(x)$ respectively. We may define

$$* : \mathbb{F}_{q^2}[x]/(x^m - 1) \times \mathbb{F}_{q^2}[x]/(x^m - 1) \to \mathbb{F}_q$$
$$(p(x), q(x)) \qquad\qquad \mapsto p(x) * q(x) := \langle v, w \rangle$$

Proposition 2. *Let $(m, q) = 1$. If C is a principal D_{2m}-code generated by $p(x)$ and $C^{\perp}$ is a principal D_{2m}-code generated by $q(x)$, then*

$$p(x) * q(x) = 0, \ \ p(x) * \overline{q}(x^{m-1}) = 0,$$

$$\overline{p}(x^{m-1}) * q(x) = 0, \ \ \overline{p}(x^{m-1}) * \overline{q}(x^{m-1}) = 0.$$

The same holds with all the shifts of $p(x)$ and $\overline{p}(x^{m-1})$.

Proof. This is clear from the definition of $*$.

Remark 6. At least two questions stand open in this context: the conditions in Proposition 2 are only necessary. It would be very interesting to find sufficient conditions for a polynomial $q(x)$ to be a generator of the dual. We may add the orthogonality with all the shifts of $p(x)$ and $\overline{p}(x^{m-1})$, but this would still be not enough. A polynomial $q(x)$ satisfying all these relations would generate a subcode of $C^{\perp}$, but not necessarily the whole dual. In fact, there is an argument on the dimension missing. Secondly, it would be nice to give some relations with the usual product of polynomials (as in the cyclic codes case) and not with the $*$ product.

For dihedral codes over fields of characteristic 2, a nice relation holds.

Proposition 3. *If q is a power of 2, then $\langle \overline{p}(x^{m-1}) \rangle_{\mathbb{F}_q} \subseteq \langle p(x) \rangle_{\mathbb{F}_q}^{\perp}$. In particular, the code generated by $p(x)$ is contained in $\langle p(x) \rangle_{\mathbb{F}_q} + \langle p(x) \rangle_{\mathbb{F}_q}^{\perp}$.*

Proof. Recall that if $p(x)$ corresponds to the vector

$$v = (a_0, b_0, a_1, b_1, \ldots, a_{m-1}, b_{m-1}),$$

then $\bar{p}(x^{m-1})$ corresponds to the vector

$$w = (b_0, a_0, b_{m-1}, a_{m-1}, \ldots, b_1, a_1),$$

so that

$$\langle v, w \rangle = 2(a_0 b_0 + a_1 b_{m-1} + a_{m-1} b_1 + \ldots) = 0$$

in any field of characteristic 2. Clearly, the same argument applies to $x^i p(x)$.

Remark 7. In many examples, we get the equality $\langle \bar{p}(x^{m-1}) \rangle_{\mathbb{F}_q} = \langle p(x) \rangle_{\mathbb{F}_q}^{\perp}$. However, we could not find a general property of $p(x)$ which guarantees it. Again, there is an argument on the dimension missing.

3.3   Minimum Distance Bounds

Let $(m, q) = 1$, t be the order of q^2 modulo m and ω be a primitive m-th root of unity in $\mathbb{F}_{q^{2t}}$. If some consecutive powers of ω are roots of both $p(x)$ and $\bar{p}(x^{m-1})$, then a BCH bound can be proved for the code generated by $p(x)$ and $\bar{p}(x^{m-1})$.

Theorem 2 (BCH bound for principal dihedral codes). *Let C be a principal dihedral code generated by $p(x)$ and $2 \leq \delta \leq m$. If $\delta - 1$ consecutive powers of ω are roots of both $p(x)$ and $\bar{p}(x^{m-1})$, then C has minimum distance at least δ.*

Proof. A codeword $c(x)$ of the folded $C \subseteq \mathbb{F}_{q^2}^m$ is of the form

$$c(x) = t_1(x)p(x) + t_2(x)\bar{p}(x^{m-1}),$$

for $t_1(x), t_2(x) \in \mathbb{F}_q[x]$. As $\delta - 1$ consecutive powers of ω are roots of both $p(x)$ and $\bar{p}(x^{m-1})$, we have $c(x) = c'(x)g(x)$ where $c'(x) \in \mathbb{F}_{q^2}[x]$ and

$$g(x) = \mathrm{lcm}\{M_{\omega^b}(x), M_{\omega^{b+1}}(x), \ldots, M_{\omega^{b+\delta-2}}(x)\},$$

where $M_{\omega^i}(x)$ is the minimal polynomial of ω^i over $\mathbb{F}_{q^2}$. It follows that the folded C is a subcode of the BCH code generated by $g(x)$, which has minimum distance at least δ by the classical BCH bound. Since a nonzero coordinate in a codeword of the folded C corresponds to at least a nonzero coordinate of the unfolded codeword in C, the minimum distance of C is at least δ.

Remark 8. As the proof of the theorem shows, the argument relies on the fact that the folding of C is a subset of a cyclic code. Any bound involving zeros of cyclic codes such as Hartmann-Tzeng bound (see [18, Theorem 4.5.6]) may be applied. However, BCH bound is simpler and leads to easier definition of codes with prescribed minimum distance, which is actually the aim of the paper.

Let r be defined as in Remark 5. For many applications, it is suitable to consider codes with a prescribed minimum distance. This can be achieved by imposing that $\delta - 1$ consecutive powers of ω, say $\omega^b, \omega^{b+1}, \ldots, \omega^{b+\delta-2}$, together with their inverse and their r-th powers, are roots of $p(x)$, which guarantees that the code generated has minimum distance at least δ.

Definition 7. *Let $(m, q) = 1$ and $2 \leq \delta \leq m$. A dihedral code $\mathcal{C} \subseteq \mathbb{F}_q^{2m}$ is a BCH-dihedral code of prescribed minimum distance δ if there exists an integer b such that its generator is*

$$p(x) = \mathrm{lcm} \left\{ \begin{array}{c} M_{\omega^b}(x), M_{\omega^{b+1}}(x), \ldots, M_{\omega^{b+\delta-2}}(x) \\ M_{\omega^{-b}}(x), M_{\omega^{-b-1}}(x), \ldots, M_{\omega^{-b-\delta+2}}(x) \\ M_{\omega^{br}}(x), M_{\omega^{br+r}}(x), \ldots, M_{\omega^{br+\delta r-2r}}(x) \end{array} \right\}$$

where $r = (m'm - 1)/q$, with m' being the inverse of m modulo q, and $M_{\omega^i}(x)$ is the minimal polynomial of ω^i over $\mathbb{F}_{q^2}$.

Remark 9. The definition above guarantees to have minimum distance at least δ. Anyway, it may probably be improved by analysing the relations between the cyclotomic cosets of the different roots. This reveals to be simpler in the binary case, that we will consider in the next subsection.

3.4   Binary Case

Let us consider now D_{2m}-codes over $\mathbb{F}_2$, with $m \geq 3$ odd. The binary case is particularly interesting, since $\nu^{2+1} - 1 = 0$. In this case

$$\overline{p}(x^{m-1}) = \nu p(x^{(m-1)/2})^2,$$

so that if $Z(p)$ is the set of zeros of $p(x)$, then $Z(p)^{2/(m-1)}$ is the set of zeros of $\overline{p}(x^{m-1})$. In this case, we are considering $p(x)$ and $\overline{p}(x^{m-1})$ as polynomials in $\mathbb{F}_4[x]$ and not in the quotient ring.

We consider an m-th root of unity ω in $\mathbb{F}_{4^t}$, where t is the order of 4 modulo m. The irreducible divisors of $x^m - 1$ are associated to the cyclotomic cosets $C_i = \{i, 4i \bmod m, 4^2 i \bmod m, \ldots\}$ (this is classical in the theory of cyclic codes - see for example [18]): actually, if $M_{\omega^i}(x)$ is the polynomial associated to C_i (which is the minimal polynomial of ω^i), its zeros are $Z(M_{\omega^i}(x)) = \{\omega^j \mid j \in C_i\}$.

Proposition 4. *The following conditions are equivalent:*

a) $Z(M_{\omega^i})^{(m-1)/2} = Z(M_{\omega^i})$ *for all $i \in \{0, \ldots, m-1\}$;*
b) $\frac{m-1}{2} C_i = C_i$ *for all $i \in \{0, \ldots, m-1\}$;*
c) *there exists an integer s such that $2^{2s+1} = -1 \bmod m$.*

If m is prime, then a), b) *and* c) *are equivalent to*

d) $s_2(m) \equiv 2 \bmod 4$, *where $s_2(m)$ is the order of 2 modulo m.*

Proof. a)⇔b): if $Z(M_{\omega^i})^{(m-1)/2} = Z(M_{\omega^i})$, then there exists $j \in C_i$ such that $\omega^{i(m-1)/2} = \omega^j$, which means that the class C_i is sent to C_i by multiplying by $\frac{m-1}{2}$. The vice versa is trivial.

b)⇒c): since $\frac{m-1}{2}C_1 = C_1$, there exists s such that $\frac{m-1}{2} = 4^s \bmod m$. Then $2^{2s+1} = -1 \bmod m$.

c)⇒b): $2^{2s+1} = -1 \bmod m$ implies $((m,2) = 1$ so that 2 is invertible) that $4^s = \frac{m-1}{2} \bmod m$. This means that for all $i \in \{0, \ldots, m-1\}$, we have $\frac{m-1}{2}i = 4^s i \in C_i$, which implies $\frac{m-1}{2}C_i = C_i$.

c)⇒d): Since $2^{4s+2} = 1 \bmod m$ and $2^{2s+1} = -1 \bmod m$, then $s_2(m)$ divides $2(2s+1)$ and $s_2(m)$ does not divide $2s+1$. So 2 divides $s_2(m)$. If 4 divides $s_2(m)$, then 4 divides $4s+2$, which is not true. So $s_2(m) \equiv 2 \bmod 4$.

d)⇒c): If $s_2(x) = 4s+2$, then 2^{2s+1} is a root of $x^2 - 1 \in \mathbb{F}_m[x]$, which has only two solutions. The only possible solution in this case is -1 (otherwise the order of 2 would be smaller than $4s+2$).

Remark 10. The set of primes $\mathcal{P} := \{m \mid s_2(m) \equiv 2 \bmod 4\} = \{3, 11, 19, 43, \ldots\}$ is infinite (its density in the set of primes is $7/24$ [22]).

Theorem 3. *If there exists an integer s such that $2^{2s+1} = -1 \bmod m$ (in particular if m is prime and $s_2(m) \equiv 2 \bmod 4$), then, for all integers $\delta \geq 2$ and $b \geq 0$, the binary D_{2m}-code generated by*

$$p(x) = \text{lcm}\{M_{\omega^b}(x), M_{\omega^{b+1}}(x), \ldots, M_{\omega^{b+\delta-2}}(x)\}$$

is a principal BCH-dihedral code with minimum distance $d \geq \delta$ and dimension $k \geq m - \deg p(x)$.

Proof. It follows from the fact that, in this case, $p(x)$ divides $\overline{p}(x^{m-1})$: actually, all roots of $p(x)$ are roots of $\overline{p}(x^{m-1})$ (as polynomial in $\mathbb{F}_4[x]$) and $p(x)$ divides $x^m - 1$.

Remark 11. Theorem 3 allows to construct binary dihedral codes with prescribed minimum distance and with a lower bound on their dimensions. With MAGMA we did some calculations and we found some codes with the best-known minimum distance for their dimension (see [17]). For example:

– the D_{22}-code generated by

$$p(x) = x^5 + \nu x^4 + x^3 + x^2 + \nu^2 x + 1,$$

   which is a $[22, 12, 6]$ code;
– the D_{66}-code generated by

$$p(x) = x^{15} + \nu x^{14} + x^{13} + x^{11} + x^{10} + \nu^2 x^9 + \nu^2 x^8 +$$

$$+\nu x^7 + \nu x^6 + x^5 + x^4 + x^2 + \nu^2 x + 1,$$

   which is a $[66, 33, 12]$ code;

- the D_{86}-code generated by

$$p(x) = x^{21} + \nu x^{20} + \nu x^{18} + \nu x^{17} + \nu x^{16} + x^{15} + \nu^2 x^{11} + \nu x^{10} +$$

$$+ x^6 + \nu^2 x^5 + \nu^2 x^4 + \nu^2 x^3 + \nu^2 x + 1,$$

which is a $[86, 44, 15]$ code;
- the D_{86}-code generated by

$$p(x) = x^7 + x^6 + \nu x^5 + \nu^2 x^2 + x + 1,$$

which is a $[86, 72, 5]$ code.

Note that the dimension is always $2(m - \deg p(x))$.

4   Open Problems

In the paper we defined dihedral codes with prescribed minimum distance and dimension. However, it would be interesting to prove better bounds on the dimension and to give a construction allowing to control it. In particular, an open problem is the following.

Problem 1. When does equality hold in Corollary 1? Can the bound be improved by adding some conditions on $p(x)$?

Related to that, there is also the problem of a canonical generator. Actually, in the theory of BCH codes we can read the dimension from the degree of the generator polynomial (the one of lowest degree). It does not seem to exist an analogue for dihedral codes. About dual codes, many questions stand open. The main one is about the relation between the generators of code. Another important problem, related to the use of dihedral codes in cryptography is the following.

Problem 2. Is there any efficient decoding algorithm for dihedral codes, based on the algebraic structure proved in the paper?

Finally, it would be interesting to extend the results to other group codes, at least in the checkable case.

Acknowledgements. The authors are grateful to G.N. Alfarano, P. Moree and A. Neri for the fruitful discussion about the paper. Moreover, they would like to thank all reviewers for their insightful comments which led to an improvement of the paper.

References

1. Barbier, M., Chabot, C., Quintin, G.: On quasi-cyclic codes as a generalization of cyclic codes. Finite Fields Appl. **18**(5), 904–919 (2012)

2. Bazzi, L.M.J., Mitter, S.K.: Some randomized code constructions from group actions. IEEE Trans. Inform. Theory **52**, 3210–3219 (2006)
3. Berger, T.P., Cayrel, P.-L., Gaborit, P., Otmani, A.: Reducing key length of the McEliece cryptosystem. In: Preneel, B. (ed.) AFRICACRYPT 2009. LNCS, vol. 5580, pp. 77–97. Springer, Heidelberg (2009). https://doi.org/10.1007/978-3-642-02384-2_6
4. Berman, S.D.: On the theory of group codes. Kibernetika **3**, 31–39 (1967)
5. Bernhardt, F., Landrock, P., Manz, O.: The extended Golay codes considered as ideals. J. Comb. Theory Ser. A **55**, 235–246 (1990)
6. Borello, M., de la Cruz, J., Willems, W.: On checkable codes in group algebras arXiv: 1901.10979 (2019)
7. Borello, M., Willems, W.: Group codes over fields are asymptotically good. Finite Fields Appl. **68**, 101738 (2020)
8. Borello, M., Willems, W.: On the algebraic structure of quasi group codes arXiv: 1912.09167 (2019)
9. Cao, Y., Cao, Y., Fu, F.W.: Concatenated structure of left dihedral codes. Finite Fields Appl. **38**, 93–115 (2016)
10. Charpin, P.: Une généralisation de la construction de Berman des codes de Reed-Muller p-aire. Comm. Algebra **16**, 2231–2246 (1988)
11. Claro, E.J.G., Recillas, H.T.: On the dimension of ideals in group algebras, and group codes. J. Algebra Appl. (2020, to appear)
12. Conway, J.H., Lomonaco Jr., S.J., Sloane, N.J.A.: A [45, 13] code with minimal distance 16. Discret. Math. **83**, 213–217 (1990)
13. Elia, M., Gorla, E.: Computing the dimension of ideals in group algebras, with an application to coding theory. J. Algebra Numb. Theory Appl. **45**(1), 13–28 (2020)
14. Faugère, J.-C., Otmani, A., Perret, L., Tillich, J.-P.: Algebraic cryptanalysis of McEliece variants with compact keys. In: Gilbert, H. (ed.) EUROCRYPT 2010. LNCS, vol. 6110, pp. 279–298. Springer, Heidelberg (2010). https://doi.org/10.1007/978-3-642-13190-5_14
15. vom Felde, A.: A new presentation of Cheng-Sloane's [32, 17, 8]-code. Arch. Math. **60**, 508–511 (1993)
16. Güneri, C., Özdemir, F., Solé, P.: On the additive cyclic structure of quasi-cyclic codes. Discret. Math. **341**(10), 2735–2741 (2018)
17. Grassl, M.: Codetables. http://www.codetables.de/
18. Huffman, W., Pless, V.: Fundamentals of Error-Correcting Codes. Cambridge University Press, Cambridge (2003)
19. Jitman, S., Ling, S., Liu, H., Xie, X.: Checkable codes from group rings arXiv: 1012.5498v1 (2010)
20. McEliece, R.J.: A public-key cryptosystem based on algebraic coding theory. DSN Progr. Rep. **42–44**, 114–116 (1978)
21. McLoughlin, I., Hurley, T.: A group ring construction of the extended binary Golay code. IEEE Trans. Inform. Theory **54**, 4381–4383 (2008)
22. Moree, P.: On the divisors of $a^k + b^k$. Acta Arith. **80**(3), 197–212 (1997)
23. Morita, K.: Duality for modules and its applications to the theory of rings with minimum condition. Sci. Rep. Tokyo Kyoiku Daigaku, Sect. A **6**(150), 83–142 (1958)
24. Niederreiter, H.: Knapsack-type cryptosystems and algebraic coding theory. Prob. Control Inf. Theory. Problemy Upravlenija i Teorii Informacii **15**, 159–166 (1986)
25. Passman, D.S.: Observations on group rings. Comm. Algebra **5**, 1119–1162 (1977)

Sequences

Recursion Polynomials of Unfolded Sequences

Ana I. Gomez[1], Domingo Gomez-Perez[1(✉)], and Andrew Tirkel[2]

[1] Department of Matemáticas, Estadística y Computación,
Universidad de Cantabria, Santander, Spain
{gomezperezai,gomezd}@unican.es
[2] Scientific Technology, 10 Marion Street, Brighton, Vic 3186, Australia
atirkel@bigpond.net.au
http://grupos.unican.es/amac

Abstract. Watermarking digital media is one of the important challenges for information hiding. Not only the watermark must be resistant to noise and against attempts of modification, legitimate users should not be aware that it is embedded in the media. One of the techniques for watermarking is using an special variant of spread-spectrum technique, called frequency hopping. It requires ensembles of periodic binary sequences with low off-peak autocorrelation and cross-correlation. Unfortunately, they are quite rare and difficult to find. The small Kasami, Kamaletdinov, and Extended Rational Cycle constructions are versatile, because they can also be converted into Costas-like arrays for frequency hopping. We study the implementation of such ensembles using linear feedback shift registers. This permits an efficient generation of sequences and arrays in real time in FPGAs. Such an implementation requires minimal memory usage and permits dynamic updating of sequences or arrays.

The aim of our work was to broaden current knowledge of sets of sequences with low correlation studying their implementation using linear feedback shift registers. A remarkable feature of these families is their similarities in terms of implementation and it may open new way to characterize sequences with low correlation, making it easier to generate them. It also validates a conjecture made by Moreno and Tirkel about arrays constructed using the method of composition.

Keywords: Periodic sequences · Multidimensional arrays · Watermarking

1 Introduction

Digital media has became a widely used product in everyday life. The availability of electronic devices, like computers and smartphones, makes possible large-scale

Supported by Consejería de Universidades e Investigación, Medio Ambiente y Política Social, Gobierno de Cantabria (ref. VP34).

J. C. Bajard and A. Topuzoğlu (Eds.): WAIFI 2020, LNCS 12542, pp. 163–173, 2021.
https://doi.org/10.1007/978-3-030-68869-1_9

distribution of digital content without proper authorization from content producers. This situation has created a need for finding ways of hiding copyright messages or serial numbers in order to trace copyright violators. Several companies decided to fund the Digital Watermarking Alliance for raising awareness and promote the adoption of digital watermarking.

There are several techniques that this consortium plan to standardize, and one proposed method to hide information in digital media is a variant of spread-spectrum techniques using ensembles of periodic sequences with low off-peak autocorrelation and low cross-correlation [12]. This makes sets of arrays with low correlations find applications in watermarking of images, audio, video, and multimedia; but they are also prized in radar and communications, because of their efficiency and noise immunity. Known ensembles of sequences, such as the small Kasami set [11], are optimal with respect to the Sidelnikov correlation bound, but their linear complexity is logarithmic in the length of the sequences, so prone to cryptanalytic attacks. Other optimal ensembles of binary sequences are Kamaletdinov ensembles of sequences [10], discovered independently by Moreno and Tirkel among other families of sequences unfolded from arrays constructed by the Extended Rational Cycle (ERC) [16]. These sequences have lengths whose factors are relatively prime, so they can be folded into two-dimensional arrays using the Chinese remainder theorem (CRT) [9]. They consist of cyclic shifts of a pseudonoise or constant column [19] and can all be generated using the *composition method* [18]. The idea behind this procedure is to build arrays using shifted versions of the same pseudonoise sequence, by means of a *shift array* or *shift sequence*. This method is very flexible and it allows also to generate higher dimensional arrays [16]. A similar family of sequences with good correlation properties are given by the interleaved sequences [8], but we remark that the definition is different and so is the theory to generated by them. While both constructions make use of the method of composition, and the concepts of shift sequence and Trace function, they are quite distinct. The constructions discussed here utilise families of novel shift sequences with low auto and cross hit correlation, together with a solitary pseudonoise column. These constructions yield sequences of length $p(p + 1)$ and $p(p - 1)$ [14] and multidimensional multi-periodic arrays [16,19]. By contrast, interleaved sequences [8] use the composition of a solitary shift sequence with ingeniously chosen column sequences. This construction is limited to sequence lengths $(2^n - 1)^2$. The construction is single periodic because of the choice of the shift sequence [5] and only two such shift sequences are available: exponential Welch and the folded m sequence introduced by Baumert and Games, see [6]. Moreover, the shift arrays used in the sequences can be converted into Costas-like arrays with bounded auto- and cross-hit correlations [17]. Apart from watermarking, such ensembles are useful in multiple access frequency/time hopping systems for UWB ranging, sonar, and wireless communications.

An important aspect which has been little discussed is implementation. Although all known constructions can be easily implemented in a computer, the challenge is to do it in low-resource devices. Linear Feedback Shift Registers (LFSRs) provide the most common technique for generating sequences. However, Kamaletdinov ensembles and ERC families require quite large LFSRs. Leukhin and Tirkel [14] proposed an implementation using cascade LFSRs and then asked for a general formula for the length of the LFSRs involved. This paper presents formulas for that parameter. For certain array sizes, this allows an efficient generation of sequences and arrays in real time in FPGAs. Such an implementation requires minimal memory usage and permits dynamic updating of sequences or arrays.

Moreover, the factorization of the minimal polynomials of these LFSRs follows a certain pattern. This provides a way to unify the above sequences and constructions and brings order to apparently haphazard discoveries. A challenging area in the field of finding families of low correlation sequences is to characterize properties of these sets like linear complexity. This is still not widely understood and this research is a step forward to close this gap. Our results also validate a conjecture made by Moreno and Tirkel about arrays constructed using the method of composition, which states the value of the linear complexity of certain families of sequences generated by the composition method (this is enunciated in conjecture 1). These results build on [14], where empirical data suggested that such unification should be possible. In turn, Leukhin and Tirkel [14] drew attention upon the pioneering works [1,4], which analysed the cycle lengths of reducible polynomials and, most importantly, those containing repeated factors. In order to understand the unified constructions, we first study the nature of the most common column sequence employed by the method of composition: the Legendre sequence which exists for every prime number. Ding et al. [3] calculated the linear complexity of the binary Legendre sequence and its minimal polynomial. We extend this result, giving the number of factors as well as their degree. Explicitly, the factors are those of cyclotomic polynomials, so similar algorithms as those by Tuxanidy and Wang [20] for odd characteristic could be applied. We leave the development of such algorithms as an open problem.

The paper is organized as follows: Sect. 2 introduces cascade LFSRs and shows how a recursion polynomial (or minimal polynomial) for a Legendre sequence factors into lower degree polynomials. Section 3 analyses the minimal polynomials for arrays generated by the method of composition using the Legendre sequence as column. By default, it also provides the minimal polynomials for m-sequence columns, a much simpler case. Section 4 discusses how the new theory is consistent with and validates the empirical findings in [14].

2   Cascade LFSRs and Legendre Sequence

Throughout the rest of the paper, we assume that the reader is familiar with the theory of LFSRs. We recommend consulting the work by Birdsall and Ristenbatt [1].

Implementation of sequences is an important and difficult issue. Although any sequence can be generated by an LFSR, it is not always efficient because its length can be close to that of the sequence.

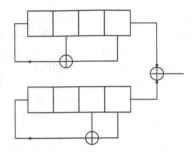

Fig. 1. A cascade with two LFSRs

There is an alternative to the naive implementation using LFSRs, called *cascade LFSRs*. The idea is to speed up the generation combining the output of several LFSRs by a XOR gate, as shown in Fig. 1. It is even more convenient when some of the LFSRs give decimated sequences of another. In this case, some memory can be saved.

Finding such a representation for a sequence is equivalent to finding factors of its minimal polynomial. In this paper, we focus on nontrivial factors, i.e. irreducible factors of degree greater than one. These are the ones which matter most, because the factor $x + 1$ represents a sequence inversion. Although there are efficient algorithms to factor polynomials with binary coefficients, our aim is deriving formulas depending directly on the parameters of the sequence.

Now, we recall that The Legendre sequence (s_i) with respect to the prime p is defined, for $0 \leq i < p$, by

$$s_i = \begin{cases} \left(1 + \left(\frac{i}{p}\right)\right)/2, & \text{if } \gcd(i,p) = 1; \\ 0, & \text{otherwise}; \end{cases} \tag{1}$$

where $\left(\frac{i}{p}\right)$ is the Legendre symbol. A binary Legendre sequence exists for all odd prime length and its correlation is perfect if $p = 3 \bmod 4$, which makes it very versatile and this is the reason it is used in the method of composition. Ding et al. proved the following result regarding its minimal polynomial.

Lemma 1 (Theorem 2 in [3]). *Let (s_i) be the Legendre sequence with respect to the prime p and $m(x)$ its minimal polynomial. We introduce the following additional elements:*

- $\mathbb{F}_2 = \{0, 1\}$, *the finite field of two elements*
- β, *a primitive root over an extension of $\mathbb{F}_2$ of order p,*
- $q(x) = \prod\{x + \beta^i \mid 0 \leq i < p, \left(\frac{i}{p}\right) = 1\}$,

Then,

- $m(x) = q(x)(x + 1)$, *if $p \equiv -1 \bmod 8$*
- $m(x) = q(x)$, *if $p \equiv 1 \bmod 8$*
- $m(x) = x^p + 1$, *if $p \equiv 3 \bmod 8$*

$-\ m(x) = (x^p + 1)/(x + 1)$, if $p \equiv 5 \bmod 8$

Next lemma give some properties of the factorization of $m(x)$.

Lemma 2 (Theorem 2.47 in [15]). *Let $\mathbb{F}_2[x]$ be the ring of polynomials with coefficients in $\mathbb{F}_2$. For a prime $p > 2$, the irreducible factors of $(x^p + 1)/(x + 1)$ over $\mathbb{F}_2[x]$ have all degree d, the minimal positive integer such that $2^d - 1$ is divisible by p. In particular, since $m(x)$ divides $x^p + 1$, all of its irreducible factors have degree d.*

A *decimation* of a p-periodic sequence (u_i) is a sequence (v_i) defined by $v_i = u_{\alpha i}$, where α is a positive integer. For the Legendre sequence, its cascade representation can be given by decimations of p-periodic sequences. This is a consequence of the trace representation of the Legendre sequence, which can be found in [13].

Lemma 3 (Theorems 2 and 4 in [13]). *Let $\mathbb{F}_p$ be the field with p elements. There exists a p-periodic sequence (u_i) such that the minimal polynomial of the Legendre sequence (s_i) can be expressed as the product of the minimal polynomials of decimations of (u_i) by quadratic residues of $\mathbb{F}_p$, times $x+1$ if $p \equiv -1, 3 \bmod 8$. All these minimal polynomials have degree d, the minimal positive integer such that $2^d - 1$ is divisible by p.*

For convinience of the reader, a proof is given here.

Proof. We denote by $\mathbb{F}_{2^d}$ the finite field with 2^d elements. By Lemma 2, all factors of the polynomial $(x^p + 1)/(x + 1)$ have degree d and are irreducible. Indeed, any LFSR (u_i) which has as minimal polynomial one of these factors is of the form $u_i = \mathrm{Tr}(\alpha^i)$, for some $\alpha \in \mathbb{F}_{2^d}$, $\alpha^p = 1$, where Tr is the trace function. If α is a generator of the multiplicative group of elements of order p, i.e. a primitive root of order p, any other LFSR must be a decimation of (u_i). Indeed, if $v_i = \mathrm{Tr}(\beta^i)$ and there exists g such that $\alpha^g = \beta$,

$$v_i = \mathrm{Tr}(\alpha^{gi}) = u_{gi \bmod p}.$$

The minimal polynomial of the Legendre sequence must be a product of these irreducible factors, each of them defining a decimation of (u_i). The fact that these decimations are obtained through quadratic residues of the finite field $\mathbb{F}_p$ is a consequence of Lemma 1. This concludes the proof. □

Example 1. We calculate the occurring LFSRs for $p = 73$. In this case, we get $2^9 = 512 \equiv 1 \bmod 73$, so $d = 9$. The factorization of the minimal polynomial of the Legendre sequence with respect to prime p is exactly

$$m(x) = (x^9 + x^4 + x^2 + x + 1)(x^9 + x^6 + x^5 + x^2 + 1)$$
$$(x^9 + x^7 + x^4 + x^3 + 1)(x^9 + x^8 + x^7 + x^5 + 1).$$

For general values of p, notice that we know exactly the linear complexity of the Legendre sequence $L(s_i)$, and that (s_i) can be represented in cascade LFSRs using:

– $(p-1)/(2d)$ nontrivial LFSRs, if $p \equiv -1, 1 \bmod 8$
– $(p-1)/(d)$ nontrivial LFSRs, otherwise

By Lemma 1, the output sequence has to be XORed with the constant sequence of ones if $p \equiv 3, 7 \bmod 8$.

3   Composition Method

The composition method builds a two-dimensional array from a shift array and a column. The output columns are cyclic shifts (as indicated by the shift array) of the input one.

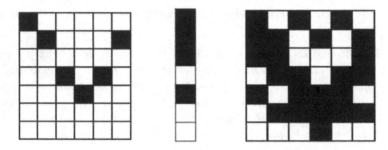

Fig. 2. Graphical example of the construction of a two-dimensional array using as column a Legendre sequence with respect to the prime 7. The doubly periodic shift sequence is $[0, 1, 3, 4, 3, 1]$ and is represented as the left two-dimensional array: the shifts for each output column are the black squares. The output is the array on the right.

This procedure admits a nice graphical representation: the shift array gets a black square in position (i, j) if the column of index i is to be shifted j positions. Figure 2 shows an example with a shift array belonging to family Kamaletdinov 1 and a Legendre sequence with respect to the prime 7 as input column. If the numbers of rows and columns are coprime, it is possible to transform the two-dimensional array into a sequence using the Chinese remainder theorem.

The shift array of $T \times N$ admits a representation as a sequence of integers between 0 and $T - 1$, which are the shifts. So, in a similar vein, consider a T-periodic binary sequence (u_i) and an N-periodic sequence of shifts (y_i) with $\gcd(N, T) = 1$. The resulting sequence using the Chinese remainder theorem is

$$S_i = u_{(i+y_i \bmod N) \bmod T}, \quad 0 \leq i < NT. \tag{2}$$

Next result gives a lower bound for the linear complexity of (S_i), which is generated by the composition method using (u_i) and (y_i).

Theorem 1. *Let (u_i) be a binary sequence of period T and (y_i) an N-periodic sequence of shifts. We define the following NT-periodic sequence:*

$$Y_i = \begin{cases} 1, & \text{if } \exists l, \quad i \equiv (l - (l + y_l)AN) \bmod NT, \ l = 0, \dots, N-1; \\ 0, & \text{otherwise;} \end{cases} \tag{3}$$

where A is the modular inverse of N modulo T, i.e. $AN \equiv 1 \mod T$. Then,

$$L(S_i) \geq NL(u_i) + L(Y_i) - NT.$$

Proof. We consider the generating polynomial associated with (S_i) (Eq. (2)):

$$S(x) = \sum_{i=0}^{NT-1} S_i x^i.$$

It is well known that the linear complexity of that sequence is

$$NT - \deg(\gcd(S(x), x^{TN} - 1)).$$

In order to calculate the greatest common divisor of these two polynomials, we denote by A an integer such that $AN \equiv 1 \mod T$.

$$\gcd\left(S(x), x^{TN} - 1\right) = \gcd\left(\sum_{i=0}^{NT-1} S_i x^i, x^{TN} - 1\right)$$

$$= \gcd\left(\sum_{l=0}^{N-1}\sum_{j=0}^{T-1} S_{Nj+l} x^{Nj+l}, x^{TN} - 1\right) = \gcd\left(\sum_{l=0}^{N-1}\sum_{j=0}^{T-1} u_{Nj+l+y_l} x^{Nj+l}, x^{TN} - 1\right)$$

$$= \gcd\left(u\left(x^N\right)\sum_{l=0}^{N-1} x^{(l-(l+y_l)AN) \mod NT}, x^{TN} - 1\right),$$

whose degree is not larger than

$$\deg\left(\gcd\left(\sum_{l=0}^{N-1} x^{(l-(l+y_l)AN) \mod NT}, x^{TN} - 1\right)\right) + \deg\left(\gcd\left(u(x^N), x^{TN} - 1\right)\right)$$

$$= \deg\left(\gcd\left(\sum_{l=0}^{N-1} x^{(l-(l+y_l)AN) \mod NT}, x^{TN} - 1\right)\right) + N \deg\left(\gcd\left(u(x), x^T - 1\right)\right).$$

Using the relation between the generating function and the linear complexity, we get the result. □

This is, to our knowledge, the first result that relates the linear complexity of the column and doubly periodic shift sequences to the linear complexity generated by the composition method and the Chinese remainder theorem.

Next, we give a formula to calculate a multiple of the minimal polynomial of the unfolded sequences presented by Leukhin and Tirkel [14]. Indeed, if a plausible conjecture holds true, we can give the LFSRs in cascade representation of the unfolded sequences, up to multiplicities.

We define the following operation: given two polynomials $f, g \in \mathbb{F}_2[x]$, $f \odot g$ is the monic polynomial defined by

$$f \odot g = \prod_{f(\alpha)=0} \prod_{g(\beta)=0} (x - \alpha\beta),$$

where the products run over all roots of f and g over a closed extension of $\mathbb{F}_2$, see [20] for more about this operation.

Theorem 2. *Take a column sequence of length p whose minimal polynomial is $M(x)$ and a doubly periodic shift sequence of length T with $\gcd(T, p) = 1$. The minimal polynomial of any sequence unfolded using the Chinese remainder theorem from an array output by the composition method is a divisor of $M(x) \odot (x^T + 1)$, if $(x^p + 1)/M(x)$ is not divisible by $x + 1$. Otherwise, the minimal polynomial is a divisor of the product of $M(x) \odot ((x^T + 1)/(x + 1))$ and $(x + 1)$.*

As a consequence, sequences coming from unfolded arrays generated by the composition method can be represented as a set of cascade LFSRs. All these LFSRs are decimations of a single linear generator of degree d, the minimum integer satisfying

$$2^d \equiv 1 \bmod pT', \quad \gcd(T', 2) = 1, \quad T = 2^f T',$$

and the multiplicity of the factors is less than 2^f.

Proof. Given $M(x)$, we define two sets:

$$\{\alpha\beta \mid M(\alpha) = 0, \quad \beta^T = 1\}, \tag{4}$$

$$\{\alpha\beta \mid M(\alpha) = 0, \quad \beta^T = 1, \quad \beta \neq 1\} \cup \{1\}. \tag{5}$$

A proof that all roots of the minimal polynomial of the unfolded array are in one of these sets can be found in [16, Lemma 5.1]. For the second statement, note that the generated sequence has period pT, which implies that the minimal polynomial is $x^{pT} + 1$, whose roots are powers of a primitive root of order $2^d - 1$ and occur with multiplicity at most 2^f. This finishes the proof. □

As an example, let us apply the theorem above to the array generated in Fig. 2. In that case, $p = 7$ and the period of the doubly periodic shift sequence is $T = 6$, so the length of the LFSRs in the cascade representation is at most $d = 6$, which is the minimum integer such that $2^d \equiv 1 \bmod 21$.

We remark that all the appearing LFSRs are decimations of the same - LFSR, but not all proper necessarily. That is why the examples in Eqs. 16 and 18 calculated by Leukhin and Tirkel involve factors with different degrees.

Computer experiments show that the roots of the minimal polynomial are those in either Eq. (4) or (5). Indeed, it is straightforward to prove this for the shift arrays defined by ERC, Family A and Kamaletdinov 1 and 2 if the following conjecture holds [7].

Conjecture 1. The sequences generated by the composition method with shift arrays defined by Kamaletdinov families have maximal linear complexity.

Example 2. Let us go through the example in [14, Fig. 7]. Take $p = 7$ and a doubly periodic shift sequence of length $T = 6$, so $6 = 2T'$ and $T' = 3$. The degree d can be calculated directly:

$$2^6 = 64 \equiv 1 \bmod pT'.$$

Notice that the factor multiplicity equals 2. In this case, Conjecture 1 holds and the roots of the minimal polynomial are given exactly by Eq. (4). Using [2, p. 119], it is possible to calculate a polynomial whose roots are exactly (4) or (5).

We finish this section with a corollary on the length of the LFSRs in the cascade representation when the Legendre sequence is used as column.

Corollary 1. *Under the conditions of Theorem 2 and for the cascade representation defined in the theorem, using as input column the Legendre sequence with respect to the prime p, all nontrivial LFSRs of the resulting cascade representation have degree d, if $T' = 1$. Otherwise, write ϕ for the Euler totient function and, for an integer $D > 1$, let d_D be the minimum positive integer such that $2^{d_D} \equiv 1 \bmod pD$. One of the following cases holds:*

- *if $p \equiv -1 \bmod 8$, there are $\phi(pD)/(2d_D)$ nontrivial LFRSs of length d_D for each divisor D of T'*
- *if $p \equiv 3 \bmod 8$, there are $\phi(pD)/d_D$ nontrivial LFRSs of length d_D for each divisor D of T'*
- *if $p \equiv 1 \bmod 8$, there are $\phi(pD)/(2d_D)$ nontrivial LFRSs of length d_D for each divisor D of T', except possibly for $D = 1$.*
- *if $p \equiv 5 \bmod 8$, there are $\phi(pD)/d_D$ nontrivial LFRSs of length d_D for each divisor D of T', except possibly for $D = 1$.*

Proof. If the sequences have maximal linear complexity, the minimal polynomial must have as many roots as possible, i.e. the set of roots must be the one defined in either Eq. (4) or (5). Now, the result is an immediate application of Theorem 2 and the factorization of $x^{pT'} + 1$ by cyclotomic polynomials. The number of - LFSRs for the case $D = 1$ is deduced in Example 1 and can be found, in general, in [15]. □

Example 3. For $p = 23$, Leukhin and Tirkel [14] give the factorization for the minimal polynomial of the sequence generated using the Extended Rational Cycle. The parameters, in that case, are $T = 24$, $T' = 3$, and $2^f = 8$. Using our notation, we have $D_1 = 11$ and $D_3 = 22$, so, applying the formula, there is only one factor of degree 11 and another of degree 22.

Interestingly enough, Theorem 2 states that the multiplicity is less than 8. In this case, the factor of degree 11 has multiplicity 7 and the other has multiplicity 8.

4   Conclusions

This paper shows that the recursion polynomial of the Legendre sequence is the product of specific irreducible polynomials. Consequently, we compute the recursion polynomials of sequences and arrays constructed by the method of composition using the Legendre sequence as input column.

This validates empirical findings and conjectures by Leukhin, Moreno, and Tirkel. It also shows that apparently unrelated constructions by Kamaletdinov, Moreno, and Tirkel can be unified under these results.

We leave two open problems: the first one is to develop similar algorithms as those in [20]. This would recover explicitly the factors of the minimal polynomial of the sequence. Although there are tables with the factorization of $X^{pT} + 1$ for some values of pT, it would still be interesting to obtain faster algorithms.

The second problem is to find an explicit formula for the multiplicity of the different factors. Computer experiments show some regularities, for example, there are always factors with maximal multiplicity, i.e. 2^f. Again, it is possible to compute the multiplicity efficiently.

As a final remark, the ideas outlayed here apply if the column is replaced by any other sequence. However, only the Legendre sequence is presented because it provides the most interesting case, due to its applications.

References

1. Birdsall, T.G., Ristenbatt, M.P.: Introduction to linear shift-register generated sequences. Technical report, The University of Michigan (1958)
2. Brawley, J.V., Carlitz, L.: Irreducibles and the composed product for polynomials over a finite field. Discrete Math. **65**(2), 115–139 (1987). https://doi.org/10.1016/0012-365X(87)90135-X
3. Ding, C., Hesseseth, T., Shan, W.: On the linear complexity of Legendre sequences. IEEE Trans. Inf. Theory **44**(3), 1276–1278 (1998)
4. Elspas, B.: The theory of autonomous linear sequential networks. IRE Trans. Circ. Theory **6**(1), 45–60 (1959)
5. Games, R.A.: Crosscorrelation of M-sequences and GMW-sequences with the same primitive polynomial. Discrete Appl. Math. **12**(2), 139–146 (1985). https://doi.org/10.1016/0166-218x(85)90067-8
6. Golomb, S.W., Gong, G.: Signal Design For Good Correlation: For Wireless Communication, Cryptography, and Radar. Cambridge University Press, Cambridge (2005)
7. Gomez-Perez, D., Høholdt, T., Moreno, O., Rubio, I.: Linear complexity for multidimensional arrays-a numerical invariant. In: Proceedings of the IEEE International Symposium on Information Theory (ISIT 2015). IEEE (2015)
8. Gong, G.: Theory and applications of Q-ary interleaved sequences. IEEE Trans. Inf. Theory **41**(2), 400–411 (1995)
9. Gyorfi, L., Massey, J.L., et al.: Constructions of binary constant-weight cyclic codes and cyclically permutable codes. IEEE Trans. Inf. Theory **38**(3), 940–949 (1992)
10. Kamaletdinov, B.: Optimal sets of binary sequences. Probl. Peredachi Informatsii **32**(2), 39–44 (1996)
11. Kasami, T.: Weight distribution formula for some class of cyclic codes. Coordinated Science Laboratory Report no. R-285 (1966)
12. Katzenbeisser, S., Petitcolas, F.: Information Hiding Techniques for Steganography and Digital Watermarking. Artech House, Norwood (2000)
13. Kim, J.H., Song, H.Y.: Trace representation of Legendre sequences. Des. Codes Crypt. **24**(3), 343–348 (2001)
14. Leukhin, A., Tirkel, A.: Ensembles of sequences and arrays. In: 2015 Seventh International Workshop on Signal Design and its Applications in Communications (IWSDA), pp. 5–9. IEEE (2015)
15. Lidl, R., Niederreiter, H.: Introduction to Finite Fields and Their Applications. Cambridge University Press, Cambridge (1994). https://doi.org/10.1017/CBO9781139172769

16. Tirkel, A., Gomez-Perez, D., Gomez, A.I.: Arrays composed from the extended rational cycle. Adv. Math. Commun. **11**(2), 313–327 (2017)
17. Tirkel, A., Gomez-Perez, D., Gomez, A.I.: Large families of sequences for CDMA, frequency hopping, and UWB. Cryptogr. Commun. **12**(1), 389–403 (2020)
18. Tirkel, A., Osborne, C., Hall, T.: Steganography-applications of coding theory. In: IEEE Information Theory Workshop, Svalbard, Norway, pp. 57–59 (1997)
19. Tirkel, A.Z., Hall, T.E.: Matrix construction using cyclic shifts of a column. In: Proceedings of the International Symposium on Information Theory, ISIT 2005, pp. 2050–2054. IEEE (2005)
20. Tuxanidy, A., Wang, Q.: Composed products and factors of cyclotomic polynomials over finite fields. Designs Codes Cryptogr. **69**, 1–29 (2013)

Finding Linearly Generated Subsequences

Claude Gravel[1,2]($\boxtimes$) (iD), Daniel Panario[3] (iD), and Bastien Rigault[1]

[1] National Institute of Informatics, Tokyo, Japan
claudegravel1980@gmail.com, rgaultb@gmail.com
[2] EAGLYS Inc., Tokyo, Japan
c_gravel@eaglys.co.jp
[3] School of Mathematics and Statistics, Carleton University, Ottawa, Canada
daniel@math.carleton.ca

Abstract. We develop a new algorithm to compute determinants of all possible Hankel matrices made up from a given finite length sequence over a finite field. Our algorithm fits within the dynamic programming paradigm by exploiting new recursive relations on the determinants of Hankel matrices together with new observations concerning the distribution of zero determinants among the possible matrix sizes allowed by the length of the original sequence. The algorithm can be used to isolate *very* efficiently linear shift feedback registers hidden in strings with random prefix and random postfix for instance and, therefore, recovering the shortest generating vector. Our new mathematical identities can be used also in any other situations involving determinants of Hankel matrices. We also implement a parallel version of our algorithm. We compare our results empirically with the trivial algorithm which consists of computing determinants for each possible Hankel matrices made up from a given finite length sequence. Our new accelerated approach on a single processor is faster than the trivial algorithm on 160 processors for input sequences of length 16384 for instance.

Keywords: Generating polynomial · Quotient-difference tables · Linear algebra over finite fields · Hankel matrices · Linear shift feedback registers · Pattern substrings · Berlekamp-Massey algorithm

1 Notation, Facts and Definitions

Let q be a prime power, $n > 0$, and $\mathbf{x} = (x_i)_{i=0}^{n-1} \in \mathbb{F}_q^n$. For integers $1 \le j \le \lceil \frac{n}{2} \rceil$ and $j - 1 \le i < n - j + 1$, define the matrix $\mathbf{X}_{i,j}$ by

$$\mathbf{X}_{i,j} = \begin{pmatrix} x_i & \cdots & x_{i+j-2} & x_{i+j-1} \\ x_{i-1} & \cdots & x_{i+j-3} & x_{i+j-2} \\ \vdots & \ddots & \vdots & \vdots \\ x_{i-j+1} & \cdots & x_{i-1} & x_i \end{pmatrix}. \tag{1}$$

D. Panario—The second author is partially funded by NSERC of Canada.

By convention, we let $\mathbf{X}_{i,0} = 1$ for $0 \leq i < n$. Every matrix $\mathbf{X}_{i,j}$ is a Hankel matrix.

Hankel matrices have a large number of applications in applied mathematics. In this paper we are interested in Hankel matrices over finite fields. We explore the well known connection of Hankel matrices and sequences over finite fields; for an introductory explanation see Sect. 8.6 in [13]. Hankel matrices are strongly connected to coprime polynomials over finite fields. Indeed, the probability of two monic polynomials of positive degree n over $\mathbb{F}_q$ to be relatively prime is the same as the uniform probability for an $n \times n$ Hankel matrix over $\mathbb{F}_q$ be nonsingular [6,7]. Elkies [5] studies the probability of Hankel matrices over finite fields be nonsingular when independent biased entries are used for the matrix. An algorithm to generate a class of Hankel matrices called superregular (that are related to MDS codes) is given in [16]. Finally, we point out that several results and applications of Hankel matrices over finite fields are given in the Handbook of Finite Fields [14]. In particular, see Sect. 13.2 for enumeration and classical results, Sect. 14.8 for connections to (t, m, s)-nets, and Sect. 16.7 for hardware arithmetic for matrices over finite fields. In this paper, we give a new algorithm to compute determinants of all possible Hankel matrices made up from a given finite length sequence over a finite field.

We denote by $d_{i,j}$ the determinant of $\mathbf{X}_{i,j}$. By definition of $\mathbf{X}_{i,j}$, we have for all i

$$d_{i,0} = 1, \quad d_{i,1} = x_i, \quad d_{i,j} = \det \mathbf{X}_{i,j}.$$

For convenience, let $h = \lceil n/2 \rceil$. We use the determinants to form a quotient-difference table [10,17]. If h is odd, the determinants form a triangle:

$$
\begin{array}{llllllllll}
0: & 1 & 1 & \dots & 1 & 1 & 1 & \dots & & 1 & 1 \\
1: & x_0 & x_1 & \dots & \dots & x_h & \dots & \dots & x_{n-2} & x_{n-1} \\
2: & & d_{1,2} & d_{2,2} & \dots & \dots & \dots & d_{n-3,2} & d_{n-2,2} \\
3: & & & d_{2,3} & \dots & \dots & \dots & d_{n-3,3} \\
4: & & & & \ddots & \vdots & \reflectbox{$\ddots$} \\
\vdots & & & & \ddots & \vdots & \reflectbox{$\ddots$} \\
h: & & & & & d_{h,h}
\end{array}
$$

If n is even, then the triangle is truncated at the hth level where there are two elements $d_{h,h}$ and $d_{h,h+1}$. We observe that i refers to columns and j refers to rows.

For integers i_0, i_1, j_0, j_1 such that $i_1 > i_0$, $h \geq j_1 - j_0 > 0$, consider the set $S(i_0, i_1, j_0, j_1) = \{(i, j) \in \mathbb{N} \times \mathbb{N} \mid i_0 \leq i \leq i_1, \quad j_0 \leq j \leq j_1, \quad j_1 - 1 \leq i < n - j_1 + 1\}$. We observe that S is nonempty and may have a k-side polygonal shape with $3 \leq k \leq 6$. Section 4 gives two examples, one with $n = 32$, $k = 6$, and one with $n = 81$, $k = 4$. We see that the tip of the triangular table from the example with $n = 32$ has length two, and therefore it yields to an hexagonal case. For a detailed explanation, see Sect. 4. If S falls entirely inside the table, then $k = 4$ necessarily, that is, we have a square of zeros. If S overlaps with the edges of the triangular table, then k may be different than 4. We use ∂S to denote the boundary of S. We prove in this work that zeros in a difference table are *always* distributed or grouped according to S.

Our goal is to design a dynamic programming algorithm to fill the table that requires the least number of determinant evaluations. More precisely, if we know the first $j - 1$ rows of the table, then we want to compute determinants for the jth row by using the least possible number of rows above the jth. In Sect. 2, we establish relations among determinants $d_{i,j}$'s no matter how $\mathbf{x}$ is generated. Our results amplify any linear patterns that could be used to generate the coordinates of $\mathbf{x}$. We show that any run of zeros in the table automatically implies a run of zeros exactly below the former so that we obtain a square of zeros. Moreover, we prove identities, that we call *cross shape identities*, relating determinants $d_{i,j}$'s located on a cross as explained later; those identities are based on Sylvester's identities, generalized by Bareiss [2], as well as Dogson's identity [1, 11].

It would be possible to avoid the evaluations of determinants of matrices by generalizing determinantal identities given in Conjecture 1 from Sect. 2. More precisely, in a true random sequence of length n, the expected length of the maximum run of zeros is $O(\log_2 n)$. Therefore using the recursive nature of determinants, and especially determinants of Hankel matrices, we conjecture that the evaluations of determinants of matrices larger than about $O(\log_2 n)$ are not required to complete the table above which would lead to a linear time algorithm to locate the linear subsequence. Our algorithm may also be used as a statistical test to determine linearity in a pseudo-random sequence.

In Section 3, we apply results from Sect. 2 to the case of a sequence $\mathbf{x} = (x_i)_{i=1}^n$ that contains a linearly shifted and fed back subsequence.

Definition 1. *Using* $\mathbf{x} = (x_i)_{i=0}^{n-1}$ *as above, let* $c = (c_0, \ldots, c_{d-1}) \in \mathbb{F}_q^d$ *with* $c_{d-1} = 1$ *and* $d < n - 1$. *The sequence* $\mathbf{x}$ *contains a* linear subsequence *if there are integers* s *and* t *with* $d \leq s \leq t < n$ *such that for all* $s \leq \ell \leq t$ *we have*

$$\sum_{i=0}^{d-1} c_i x_{\ell-d+i} = 0.$$

Indeed one of our motivations is to identify the indices s and t as well as to find the generating vector c. This relates our work to the Berlekamp-Massey algorithm. As shown later our method does not assume any upper bound on the length of c or equivalently on the degree of the generating polynomial in the framework of Berlekamp-Massey.

Given a prime power q, $d > 0$, and a sequence $\mathbf{x} = (x_i)_{i=0}^\infty$ with $x_i \in \mathbb{F}_q$, the Berlekamp-Massey algorithm is an iterative algorithm that finds the shortest linear feedback shift register (LFSR) that generates $\mathbf{x}$. A register of size d over $\mathbb{F}_q$ is an element of $\mathbb{F}_q^d$. More precisely, an LFSR consists of an initial register $(x_0, x_1, \ldots, x_{d-1}) \in \mathbb{F}_q^d$, a non zero vector $c = (c_0, \ldots, c_{d-1}) \in \mathbb{F}_q^d$ such that for $i \geq 0$

$$(x_i, x_{i+1}, \ldots, x_{i+d-2}, x_{i+d-1}) \longrightarrow (x_{i+1}, x_{i+2}, \ldots, x_{i+d-1}, \sum_{j=0}^{d-1} c_j x_{i+j}). \qquad (2)$$

The arrow in Eq. (2) expresses the transition. The state of a system at a point in time is the content of the register. In Eq. (2), the system transits from the state $(x_i, x_{i+1}, \ldots, x_{i+d-2}, x_{i+d-1})$ to $(x_{i+1}, x_{i+2}, \ldots, x_{i+d-1}, x_{i+d})$ where x_{i+d} is a given as linear combination of the previous x_i's. The content of the register at time $i + d - 1$ is being fed back into the right end of it through the linear combination $\sum_{j=0}^{d-1} c_j x_{i+j}$. At time $i + d$, the register is updated to $(x_{i+1}, x_{i+2}, \ldots, x_{i+d-1}, x_{i+d})$ where $x_{i+d} = c_0 x_i + c_1 x_{i+1} + \cdots + c_{d-1} x_{i+d-1}$.

For more information on the Berlekamp-Massey algorithm, see [3] where interesting connections between this algorithm and the extended Euclidean algorithm are given. LaMacchia and Odlyzko [12] also review how Berlekamp-Massey algorithm is used in the Wiedemann algorithm to find linear recurrences over finite fields and also show interesting connections to determinants of Hankel matrices. For more information on LFSR sequences, see [8,9].

We conclude the section giving the structure of the paper. Section 2 gives several theoretical relations among Hankel determinants that are crucial in this paper. Those relations are used in Sect. 3 where we provide our algorithm to compute all determinants from Hankel matrices. Illustrative examples over $\mathbb{F}_2$ are given in Sect. 4. Due to the lack of space, experimental runs of our algorithm against a standard method of computation are given in Appendix 5. We compare our results empirically with the trivial algorithm which consists of computing determinants for each possible Hankel matrices made up from a given finite length sequence. Our new accelerated approach on a single processor is faster than the trivial algorithm on 160 processors for input sequences of length 16384 for instance as shown in Sect. 5.

2   Relations Among Hankel Determinants

In this section, we derive useful results to allow the computations of $d_{i,j}$ without actually computing explicitly or directly determinants of size j and instead using determinants $d_{i,j'}$ with $j < j'$. Then in Sect. 3, we fill the triangular table using a dynamic programming approach. Before that, let us recall one of the results from [2] applied to Hankel matrices and adapted to our notation. If i, j are such that $i_0 < i < i_1$, $j_0 \le j \le j_0 + (i_1 - i_0 - 1)$, and with the convention that $d_{i,0} = 1$, $d_{i,1} = x_i$, then

$$d_{i,j} d_{i,j_0-1}^{j-j_0} = \det \begin{pmatrix} d_{i,j_0} & \cdots & d_{i+j-j_0,j_0} \\ \vdots & \ddots & \vdots \\ d_{i-(j-j_0),j_0} & \cdots & d_{i,j_0} \end{pmatrix}. \tag{3}$$

Equation (3) is called a jth-step integer preserving identity in [2]. We call an identity like in Eq. (3) a cross shape identity because $d_{i,j}$, d_{i,j_0} and d_{i,j_0-1} are located on the vertical part of a cross, and the other non-diagonal elements of the matrix are located on the horizontal part of the aforementioned cross. A visual representation of Eq. (3) is as follow:

$$
\begin{array}{llllllll}
0 & : 1 \;\; 1 \;\cdots & 1 & \cdots & 1 & \cdots & 1 & \cdots \;\; 1 \quad 1 \\
1 & : x_0 \;\; x_1 \cdots & x_{i-j+j_0} & \cdots & x_i & \cdots & x_{i+j-j_0} & \cdots \; x_{n-2} \;\; x_{n-1} \\
\vdots & \quad\; \ddots & \vdots & & \vdots & & \vdots & \quad\;\; \iddots
\end{array}
$$

$$
\begin{array}{ll}
j_0 - 1 : & \ddots \qquad\qquad\quad d_{i,j_0-1} \\
j_0 \quad\;\; : & d_{i-j+j_0,j_0} \cdots \; d_{i,j_0} \; \cdots \; d_{i+j-j_0,j_0} \\
\vdots \\
\\
j \quad\;\; : & \qquad\qquad\qquad d_{i,j} \\
\vdots
\end{array}
$$

We come back to Eq. (3) at the end of this section with a brief explanation of its proof. Eq. (3) remains valid even if $d_{i,j_0-1} = 0$ as pointed in [2].

Theorem 1. *Let* $i_0 < i < i_1$, *and* j_0 *be such that* $d_{i,j_0} \neq 0$, $d_{i,j_0+1} = 0$, $d_{i_0,j_0+1} \neq 0$, $d_{i_1,j_0+1} \neq 0$. *Then with* $j_1 = j_0 + (i_1 - i_0)$ *and* $S(i_0, i_1, j_0, j_1)$ *non-empty, we have*

$$
\begin{aligned}
d_{i,j} = 0 \quad &\text{for all } (i,j) \in S(i_0, i_1, j_0, j_1), \\
d_{i,j} \neq 0 \quad &\text{for all } (i,j) \in \partial S(i_0, i_1, j_0, j_1).
\end{aligned}
$$

Proof. Without loss of generality, assume that S falls entirely inside the table with left and right boundaries at (i_0, j_0) and (i_1, j_0), respectively, and with upper and lower boundaries at (i_0, j_0) and (i_0, j_1), respectively. To fall entirely inside the table, one must have $2(i_0+1) - i_1 \geq 0$ so that the Hankel matrix $\mathbf{X}_{i_0, i_1 - i_0 - 1}$ is properly defined; the number of consecutive zeros on level j_0 that occur between i_0 and i_1 is $i_1 - i_0 - 1$.

Fix i such that $i_0 < i < i_1$ and let $j_0 \leq w \leq j_0 + (i_1 - i_0 - 1)$. Then using Eq. (3) with $w = j_0 + 1$ as the basis for induction, we obtain that $d_{i,j_0+1} = 0$, that is, we obtain the second row of zeros below the first one. For the inductive step, assume that $d_{i,w'} = 0$ for $j_0 \leq w' < w$, and rewrite Eq. (3) as

$$
d_{i,w} = d_{i,j_0-1}^{j_0-w} \det \begin{pmatrix} d_{i,j_0} & \cdots & d_{i+w-j_0,j_0} \\ \vdots & \ddots & \vdots \\ d_{i+j_0-w,j_0} & \cdots & d_{i,j_0} \end{pmatrix}.
$$

Therefore at least one row of the previous matrix is made only of zeros which implies the desired result.

We remark that Theorem 1 does not depend on the input sequence, and it is solely a property of determinants for Hankel matrices. If for instance the input sequence is chosen entirely at random with independent identically unbiased distributed Bernoulli random variables, then the biggest squares have average side length $O(\log_2 n)$ which is the expected length of the longest run of zeros in a random sequence of Bernoulli random variables with length n.

Given a square matrix $\mathbf{X}$ of size $\ell \times \ell$, we consider its sub-matrix $\mathbf{C}$ of size $(\ell - 2) \times (\ell - 2)$ located in the center $\mathbf{X}$, and its 4 sub-matrices $\mathbf{N}$, $\mathbf{S}$, $\mathbf{E}$ and $\mathbf{W}$ of size $(\ell - 1) \times (\ell - 1)$ located in the top left, bottom right, top right and bottom left of $\mathbf{X}$, respectively. In other words let

$$
\mathbf{X} = \begin{pmatrix} x_{1,1} & \cdots & x_{1,\ell} \\ \vdots & \mathbf{C} & \vdots \\ x_{\ell,1} & \cdots & x_{\ell,\ell} \end{pmatrix}
$$

$$
= \begin{pmatrix} \mathbf{N} & x_{1,\ell} \\ & \vdots \\ x_{\ell,1} & \cdots & x_{\ell,\ell} \end{pmatrix} = \begin{pmatrix} x_{1,1} & \cdots & x_{1,\ell} \\ \vdots & \mathbf{S} \\ x_{\ell,1} & \end{pmatrix}
$$

$$
= \begin{pmatrix} x_{1,1} & \cdots & x_{1,\ell} \\ \mathbf{W} & \vdots \\ & x_{\ell,\ell} \end{pmatrix} = \begin{pmatrix} x_{1,1} & \mathbf{E} \\ \vdots & \\ x_{\ell,1} & \cdots & x_{\ell,\ell} \end{pmatrix}.
$$

Then we have Dodgson's identity (see [1], or page 29 of [11]):

$$
\det(\mathbf{X}) \det(\mathbf{C}) = \det(\mathbf{N}) \det(\mathbf{S}) - \det(\mathbf{E}) \det(\mathbf{W}). \tag{4}
$$

If the entry $x_{\ell,\ell}$ is an unknown and all other elements of $\mathbf{X}$ are known, then, for some $\alpha, \beta \in \mathbb{F}_q$, we have that

$$
\big(x_{\ell,\ell} \det(\mathbf{N}) + \alpha\big) \det(\mathbf{C}) = \det(\mathbf{N})\big(x_{\ell,\ell} \det(\mathbf{C}) + \beta\big) - \det(\mathbf{E}) \det(\mathbf{W}). \tag{5}
$$

Equation (5) implies that $x_{\ell,\ell}$ cannot be determined if $\det(\mathbf{N}) = 0$ or $\det(\mathbf{C}) = 0$. This simply implies that $x_{\ell,\ell}$ cannot be determined from a determinantal equation of the type obtained by Dodgson's identity.

We now derive a useful identity using Eq. (4) which can also be proved using results from [2].

Proposition 1 (North-South-East-West). *For all (i,j) such that $i-j+1 \geq 0$, and $2 \leq j \leq \lceil n/2 \rceil$ the following identity is true:*

$$
d_{i,j}d_{i,j-2} = d_{i,j-1}^2 - d_{i+1,j-1}d_{i-1,j-1}.
$$

Proof. Apply Eq. (4) on the matrix $\mathbf{X}_{i,j}$ given from (1) where $\det(\mathbf{W}) = d_{i-1,j-1}$, $\det(\mathbf{E}) = d_{i+1,j-1}$, $\det(\mathbf{N}) = d_{i,j-1}$, $\det(\mathbf{S}) = d_{i,j-1}$, and $\det(\mathbf{C}) = d_{i,j-2}$.

We observe that Proposition 1 is reminiscent to the North-South-East-West identity [18] for quotient-difference table. Proposition 1 is similar to the 1st-order step integer preserving relation from [2] with a much easier proof. The condition $d_{i,j-2} \neq 0$ is not required as explained in [2], or as it follows directly from Eq. (4), but it matters for our dynamic programming method since we cannot determine

$d_{i,j}$ if $d_{i,j-2} = 0$ using the table information from the $(j-1)$th and $(j-2)$th rows.

In order to accelerate the computation of determinants within a dynamical programming approach, we must ensure that $d_{i,j_0-1} \neq 0$ from Eq. (3). For that we have the next theorem.

Theorem 2. *For all (i,j) such that $i - j + 1 \geq 0$, and $j_0 \leq j \leq \lceil n/2 \rceil$, if*

$$d_{i,j_0-1} \neq 0, \quad d_{i,k} = 0 \text{ for } j_0 \leq k \leq j - 1,$$

then

$$d_{i,j} = d_{i,j_0-1}^{j_0-j} \det \begin{pmatrix} d_{i,j_0} & \cdots & d_{i+j-j_0,j_0} \\ \vdots & \ddots & \vdots \\ d_{i-(j-j_0),j_0} & \cdots & d_{i,j_0} \end{pmatrix}.$$

Proof. Before starting, for a fixed position i and for any size j_0, we observe that the value $j - j_0$ expresses the depth of singularity, that is, the number of zeros below the non-zero cell indexed by (i, j_0-1). The depth of singularity also relates to the concentration of zeros aligned horizontally around the cell $(i, j_0 - 1)$. By concentration of zeros, we mean the length of a run of consecutive zeros.

$(j = j_0 + 1)$**th step:** Suppose that $d_{i,j_0-1} \neq 0$ and $d_{i,j_0} = 0$. If $d_{i,j_0+1} = 0$, then there is at least one zero to the left or to the right of (i, j_0) or both. Indeed Proposition 1 entails that $d_{i,j_0+1} d_{i,j_0-1} = d_{i,j_0}^2 - d_{i+1,j_0} d_{i-1,j_0}$ which, in this case, is equivalent to $d_{i,j_0+1} d_{i,j_0-1} = -d_{i+1,j_0} d_{i-1,j_0}$ from which we infer that either $d_{i-1,j_0} = 0$ or $d_{i+1,j_0} = 0$ whenever $d_{i,j_0+1} = 0$. So there is qualitatively speaking a small concentration of zeros aligned horizontally around the cell (i, j_0).

$(j = j_0 + 2)$**th step:** Now suppose that $d_{i,j_0-1} \neq 0$, $d_{i,j_0} = d_{i,j_0+1} = 0$ and write

$$0 = d_{i,j_0+1} d_{i,j_0-1}^2 = \det \begin{pmatrix} 0 & d_{i+1,j_0} & d_{i+2,j_0} \\ d_{i-1,j_0} & 0 & d_{i+1,j_0} \\ d_{i-2,j_0} & d_{i-1,j_0} & 0 \end{pmatrix}.$$

By the $(j = j_0 + 1)$th-step, if $d_{i+1,j_0} = 0$, then $0 = d_{i,j_0+1} d_{i,j_0-1}^2 = d_{i-1,j_0}^2 d_{i+2,j_0}$ from which either $d_{i-1,j_0} = 0$ or $d_{i+2,j_0} = 0$; if $d_{i-1,j_0} = 0$, then $0 = d_{i,j_0+1} d_{i,j_0-1}^2 = d_{i+1,j_0}^2 d_{i-2,j_0}$ from which either $d_{i+1,j_0} = 0$ or $d_{i-2,j_0} = 0$. Therefore we conclude that $d_{i-1,j_0} = d_{i+1,j_0} = 0$ as well. The horizontal part of the cross contains therefore a higher concentration of zeros around d_{i,j_0} with respect to the previous step. We cannot conclude at this moment that $d_{i+1,j_0} = 0 = d_{i+2,j_0}$ or $d_{i-1,j_0} = 0 = d_{i-2,j_0}$ without further adding deeper singularities.

$(j = j_0 + 3)$**th step:** Now suppose that $d_{i,j_0-1} \neq 0$ and $d_{i,j_0} = d_{i,j_0+1} = d_{i,j_0+2} = 0$ and write

$$0 = d_{i,j_0+2} d_{i,j_0-1}^3 = \det \begin{pmatrix} 0 & d_{i+1,j_0} & d_{i+2,j_0} & d_{i+3,j_0} \\ d_{i-1,j_0} & 0 & d_{i+1,j_0} & d_{i+2,j_0} \\ d_{i-2,j_0} & d_{i-1,j_0} & 0 & d_{i+1,j_0} \\ d_{i-3,j_0} & d_{i-2,j_0} & d_{i-1,j_0} & 0 \end{pmatrix}. \tag{6}$$

From the $(j = j_0 + 2)$th step, $d_{i-1,j_0} = 0 = d_{i+1,j_0}$, and Eq. (6) is equivalent to

$$0 = d_{i,j_0+2}d_{i,j_0-1}^3 = \det \begin{pmatrix} 0 & 0 & d_{i+2,j_0} & d_{i+3,j_0} \\ 0 & 0 & 0 & d_{i+2,j_0} \\ d_{i-2,j_0} & 0 & 0 & 0 \\ d_{i-3,j_0} & d_{i-2,j_0} & 0 & 0 \end{pmatrix}$$

$$= d_{i+2,j_0}^2 d_{i-2,j_0}^2,$$

so that either $d_{i-2,j_0} = 0$ or $d_{i+2,j_0} = 0$. With the knowledge of the $(j_0 + 2)$th-step, we can safely conclude that either $d_{i+1,j_0} = 0 = d_{i+2,j_0}$ or $d_{i-2,j_0} = 0 = d_{i-1,j_0}$. Hence the concentration of zeros increases on the j_0th row with respect to the previous steps. We observe that the previous determinant has at least one row with 3 consecutive zeros. The position of a run of zeros from one row to the following is shifted cyclically by one position.

The process stops when we can no longer add deeper singularity, that is, we stop for the smallest index $j > j_0$ such that $d_{i,j} \neq 0$. When such index j is found, then we can no longer deduce zero determinants on the horizontal part of the cross.

jth step: Assume that $d_{i,j_0-1} \neq 0$ and $d_{i,k} = 0$ for $j_0 \leq k \leq j - 1$, and now assume $d_{i,j} \neq 0$. The matrix to consider at this step has size $(j + 1) \times (j + 1)$. Thus at this current jth-step, we can find $d_{i,j}d_{i,j_0-1}^{j-j_0} \neq 0$. We observe that we hit the boundaries of a square of zeros. At the following $(j + 1)$th step, all rows would contain at least 2 non-zero elements or equivalently there would not be a row with at least $j + 1$ consecutive zeros. Thus $d_{i,j+1}d_{i,j_0-1}^{j+1-j_0}$ would be the sum of at least two products and it would become impossible to correctly deduce the values of the determinants.

Now we work for our next result, a partially proved conjecture. Given valid indices i, j for the column and the row of the triangular table of determinants, and $2k \leq j$ let $\mathbf{G}_{i,j,k}$ be a matrix of size $(k + 1) \times (k + 1)$ defined as

$$\mathbf{G}_{i,j,k} = \begin{pmatrix} d_{i,j-2k} & d_{i+1,j-2k+1} & \cdots & d_{i+k,j-k} \\ d_{i-1,j-2k+1} & d_{i,j-2k+2} & \cdots & d_{i+k-1,j-k+1} \\ \vdots & \vdots & \ddots & \vdots \\ d_{i-k,j-k} & d_{i-k+1,j-k+1} & \cdots & d_{i,j} \end{pmatrix}. \tag{7}$$

In other words, for $0 \leq r, c \leq k$, the entry of $\mathbf{G}_{i,j,k}$ located on the rth row and cth column is given by $d_{i-r+c,j-2k+r+c}$. The pair $(i-r+c, j-2k+r+c)$ indexing an element of $\mathbf{G}_{i,j,k}$ is the intersection of two perpendicular lines. The intersection of a group of k parallel lines intersecting perpendicularly another group of k parallel lines as it might be easier to see with the following representation by drawing k lines with slope $\pi/4$ and separated at distance $\sqrt{2}$ intersecting k other lines with slope $3\pi/4$ also at distance $\sqrt{2}$ of each other:

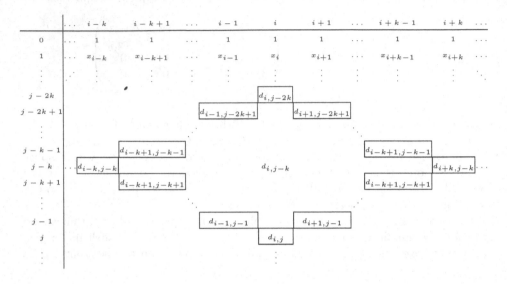

If the information about the determinants $d_{i,j'}$, $0 \leq j' \leq j-1$, is known, then we might hope to solve a determinantal equation like $\det(\mathbf{G}_{i,j,k}) = g$ for some $g \in \mathbb{F}_q$ for the unknown $d_{i,j}$ located in the bottom right corner of $\mathbf{G}_{i,j,k}$.

We may sometimes abuse the language to denote the index (i,j) or the value indexed by (i,j) which is $d_{i,j}$. It is very convenient to refer to k as a radius of an ℓ_1-ball centered around $d_{i,j-k}$ or more precisely around the index $(i, j-k)$. An ℓ_1-ball is a square grid. The grid can be seen as the intersection of the two families of parallel lines and each family perpendicular to each other as mentioned previously. The indices obtained by the intersection of the two families are used to define $\mathbf{G}_{i,j,k}$. We refer the neighbourhood around $(i, j-k)$, which is the center of the grid, as the ℓ_1-ball of radius k. If k is even, then the center $(i, j-k)$ is deleted. If k is odd, then the center is part of the ball. If $\det(\mathbf{G}_{i,j,k}) = 0$, then there is a local linear dependency around $(i, j-k)$. From Dodgson's identity, $\mathbf{G}_{i,j,k-2}$ plays the role of the center which can be seen as the interior of the neighbourhood of $(i, j-k)$.

We postulate the following conjecture about the local linear dependency or more precisely about the minimum amount of information required to determine $d_{i,j}$ assuming the table is known up to the $(j-1)$th level, inclusively.

Conjecture 1. For $2 \leq k \leq 6$, $j \geq 2k$, and $j-1 \leq i \leq n-j$, where n is the length of the sequence, the smallest radius k for which $\det \mathbf{G}_{i,j,k} = 0$ is the smallest value k for which $\det \mathbf{G}_{i,j,k-1} \neq 0$. Assuming $\det \mathbf{G}_{i,j,k'} = 0$ for $2 \leq k' \leq 6$, then we have $\det \mathbf{G}_{i,j,7} = 0$ if and only $\det \mathbf{G}_{i,j,6} \neq 0$ and $d_{i,j-7} = 0$.

We recall from linear algebra that the condition $\det(\mathbf{G}_{i,j,k-1}) \neq 0$ is necessary and sufficient for the uniqueness to the solution of the linear equation $\det(\mathbf{G}_{i,j,k}) = 0$ with $d_{i,j}$ as unknown for all value of k and $j \geq 2k$.

We verified the previous conjecture by comparing $d_{i,j}$ obtained from solving the corresponding determinantal equation with the value obtained from the trivial algorithm. Given that we never found any counter-example to Conjecture 1, we decided to include its algorithmic flavour, that is Algorithm 1, in our dynamic method given in Algorithm 2.

Algorithm 1. Growing an ℓ_1-metric ball and solving for the unknown

---

Input: Integer $n > 0$, $2 \leq k \leq 7$, $j \geq 2k$, and $j - 1 \leq i \leq n - j$
1: **if** $k \leq 6$ **then**
2:      **for** $k' = 2$ to k **do**
3:          **if** $\det(\mathbf{G}_{i,j,k'-1}) \neq 0$ **then**
4:              Solve $\det(\mathbf{G}_{i,j,k'}) = 0$ with $d_{i,j}$ as unknown
5:              **Return** $d_{i,j}$
6:          **end if**
7:      **end for**
8: **else if** $k = 7$, $\det(\mathbf{G}_{i,j,6}) \neq 0$, $d_{i,j-7} = 0$ **then**
9:      Solve $\det(\mathbf{G}_{i,j,7}) = 0$ with $d_{i,j}$ as unknown
10:      **Return** $d_{i,j}$
11: **else**
12:      k out of range   // We need more research for larger radius k.
13: **end if**

We finish by briefly explaining Eq. (3) as we promised at the beginning of this section. Given a square matrix A of size $h \times h$, with $h = \lceil n/2 \rceil$, we introduce the notation from [2]

$$a_{r,c}^{(k)} = \det \begin{pmatrix} a_{0,0} & a_{0,1} & \cdots & a_{0,k-1} & a_{0,c} \\ a_{1,0} & a_{1,1} & \cdots & a_{1,k-1} & a_{1,c} \\ \vdots & \vdots & \ddots & \vdots & \vdots \\ a_{k-1,0} & a_{k-1,1} & \cdots & a_{k-1,k-1} & a_{k-1,c} \\ a_{r,0} & a_{r,1} & \cdots & a_{r,k-1} & a_{r,c} \end{pmatrix} \quad \text{for } k \leq r, c \leq h.$$

Clearly $a_{r,c}^{(k)}$ is the determinant of a $(k+1) \times (k+1)$ matrix. We observe that the principal minors of A are $a_{k,k}^{(k)}$. In [2], it is shown that

$$a_{r,c}^{(k)} = \frac{1}{\left(a_{\ell,\ell}^{(\ell-1)}\right)^{k-\ell}} \det \begin{pmatrix} a_{\ell,\ell}^{(\ell)} & \cdots & a_{\ell,k-1}^{(\ell)} & a_{\ell,c}^{(\ell)} \\ \vdots & \ddots & \vdots & \vdots \\ a_{k-1,\ell}^{(\ell)} & \cdots & a_{k-1,k-1}^{(\ell)} & a_{k-1,c}^{(\ell)} \\ a_{r,\ell}^{(\ell)} & \cdots & a_{r,k-1}^{(\ell)} & a_{r,c}^{(\ell)} \end{pmatrix} \quad \text{for } 0 < \ell < k. \quad (8)$$

So $a_{r,c}^{(k)}$ is also the determinant of a $(k-\ell+1) \times (k-\ell+1)$ matrix of determinants. The left side of Eq. (8) does not depend on ℓ. Let us concentrate on the principal minors when $r = c = k$, and substitute $\ell = j_0$ and $k = j$ in Eq. (8) to get that

$$
a_{j,j}^{(j)} = \frac{1}{\left(a_{j_0,j_0}^{(j_0-1)}\right)^{j-j_0}} \det \begin{pmatrix} a_{j_0,j_0}^{(j_0)} & \cdots & a_{j_0,j-1}^{(j_0)} & a_{j_0,j}^{(j_0)} \\ \vdots & \ddots & \vdots & \vdots \\ a_{j-1,j_0}^{(j_0)} & \cdots & a_{j-1,j-1}^{(j_0)} & a_{j-1,j}^{(j_0)} \\ a_{j,j_0}^{(j_0)} & \cdots & a_{j,j-1}^{(j_0)} & a_{j,j}^{(j_0)} \end{pmatrix} \quad \text{for } 0 < j_0 < j.
$$

Entries (determinants) inside of the previous matrix are those d_{j-j_0,j_0}'s introduced at the beginning of this section. We observe that we can shift $j - j_0$ by any quantity modulo n and therefore get Eq. (3).

3   Algorithm to Compute Determinants of Hankel Matrices over Finite Fields

In the following algorithm, the symbol j indexes the rows of the table and j is the size of the Hankel matrix $\mathbf{X}_{i,j}$ as in the introduction. The symbol i indexes the columns and is related to the position in the input vector $\mathbf{x}$ from where we build the Hankel matrix $\mathbf{X}_{i,j}$. We use the symbol M to denote the dynamic table under consideration, and $M[j][i]$ stands for $d_{i,j} = \det \mathbf{X}_{i,j}$. The table grows from top to bottom by considering the smallest possible Hankel matrices to the largest one if n is odd or the largest two if n is even. For a given matrix of size j or equivalently a given row j of M, the algorithm sweeps from the left to the right using the input vector $\mathbf{x}$ in order to consider all possible Hankel matrices of size j. We use also $M[j][\cdot]$ to refer to the jth row of the table M.

Algorithm 2. Computing determinants for all possible Hankel matrices made up from a sequence $\mathbf{x} \in \mathbb{F}_q^n$

Input: Integer $n > 0$ and vector $\mathbf{x} \in \mathbb{F}_q^n$.
Output: Triangular table M.
1: $h \leftarrow \lceil n/2 \rceil$
2: $M \leftarrow \emptyset$ // Allocate space for M with base h and width n.
3: **for** $i = 0$ to $n - 1$ **do**   // Initialize first two rows M.
4:      $M[0][i] \leftarrow 1$
5:      $M[1][i] \leftarrow x_i$
6: **end for**
7: **for** $j = 2$ to h **do**
8:      Find new squares of zeros   // Use Theorem 1 with the knowledge
                                             of rows $M[j-1][\cdot]$ and $M[j-2][\cdot]$.
9:          **for** $i = j - 1$ to $n - j$ **do**   // Loop is parallelized.
10:             **if** $M[j][i]$ has not been yet evaluated **then**
11:                 **if** $M[j-2][i] \neq 0$ **then**
12:                     Compute $M[j][i]$ using Proposition 1
13:                 **else if** Conditions for Conjecture 1 **then**
14:                     Compute $M[j][i]$ accordingly with Algorithm 1

```
15:              else if Conditions for Theorem 2 then
16:                    Compute M[j][i] accordingly
17:              else
18:                    Compute M[j][i] explicitly from its definition
19:              end if
20:          end if
21:      end for
22: end for
```

Based on the results from Sect. 2, the algorithm correctly terminates. We note that an auxiliary table can be maintained to flag entries of M that were computed or not. Given j from line (7), to find squares of zeros using $M[j-1][\cdot]$ and $M[j-2][\cdot]$, we look for consecutive non zero elements between two indices, say i_0 and i_1 (including $M[j-2][i_0] \neq 0$ and $M[j-2][i_1] \neq 0$) at level $j-2$, then check for $M[j-1][i_0] \neq 0$ followed by zero elements until $M[j-1][i_1] \neq 0$; the procedure begins with $i_0 = j - 1$ and if i_1 is found to be the right upper corner, then a square is filled, and the procedure continues from i_1 until reaching $n - j$. Once the squares are filled, the remaining elements on a given row must be evaluated. The goal is to use as few as possible knowledge from the previous rows by using Proposition 1, Theorem 2, and Conjecture 1. If none of the previous results applied, then we revert to the trivial and expensive evaluation.

We end this section by explaining briefly how to find the generating vector of a linear subsequence. Once an unusual long run of zeros is found on a row of the table, we stop the computations of determinants since actually there is no need to further complete the table. Indeed, all the knowledge we need to build the adjugate, in order to invert a Hankel matrix connecting the generating vector to a part of the original sequence, is located on the previous rows that had been already computed.

4   Illustrative Visual Examples

In this section, we give two examples illustrating our new results over $\mathbb{F}_2$. For the first example, we generate a sequence of length 32 indexed from left to right starting with index 0, ending with index 31, and which is given by

$$01010110100111010011101011101110$$

Red color represents the prefix and the postfix that are generated randomly. Green color represents the middle linear substring. The big square of zeros due to the linearity of the middle string is in blue color. The sequence is used to initialize the table so it is identical to the row indexed by 1 below. Row 0 contains only unit elements. The generating vector is $(1, 0, 1, 1) = (c_0, c_1, c_2, c_3)$. We note that as mentioned previously, $c_3 = 1$ to ensure the vector is not trivial. The leftmost index of linear subsequence is 8, that is $i_0 = 8$, and so the generated random prefix is $01010110 = x_0 x_1 \cdots x_7$. The rightmost index of the linear subsequence is 24, and the generated random postfix is $11101110 = x_{24} \cdots x_{31}$.

The middle linearly substring is given by $1001110100111010 = x_8 x_9 \cdots x_{23}$, and is generated linearly from the prefix string: $c_3 x_8 + c_2 x_7 + c_1 x_6 + c_0 x_5 = 0$ implies $x_8 = c_2 x_7 + c_1 x_6 + c_0 x_5 = x_7 + x_5 = 0 + 1 = 1$, $x_9 = x_8 + x_6 = 1 + 1 = 0$, $x_{10} = x_9 + x_7 = 0 + 0 = 0$, and so on.

Since the generating vector has length 4, then the row at which appears a long run of zeros is on the row indexed by 4. The shape of S is hexagonal, and the values of j_1 varies with those of the positional indices i. The value $j_0 = 4$.

```
 0 :1111111111111111111111111111111111
 1 :010101101001110100111010011101110
 2 : 1111111100101110010111111011101
 3 :  00100111111111111111101111011
 4 :   0100100000000000000011101111
 5 :    1111000000000000001111100
 6 :     0110000000000000100110
 7 :      11000000000000010011
 8 :       100000000000001111
 9 :        0000000000000100
10 :         00000000000010
11 :          000000000001
12 :           0000000000
13 :            00000000
14 :             000000
15 :              0000
16 :               00
```

For the second example, we generate a sequence of length 81 indexed from left to right starting with index 0, ending with index 80, and which is given by

10110000001010111011010110101100100011110101100100010101110110011001100000000100

```
 0 :1111111111111111111111111111111111111111111111111111111111111111111111111111111
 1 :10110000001010111011010110101100100011110101100100010101110110011001100000000100
 2 :111000001111100111111111111110010001001111100100011110011100110011001000000010
 3 :010000010011001000101001010111100010010101101111000100110010011111111100000001
 4 :100000010011110001100111111111111111111111111001111100100000010000000
 5 :000000111010100011111110000000000000000011011110101110000000001000000
 6 :000001011111111110000010000000000000000011111111111010000000100000
 7 :111111100000111100000000000000000000000010000000010111000000010000
 8 :001011000001001000001000000000000000000001000000001111100000001111
 9 :011110000010010000010000000000000000000001000000010010000000100
10 :100100000111100000100000000000000000000001000000010010000000010
11 :001000001111111111000000000000000000000001000000010001111111111
12 :111111110000000010000000000000000000000001000000001111101010010
13 :000001100000000100000000000000000000000001000000001110111001
14 :000011000000001000000000000000000000000001000000001001111111
15 :000110000000010000000000000000000000000111111111111100110001
16 :001100000001000000000000000000000000001111110111111111000
17 :011000000010000000000000000000000000001000011100100100
18 :110000001000000000000000000000000000100010100010011
19 :100000001000000000000000000000000001000011111111
20 :111111110000000000000000000000000001000010010011
21 :111100010000000000000000000000000111111001001
22 :001000100000000000000000000000010101111111
23 :010001000000000000000000000000001111110000
24 :111110000000000000000000000100110000
25 :001100000000000000000000000010011000
26 :011111111111111111111111111111111100
27 :101111000110100011010000001010
28 :110010001110001111100000111
29 :100100010110001011000011
30 :111111111111111111000001
31 :001110000000000100000
32 :010101000000000011111
33 :111000000000001010
34 :110000000000111
35 :100000000010
36 :00000000001
37 :000000000
38 :0000000
39 :00000
40 :000
41 :1
```

In this case the generating vector is $(1, 0, 0, 1, 1) = (c_0, c_1, c_2, c_3, c_4)$. The leftmost index of the linear subsequence is 30, and the rightmost index of linear subsequence is 50. The shape of S is a square and $i_0 = 30$, $i_1 = 50$, $j_0 = 5$, *and* $j_1 = j_0 + (i_1 - i_0) = 25$ as it can be seen as well from the visual aid. The prefix random string is given by $x_0 x_1 \cdots x_{29} = 10110000001010111011010110101$. The postfix random string is $x_{50} x_{51} \cdots x_{80} = 00101011110110011001100000000100$. The middle linear substring is given by $x_{30} \cdots x_{49} = 10010001111010110010$.

5   Conclusion and Further Work

We believe that there are still more relations to be found and to be coded in order to avoid the computation of determinants, and this is currently under study. An ultimate goal is to get rid entirely of the evaluations of large determinants by proving and generalizing Conjecture 1. How would the linear dependency vanish as the radius gets larger or synonymously how far does it propagate around the center? Can we further enlarge the radius by adding new conditions for $k > 7$?

It would be interesting to adapt our algorithm to output the generating vector and compare it to efficient implementations of the Berlekamp-Massey algorithm. We would need to stop at the level containing a long run of zeros and use the information of the row preceding this one to solve efficiently the linear system for the generating vector using adjugate matrices.

It is known that Berlekamp-Massey algorithm is virtually the same as the extended Euclidean algorithm for polynomials over finite field. Could we find a similar equivalence to our algorithm for problems involving Bezout identities that express linear dependencies among elements in fields?

Our dynamic approach can be easily adapted to multiple and combined linear feedback shift registers. Further research also includes to analyze the case of non-linear feedback shift register by linearizing the generator; more precisely, linearizing a non-linear boolean feedback function pertains to add constraints which are reflected in the determinant identities.

Appendix: Run Times and Distribution of Counts

In order to compare in practice the running times between the trivial method and our new method, we generate sequences of length n that we linearly filled. In order to compute determinants of large Hankel matrices whenever necessary, we do not use the Levinson-Durbin algorithm [4,15] that can be adapted to Hankel matrices instead of Toeplitz matrices. We created an extremely fast C/C++ low-level module to compute determinants over $\mathbb{F}_2$ in order to do not rely on any external libraries. Our module to compute determinants over $\mathbb{F}_2$ is quite faster than NTL; it however only applies to binary matrices. We recall that one of our future goals is to avoid such computation of determinants of large matrices, and only use local information. We look at typical worst-case instances when the linear subsequence is "buried" between two long random sequences serving as a prefix and a postfix. The prefix random string together with the generator vector are used to built the linear subsequence in the middle. The prefix random string must be at least as long as the length of the generating vector to be used as initial data. We also consider typical easy-case instances when there is no random postfix sequence and when the length of the random prefix sequence is the same as the generating vector.

For our accelerated dynamic algorithm, we give the distribution of counts of the number of times, with respect to the number of entries in the table, that we branch to Proposition 1, Theorems 1 or 2, Conjecture 1 or to an explicit computation (where Levinson-Durbin could be used for instance). The time to verify that the tables obtained from the trivial and our accelerated methods coincide is not taken into account; we must check this because at this time we cannot prove the validity of Conjecture 1 and/or further enhanced it. We also parallelize both the naive and accelerated algorithms. We notice that our accelerated algorithm on a single core is faster than the trivial algorithm on 160 cores for instance as shown here in the following tables. For $n = 2^{14} = 16384$, the time to run the trivial algorithm is prohibitive and we did not run the trivial algorithm for length $n = 2^{14}$. The meanings of the abbreviations in the following tables are: **Tri. S.T.** for trivial algorithm single threaded, **Tri. M.T.** for trivial algorithm multi threaded, **Acc. S.T.** for accelerated algorithm single threaded, and **Acc. M.T.** for accelerated algorithm multi threaded. Roughly speaking, a thread is a core. All threads share a unique space in memory.

EXCERPT OF RUNNING TIMES FOR $n = 4096$ (in milliseconds)

Tri. S.T.	Tri. M.T.	Acc. S.T.	Acc. M.T.
15931964.434875	793513.102411	209495.875914	79493.525290
15930582.584026	793131.331561	212010.715863	79784.405378
15931070.671137	793029.664463	211231.649798	79538.533773
≈4 h 30 min	≈14 min	≈4 min	≈1 min 33 s

EXCERPT OF RUNNING TIMES FOR $n = 16384$ (in milliseconds)

Acc. S.T.	Acc. M.T.
95397251.744779	18823959.088013
92977447.400304	18820690.826587
93372757.004548	18817978.092718
≈26 h 30 min	≈5 h 36 min

The hardware specification for the computer we used is: Intel(R) Core(TM) i7-8700 CPU @ 3.20 GHz, 160 cores, 1TB RAM. Our code is available at https:// github.com/63EA13D5/. We coded Algorithm 2 over $\mathbb{F}_2$, and a compile switch can be enable to avoid using the library NTL or to use it. For the worst-case instances, we generate the sequences using the following parameters:

1. Elements indexed from 0 to $7n/16$ inclusively are generated randomly.
2. Elements indexed from $7n/16+1$ to $9n/16$ inclusively are linearly filled using a non-trivial generating vector of length $d = n/8$. The generating vector is randomly created and the rightmost coordinate is set to the unit element in base field.
3. Elements indexed from $9n/16$ to n inclusively are generated randomly.

The ratio of the number of entries in the big square over the number of entries for the table of a given instance is about $1/16$ up to a few decimals. For each value of n, a sample of sequences is used to estimate the running time by evaluating the averages over the sample, one average for the trivial and one average for our method. For comparison, each method is applied to a sequence from the sample. The ratios of the averages of the new method by the trivial are given. We generate a sample of 1000 linearly filled vectors as described above for each value of n. Zero counts are not shown in the tables (Tables 1 and 2).

Table 1. Time complexity and distribution of counts–hard instances.

Sample size	10140		
Sequence length	4096		
Generating vector length	256		
Subsequence leftmost index	1792		
Subsequence rightmost index	2304		
Number of entries	4192256		
Average time for accelerated method (ms)	482211.336405 (i)		
Average time for trivial method (ms)	39627789.122209 (ii)		
Ratio i/ii	0.012169		
Average counts NSEW	1720642.408481		
Average counts square filling	925814.212032		
Average counts direct	60875.169231		
Average counts 2 × 2 grid	467181.981164		
Average counts 3 × 3 grid	382111.856114		
Average counts 4 × 4 grid	265305.747436		
Average counts 5 × 5 grid	169767.046746		
Average counts 6 × 6 grid	94158.297732		
Average counts 7 × 7 grid	51134.827416		
Average counts 2-cross	3772.531657	Average counts 3-cross	2480.830473
Average counts 4-cross	1473.981558	Average counts 5-cross	821.684813
Average counts 6-cross	4415.502071	Average counts 7-cross	4449.331460
Average counts 8-cross	16683.110947	Average counts 9-cross	9334.250197
Average counts 10-cross	5165.383136	Average counts 11-cross	2828.874063
Average counts 12-cross	1536.936785	Average counts 13-cross	828.868540
Average counts 14-cross	444.807101	Average counts 15-cross	237.420710
Average counts 16-cross	125.906312	Average counts 17-cross	67.308679
Average counts 18-cross	35.672189	Average counts 19-cross	18.771893
Average counts 20-cross	9.915779	Average counts 21-cross	5.115089
Average counts 22-cross	2.713708	Average counts 23-cross	1.373570
Average counts 24-cross	0.696746	Average counts 25-cross	0.388955

(*continued*)

Table 1. (*continued*)

Average counts 26-cross	0.246154	Average counts 27-cross	0.098422
Average counts 28-cross	0.054734	Average counts 29-cross	0.043590
Average counts 30-cross	0.020809	Average counts 31-cross	0.009369
Average counts 32-cross	0.006312	Average counts 256-cross	21.695464
Average counts 257-cross	53.271203	Average counts 258-cross	70.017061
Average counts 259-cross	74.697732	Average counts 260-cross	67.246943
Average counts 261-cross	56.079487	Average counts 262-cross	45.345661
Average counts 263-cross	35.773866	Average counts 264-cross	26.859369
Average counts 265-cross	20.313807	Average counts 266-cross	14.738166
Average counts 267-cross	9.678205	Average counts 268-cross	7.546055
Average counts 269-cross	5.817850	Average counts 270-cross	3.680079
Average counts 271-cross	2.569625	Average counts 272-cross	1.961440
Average counts 273-cross	1.428895	Average counts 274-cross	1.082840
Average counts 275-cross	0.923471	Average counts 276-cross	0.490335
Average counts 277-cross	0.383432	Average counts 278-cross	0.439250
Average counts 279-cross	0.247929	Average counts 280-cross	0.082840
Average counts 281-cross	0.027811	Average counts 282-cross	0.027811
Average counts 285-cross	0.028205	Average counts 286-cross	0.028205
Average counts 287-cross	0.028402	Average counts 288-cross	0.028402
Average counts 293-cross	0.028994	Average counts 294-cross	0.028994
Sum over all average counts	4192256		

Table 2. Time complexity and distribution of counts–easy instances

Sample size	10140		
Sequence length	4096		
Generating vector length	256		
Subsequence leftmost index	256		
Subsequence rightmost index	4096		
Number of entries	4192256		
Average time for accelerated method (ms)	4331.896007 (i)		
Average time for trivial method (ms)	25765730.530170 (ii)		
Ratio i/ii	0.000168		
Average counts NSEW	407852.065385		
Average counts square filling	3419936.066568		
Average counts direct	19444.707988		
Average counts 2 × 2 grid	106227.368146		
Average counts 3 × 3 grid	86949.429093		
Average counts 4 × 4 grid	60390.460256		
Average counts 5 × 5 grid	38652.743195		
Average counts 6 × 6 grid	21438.953156		
Average counts 7 × 7 grid	11646.030572		
Average counts 2-cross	3772.586193	Average counts 3-cross	2479.999310
Average counts 4-cross	1473.537081	Average counts 5-cross	821.756805
Average counts 6-cross	1341.515483	Average counts 7-cross	1185.020414
Average counts 8-cross	3878.167554	Average counts 9-cross	2161.438363
Average counts 10-cross	1191.903748	Average counts 11-cross	651.711736
Average counts 12-cross	351.648225	Average counts 13-cross	189.478107

(*continued*)

Table 2. (*continued*)

Average counts 14-cross	102.336391	Average counts 15-cross	54.392998
Average counts 16-cross	28.796746	Average counts 17-cross	15.180178
Average counts 18-cross	8.078008	Average counts 19-cross	4.166469
Average counts 20-cross	2.130868	Average counts 21-cross	1.140039
Average counts 22-cross	0.562821	Average counts 23-cross	0.254734
Average counts 24-cross	0.159073	Average counts 25-cross	0.104832
Average counts 26-cross	0.051972	Average counts 27-cross	0.035207
Average counts 28-cross	0.019428	Average counts 29-cross	0.002860
Average counts 1792-cross	0.645759	Average counts 1793-cross	0.501775
Average counts 1794-cross	0.350099	Average counts 1795-cross	0.212623
Average counts 1796-cross	0.121893	Average counts 1797-cross	0.073373
Average counts 1798-cross	0.039645	Average counts 1799-cross	0.022091
Average counts 1800-cross	0.015385	Average counts 1801-cross	0.007298
Average counts 1802-cross	0.005128	Average counts 1803-cross	0.001578
Average counts 1804-cross	0.002564	Average counts 1805-cross	0.000394
Average counts 1806-cross	0.000197	Average counts 1812-cross	0.000197
Sum over all average counts		4192256	

We observe that we are about 83 times faster on typical hard instances and about 5947 times faster on easy ones. In practice, to detect linearity or to solve backward for the generating vector, we only need to stop at the first level that contains a long run of zeros.

References

1. Abeles, F.F.: Chiò's and dodgson's determinantal identities. Linear Algebra Appl. **454**, 130–137 (2014). https://doi.org/10.1016/j.laa.2014.04.010
2. Bareiss, E.H.: Sylvester's identity and multistep integer-preserving gaussian elimination. Math. Comput. **22**(103), 565–578 (1968)
3. Blahut, R.E.: Algebraic Codes for Data Transmission. Cambridge University Press, Cambridge (2003). https://doi.org/10.1017/CBO9780511800467
4. Bojanczyk, A.W., Brent, R.P., De Hoog, F.R., Sweet, D.R.: On the stability of the bareiss and related Toeplitz factorization algorithms. SIAM J. Matrix Anal. Appl. **16**(1), 40–57 (1995). https://doi.org/10.1137/S0895479891221563
5. Elkies, N.: On finite sequences satisfying linear recursions. New York J. Math. **8**, 85–97 (2001)
6. Gao, Z., Panario, D.: Degree distribution of the greatest common divisor of polynomials over Fq. Random Struct. Algorithms **29**(1), 26–37 (2006). https://doi.org/10.1002/rsa.20093
7. García-Armas, M., Ghorpade, S.R., Ram, S.: Relatively prime polynomials and nonsingular Hankel matrices over finite fields. J. Combin. Theory Ser. A **118**(3), 819–828 (2011). https://doi.org/10.1016/j.jcta.2010.11.005
8. Golomb, S.W.: Shift Register Sequences. 3rd revised edn. World Scientific (2017). https://doi.org/10.1142/9361, https://www.worldscientific.com/doi/abs/10.1142/9361
9. Golomb, S.W., Gong, G.: Signal Design for Good Correlation: For Wireless Communication, Cryptography, and Radar. Cambridge University Press, Cambridge (2005). https://doi.org/10.1017/CBO9780511546907

10. Henrici, P.: The quotient-difference algorithm. National Bureau Stand. Appl. Math. Ser. **49**, 23–46 (1958)
11. Horn, R., Johnson, C.: Matrix Analysis. Matrix Analysis. Cambridge University Press, Cambridge (2013)
12. LaMacchia, B.A., Odlyzko, A.M.: Solving large sparse linear systems over finite fields. In: Menezes, A.J., Vanstone, S.A. (eds.) CRYPTO 1990. LNCS, vol. 537, pp. 109–133. Springer, Heidelberg (1991). https://doi.org/10.1007/3-540-38424-3_8
13. Lidl, R., Niederreiter, H.: Finite Fields. Encyclopedia of Mathematics and its Applications. Cambridge University Press, Cambridge (1996). https://doi.org/10.1017/CBO9780511525926
14. Mullen, G., Panario, D.: Handbook of Finite Fields. Discrete Mathematics and Its Applications. CRC Press, Boca Raton (2013)
15. Musicus, B.R.: Levinson and fast Choleski algorithms for Toeplitz and almost Toeplitz matrices. Technical report 538, Research Laboratory of Electronics, Massachusetts Institute of Technology (1988)
16. Raemy, I.: Superregular Hankel matrices over finite fields: an upper bound of the matrix size and a construction algorithm. Master's thesis, University of Zurich (2015)
17. Sloane, N., Plouffe, S.: The Encyclopedia of Integer Sequences. Elsevier Science, Amsterdam (1995)
18. Weisstein, E.W.: Quotient-different table. http://mathworld.wolfram.com/Quotient-DifferenceTable.html. Accessed 3 Mar 2020

Special Functions over Finite Fields

Generalization of a Class of APN Binomials to Gold-Like Functions

D. Davidova$^{(\boxtimes)}$ and N. Kaleyski

Department of Informatics, University of Bergen, 7803 NO-5020, Bergen, Norway
{Diana.Davidova,Nikolay.Kaleyski}@uib.no

Abstract. In 2008 Budaghyan, Carlet and Leander generalized a known instance of an APN function over the finite field $\mathbb{F}_{2^{12}}$ and constructed two new infinite families of APN binomials over the finite field $\mathbb{F}_{2^n}$, one for n divisible by 3, and one for n divisible by 4. By relaxing conditions, the family of APN binomials for n divisible by 3 was generalized to a family of differentially 2^t-uniform functions in 2012 by Bracken, Tan and Tan; in this sense, the binomials behave in the same way as the Gold functions. In this paper, we show that when relaxing conditions on the APN binomials for n divisible by 4, they also behave in the same way as the Gold function x^{2^s+1} (with s and n not necessarily coprime). As a counterexample, we also show that a family of APN quadrinomials obtained as a generalization of a known APN instance over $\mathbb{F}_{2^{10}}$ cannot be generalized to functions with 2^t-to-1 derivatives by relaxing conditions in a similar way.

Keywords: Almost perfect nonlinear · Boolean functions · Differential uniformity · Walsh transform · Walsh spectrum

1 Introduction

Let n, m be natural numbers. A *vectorial Boolean (n, m)-function*, or simply an *(n, m)-function*, or vectorial Boolean function, is a mapping from the n-dimensional vector space $\mathbb{F}_2^n$ over the finite field $\mathbb{F}_2 = \{0, 1\}$ to the m-dimensional vector space $\mathbb{F}_2^m$. Since the extension field $\mathbb{F}_{2^n}$ can be identified with an n-dimensional vector space over $\mathbb{F}_2$, (n, m)-functions can be seen as functions between the Galois fields $\mathbb{F}_{2^n}$ and $\mathbb{F}_{2^m}$. Vectorial Boolean functions have many applications in mathematics and computer science. In cryptography, they are the basic building blocks of block ciphers, and the choice of functions directly influences the security of the cipher. In order to construct cryptographically secure ciphers, it is necessary to understand what properties such functions need to possess in order to resist various types of cryptanalytic attacks, and to find methods for constructing functions having these desirable properties. In our work, we mostly concentrate on the case when $n = m$, i.e. when the number of input and output bits is the same. A comprehensive survey on (n, m)-functions can be found in [4,8].

© Springer Nature Switzerland AG 2021
J. C. Bajard and A. Topuzoğlu (Eds.): WAIFI 2020, LNCS 12542, pp. 195–206, 2021.
https://doi.org/10.1007/978-3-030-68869-1_11

One of the most powerful attacks against block ciphers is differential cryptanalysis, introduced by Biham and Shamir [1]. The attack is based on studying how the difference in two inputs to a function affects the difference in the corresponding outputs. The resistance to differential attacks of an (n, m)-function is measured by a property called its differential uniformity. The lower the differential uniformity, the more resistant the cryptosystem is to differential attacks. The class of almost perfect nonlinear (APN) functions is defined as the class of (n, n)-functions having the best possible differential uniformity, and thus provides optimal security against differential cryptanalysis.

Another powerful attack against block ciphers is linear cryptanalysis, introduced by Matsui [12]. The property of a function which measures the resistance to this kind of attack is called nonlinearity. The nonlinearity $\mathcal{NL}(F)$ of an (n, m)-function F is defined to be the minimum Hamming distance between any component of F and any affine $(n, 1)$-function. An upper bound on the nonlinearity of any (n, n)-function can be derived, and the class of almost bent (AB) functions is defined as the class of those functions that meet this bound with equality and therefore provide the best possible resistance to linear attacks.

Recall that the Gold functions are APN power functions over $\mathbb{F}_{2^n}$ of the form x^{2^s+1} for some natural number s satisfying $\gcd(s, n) = 1$. Relaxing the condition to $\gcd(s, n) = t$ for some positive integer t, the functions of the form $F(x) = x^{2^s+1}$ become differentially 2^t-uniform, with all their derivatives $D_a F(x) = F(x) + F(a + x)$ for $a \neq 0$ being 2^t-to-1 functions. These functions are permutations if and only if $n/\gcd(s, n) = n/t$ is odd [13], and are $(2^t + 1)$–to–1 functions otherwise. Their nonlinearity is $2^{n-1} - 2^{(n+t)/2}$ when n/t is odd, and $2^{n-1} - 2^{(n+2t)/2}$ otherwise.

In 2008, two infinite families of (n, n)-APN binomials inequivalent to power functions were introduced in [5] for values of n divisible by 3 or by 4 as generalizations of a known sporadic APN instance over $\mathbb{F}_{2^{12}}$ [11]. These were the first known infinite families of APN functions that are inequivalent to power functions. It was later shown in 2012 that the family of APN binomials for n divisible by 3 can be generalized to functions with 2^t-to-1 derivatives (for some positive integer t) with nonlinearity equal to $2^{n-1} - 2^{(n+t)/2}$ for $n + t$ even, and $2^{n-1} - 2^{(n+t-1)/2}$ for $n + t$ odd by relaxing conditions [3]. Thus, the APN binomials for n divisible by 3 behave in the same way as the Gold functions from the point of view of differential uniformity, nonlinearity and properties of the image set.

In this paper we show that the second class of APN binomials from [5] (for n divisible by 4) also behaves in the same way as the Gold functions in this respect. We note that all the constructed functions (much like the APN binomials) are quadratic, and are therefore not directly suitable for cryptographic use in practice. Nonetheless, the vast majority of known APN functions are given by a quadratic representation, but contain representatives of higher algebraic degrees in their CCZ-equivalence class. We also consider the family of APN quadrinomials constructed by generalizing a known APN instance over $\mathbb{F}_{2^{10}}$ [7] and computationally verify that they provide a counterexample to this approach, in the sense that they cannot be generalized to functions with 2^t-to-1 derivatives by relaxing conditions in a similar way for any even dimension n in the range $6 \leq n \leq 14$.

The paper is structured as follows. In Sect. 2, we recall the basic definitions and results that we use throughout our work. In Sect. 3, we compute the differential uniformity of the generalized families of binomials; an upper bound on their nonlinearity is then derived in Sect. 4. Section 5, in which we computationally show that the APN quadrinomials constructed in [7] cannot be generalized to 2^t-uniform functions over $\mathbb{F}_{2^n}$ with $6 \leq n \leq 14$, concludes the paper.

2   Preliminaries

Let n be a positive integer. Then $\mathbb{F}_{2^n}$ denotes the finite field with 2^n elements, and $\mathbb{F}_{2^n}^*$ denotes its multiplicative group. For any positive integer k dividing n, the trace function Tr_k^n is the mapping from $\mathbb{F}_{2^n}$ to $\mathbb{F}_{2^k}$ defined by $\mathrm{Tr}_k^n(x) = \sum_{i=0}^{\frac{n}{k}-1} x^{2^{ik}}$. For $k = 1$, the function $\mathrm{Tr}_1^n : \mathbb{F}_{2^n} \to \mathbb{F}_2$ is called the *absolute trace* over $\mathbb{F}_{2^n}$ and is denoted simply by $\mathrm{Tr}_n(x)$, or by $\mathrm{Tr}(x)$ if the dimension n is clear from context.

Let n and m be positive integers. An (n,m)-*function* is any function F from $\mathbb{F}_{2^n}$ to $\mathbb{F}_{2^m}$. For any (n,m)-function F and for any $a \in \mathbb{F}_{2^n}$, the function $D_a F(x) = F(x + a) + F(x)$ is called the *derivative of F in the direction a*. Let $\delta_F(a,b)$ denote the number of solutions of the equation $D_a F(x) = b$ for some $a \in \mathbb{F}_{2^n}$ and $b \in \mathbb{F}_{2^m}$. The multiset $\{\delta_F(a,b) : a \in \mathbb{F}_{2^n}^*, b \in \mathbb{F}_{2^m}\}$ is called the *differential spectrum* of F. The *differential uniformity* of F is the largest value in its differential spectrum. We say that F is *differentially δ-uniform* if its differential uniformity is at most δ. The differential uniformity of any (n,m)-function is clearly always even, since if $x \in \mathbb{F}_{2^n}$ is a solution to $D_a F(x) = b$ for some $a \in \mathbb{F}_{2^n}$ and $b \in \mathbb{F}_{2^m}$, then so is $x + a$. The lowest possible differential uniformity of any function is thus 2. A function with differential uniformity equal to 2 is called *almost perfect nonlinear (APN)*. Since a low differential uniformity corresponds to a strong resistance to differential cryptanalysis, APN functions provide optimal security against this type of attack.

A *component function* of an (n,m)-function F is any function of the form $x \mapsto \mathrm{Tr}_m(cF(x))$ for $c \in \mathbb{F}_{2^m}^*$. The component functions are clearly $(n,1)$-functions. The nonlinearity $\mathcal{NL}(F)$ of F is the minimum Hamming distance between any component function of F and any affine $(n,1)$-function, i.e. any function $a :$ $\mathbb{F}_{2^n} \to \mathbb{F}_2$ satisfying $a(x) + a(y) + a(z) = a(x + y + z)$ for all $x, y, z \in \mathbb{F}_{2^n}$. Recall that the Hamming distance between two $(n,1)$-functions f and g is the number of inputs $x \in \mathbb{F}_{2^n}$ for which $f(x) \neq g(x)$.

An important tool for analyzing any (n,m)-function F is the so-called Walsh transform. The *Walsh transform of F* is the function $W_F : \mathbb{F}_{2^m} \times \mathbb{F}_{2^n} \to \mathbb{Z}$ defined as $W_F(a,b) = \sum_{x \in \mathbb{F}_{2^n}} (-1)^{\mathrm{Tr}_m(aF(x)) + Tr_n(bx)}$.

The nonlinearity of an (n,m)-function F can be expressed as $\mathcal{NL}(F) = 2^{n-1} - \frac{1}{2} \max_{a \in \mathbb{F}_{2^m}^*, b \in \mathbb{F}_{2^n}} |W_F(a,b)|$. The nonlinearity of any (n,n)-function is bounded from above by $2^{n-1} - 2^{(n-1)/2}$ [10]. Functions attaining this bound are called *almost bent (AB)*. Clearly, AB functions exist only for odd values of n; when n is even, functions with nonlinearity $2^{n-1} - 2^{n/2}$ are known, and it is conjectured that this value is optimal in the even case. Nonlinearity measures the resistance to linear cryptanalysis; the higher the nonlinearity, the better. Thus, AB functions provide optimal security against linear cryptanalysis when n is odd. Furthermore, all AB functions are necessarily APN [10], so that AB functions are optimal with respect to differential cryptanalysis as well.

Due to the huge number of (n, m)-functions for non-trivial values of n and m, they are typically classified up to some notion of equivalence. The most general known equivalence relation which preserves differential uniformity (and hence APN-ness) is Carlet-Charpin-Zinoviev (or CCZ) equivalence [6,9]. We say that two (n, m)-functions F and F' are *CCZ-equivalent* if there is an affine permutation $\mathcal{A}$ of $\mathbb{F}_{2^n} \times \mathbb{F}_{2^m}$ that maps the graph $\mathcal{G}(F) = \{(x, F(x)) : x \in \mathbb{F}_{2^n}\}$ of F to the graph $\mathcal{G}(F')$ of F'. A special case of CCZ-equivalence is extended affine (or EA) equivalence. We say that F and F' are *EA-equivalent* if there are affine permutations A_1 and A_2 of $\mathbb{F}_{2^m}$ and $\mathbb{F}_{2^n}$, respectively, and an affine (n, m)-function A such that $F' = A_1 \circ F \circ A_2 + A$.

In [5], Budaghyan, Carlet and Leander introduced the following two infinite families of APN binomials:

1. For $n = 3k$:

$$F_3(x) = x^{2^s+1} + w^{2^k-1}x^{2^{ik}+2^{mk+s}}, \tag{1}$$

   where s and k are positive integers such that $s \leq 4k - 1$, $\gcd(k, 3) = \gcd(s, 3k) = 1$, $i = sk \mod 3$, $m = 3 - i$ and w is a primitive element of the field $\mathbb{F}_{2^n}$.
2. For $n = 4k$:

$$F_4(x) = x^{2^s+1} + w^{2^k-1}x^{2^{ik}+2^{mk+s}}, \tag{2}$$

   where s and k are positive integers such that $s \leq 4k - 1$, $\gcd(k, 2) = \gcd(s, 2k) = 1$, $i = sk \mod 4$, $m = 4 - i$ and w is a primitive element of the field $\mathbb{F}_{2^n}$.

The first class of APN binomials (for n divisible by 3) are permutations if and only if k is odd.

As we show below, if the condition of k being odd is omitted, the binomials for n divisible by 4 are EA-equivalent to the Gold functions. Indeed, let k be even. Then $i = sk \mod 4$ is also even. If $i = 2$, then

$$F(x) = x^{2^s+1} + w^{2^k-1}x^{2^{ik}+2^{mk+s}} = x^{2^s+1} + w^{2^k-1}x^{2^{2k}+2^{2k+s}}$$
$$= x^{2^s+1} + w^{2^k-1}x^{2^{2k}(1+2^s)} = x^{2^s+1} + w^{2^k-1}(x^{2^s+1})^{2^{2k}}$$

which is EA-equivalent to x^{2^s+1} since $x \mapsto x + w^{2^k-1}x^{2^{2k}}$ is a linear permutation. Indeed, if $x + w^{2^k-1}x^{2^{2k}} = y + w^{2^k-1}y^{2^{2k}}$ and $x \neq y$, then we must have $w^{1-2^k} = $

$(x + y)^{2^{2k}-1}$ which is impossible since $2^{2k} - 1$ is a multiple of 5 under the hypothesis, whereas $2^k - 1$ is not.

In the same manner, if $i = 0$, we get

$$F(x) = x^{2^s+1} + w^{2^k-1} x^{2^{ik}+2^{mk+s}} = x^{2^s+1} + w^{2^k-1} x^{1+2^s} = x^{2^s+1}\left(1 + w^{2^k-1}\right).$$

The complete Walsh spectra of the functions F_3 and F_4 were determined in [2].

As previously mentioned, relaxing the conditions allows the functions F_3 to be generalized to a family of 2^t-differentially uniform functions in the same way as the Gold functions [3]. In this paper, we show how the family F_4 can be generalized to functions with 2^t-to-1 derivatives in a similar way. Further, we provide a counterexample to the question of whether this construction can be used to generalize any family of quadratic APN functions to a family of 2^t-uniform functions: for the family of quadrinomials from [7], we computationally verify that relaxing conditions does not lead to functions with 2^t-to-1 derivatives for $t > 1$ over $\mathbb{F}_{2^n}$ for any $6 \leq n \leq 14$.

For background on APN functions and cryptographic Boolean functions, we refer the reader to [4] or [8].

3   Differential Uniformity

In the following theorem, we show that by relaxing the condition $\gcd(s, 2k) = 1$ in (2) to $\gcd(s, 2k) = t$ for some positive integer t, we obtain functions over $\mathbb{F}_{2^{4k}}$ all of whose derivatives are 2^t-to-1 functions.

Theorem 1. *Let s, k, t be positive integers and let $n = 4k$. Let $\gcd(s, 2k) = t$, $2 \nmid k$, $i = sk \mod 4$, $m = 4 - i$, and w be a primitive element of $\mathbb{F}_{2^n}$. Then all derivatives $D_a F$ for $a \in \mathbb{F}_{2^n}^*$ of the function*

$$F(x) = wx^{2^s+1} + w^{2^k} x^{2^{ik}+2^{mk+s}} \tag{3}$$

are 2^t-to-1 functions. In particular, F is differentially 2^t-uniform.

Proof. We first show that for i even, F is EA-equivalent to x^{2^s+1}. To see this, consider two cases depending on the value of i. First, suppose $i = 2$. Then

$$F(x) = wx^{2^s+1} + w^{2^k} x^{2^{2k}+2^{2k+s}} = wx^{2^s+1} + w^{2^k}(x^{2^s+1})^{2^{2k}}$$

which is EA-equivalent to x^{2^s+1} since $x \mapsto wx + w^{2^k} x^{2^{2k}}$ is a linear permutation. Indeed, suppose that $wx + w^{2^k} x^{2^{2k}} = wy + w^{2^k} y^{2^{2k}}$ for some two distinct elements $x, y \in \mathbb{F}_{2^n}$; then $(x+y)^{2^{2k}-1} = w^{1-2^k}$ which is a contradiction since the exponent on the left-hand side is a multiple of three, while the one on the right-hand side is not. Finally, note that the derivatives of x^{2^s+1} are all 2^t-to-1 functions since $\gcd(s, 4k) = \gcd(s, 2k) = t$.

If $i = 0$, then

$$F(x) = wx^{2^s+1} + w^{2^k} x^{1+2^{4k+s}} = wx^{2^s+1} + w^{2^k} x^{1+2^s} = x^{2^s+1}\left(w + w^{2^k}\right),$$

which is EA-equivalent to x^{2^s+1} (as w is a primitive element, we have $w + w^{2^k} \neq 0$), and hence all of its derivatives are 2^t-to-1 under the conditions on s, t and k.

We now consider the case of i odd. Both possibilities for i produce functions in the same EA-equivalence class. For $i = 1$, the function (3) takes the form

$$F(x) = wx^{2^s+1} + w^{2^k} x^{2^k+2^{3k+s}}. \tag{4}$$

Consider the function F' defined by

$$F'(x) = F(x)^{2^{3k}} = \left(wx^{2^s+1} + w^{2^k} x^{2^k+2^{3k+s}}\right)^{2^{3k}} = wx^{2^{2k+s}+1} + w^{2^{3k}} x^{2^{3k}(2^s+1)}.$$

Clearly, F' is EA-equivalent to F. From the condition $ks = 1 \mod 4$ we get $k \mod 4 = s \mod 4$, i.e. $2k + s = 3s \mod 4$, hence $(2k + s)k = 3sk = 3 \mod 4$. Thus, denoting $2k + s$ by s', we get $F'(x) = wx^{2^{s'}+1} + w^{2^{-k}} x^{2^{3k}+2^{k+s'}}$, which is precisely the function from (3) for $i = 3$.

It is thus enough to prove the theorem for $i = 3$, i.e. for the function $F(x) = wx^{2^s+1} + w^{2^k} x^{2^{3k}+2^{k+s}}$.

The derivatives of F are 2^t-to-1 functions if and only if the equation $F(x) + F(x + v) = u$ has either 0 or 2^t solutions for any $u, v \in \mathbb{F}_2^n, v \neq 0$. The left-hand side of this equality takes the form

$$
\begin{aligned}
&F(x) + F(x + v) \\
&= wx^{2^s+1} + w^{2^k} x^{2^{3k}+2^{k+s}} + w(x + v)^{2^s+1} + w^{2^k}(x + v)^{2^{3k}+2^{k+s}} \\
&= wx^{2^s+1} + w^{2^k} x^{2^{3k}+2^{k+s}} + wx^{2^s+1} + wv^{2^s+1} + wx^{2^s}v + wxv^{2^s} + w^{2^k} x^{2^{3k}+2^{k+s}} \\
&\quad + w^{2^k} v^{2^{3k}+2^{k+s}} + w^{2^k} x^{2^{3k}} v^{2^{k+s}} + w^{2^k} v^{2^{3k}} x^{2^{k+s}} \\
&= wv^{2^s+1} + wx^{2^s}v + wxv^{2^s} + w^{2^k} v^{2^{3k}+2^{k+s}} + w^{2^k} x^{2^{3k}} v^{2^{k+s}} + w^{2^k} v^{2^{3k}} x^{2^{k+s}} \\
&= w^{2^k} v^{2^{3k}+2^{k+s}}\left(\left(\frac{x}{v}\right)^{2^{3k}} + \left(\frac{x}{v}\right)^{2^{k+s}}\right) + wv^{2^s+1}\left(\left(\frac{x}{v}\right)^{2^s} + \left(\frac{x}{v}\right)\right) + wv^{2^s+1} \\
&\quad + w^{2^k} v^{2^{3k}+2^{k+s}}.
\end{aligned}
$$

Dividing the last expression by wv^{2^s+1} and substituting vx for x, we get a linear expression in x:

$$a\left(x^{2^{3k}} + x^{2^{k+s}}\right) + \left(x^{2^s} + x\right) + 1 + a,$$

where $a = w^{2^k-1} v^{2^{3k}+2^{k+s}-(2^s+1)}$. So, $F(x) + F(x + v) = u$ has 0 or 2^t solutions if and only if the kernel of the linear map

$$\Delta_a(x) = a\left(x^{2^{3k}} + x^{2^{k+s}}\right) + \left(x^{2^s} + x\right)$$

has 2^t elements. Consider the equation $\Delta_a(x) = 0$. We use Dobbertin's multi-variate method and follow the computations from Theorem 2 of [5]. Let $b = a^{2^k}$ and $c = b^{2^k}$. We get that

$$\Delta_a(x) = 0 \text{ if and only if } ab\left(bc + 1\right)^{2^s+1}\left(x^{2^{2s}} + x^{2^s}\right) = 0,$$

assuming that $P(a) = c(ab + 1)^{2^s+1} + a^{2^s}(bc + 1)^{2^s+1} \neq 0$.

We now show that $bc + 1 \neq 0$. Clearly, $bc + 1 = 0$ if and only if $ab + 1 = 0$. Suppose $ab = 1$, i.e. $a^{2^k+1} = 1$. From

$$\left(2^{3k} + 2^{k+s} - (2^s + 1)\right)\left(2^k + 1\right) = (2^{2k} - 1)(2^k + 2^s) \mod (2^{4k} - 1)$$

we get

$$1 = a^{2^k+1} = \left(w^{2^k-1}v^{2^{3k}+2^{k+s}-(2^s+1)}\right)^{2^k+1} = w^{2^{2k}-1}v^{(2^{2k}-1)(2^k+2^s)}$$
$$= \left(wv^{2^k+2^s}\right)^{2^{2k}-1},$$

hence $wv^{2^k+2^s}$ is a $(2^{2k}+1)$-st power of an element from $\mathbb{F}_{2^n}$. On the other hand, from $ks = 3 \mod 4$ and $2 \nmid k$ we have that k and s are odd, and $k \neq s \mod 4$, which means that $k - s = 2p$ for some odd p. Thus, $2^k + 2^s = 2^s(2^{k-s} + 1) = 2^s(2^{2p} + 1)$. Since p is odd, we have $5 \mid 2^{2p} + 1$, and therefore $u^{2^k+2^s}$ is the fifth power of an element of the field, while $wu^{2^k+2^s}$ is not. Thus $wu^{2^k+2^s}$ is also not a $(2^{2k} + 1)$-st power. Hence, we get a contradiction, and so we must have $ab + 1 \neq 0$ and hence $bc + 1 \neq 0$. Therefore, we have

$$\Delta_a(x) = 0 \text{ if and only if } x^{2^{2s}} + x^{2^s} = 0$$

when $P(a) \neq 0$.

By the statement of Theorem 1, k is odd and $sk = 3 \mod 4$, so that s is also odd, and from $\gcd(s, 2k) = t$ it follows that $\gcd(s, 4k) = t$. Therefore the equation $x^{2^{2s}} + x^{2^s} = 0$, which is equivalent to $x^{2^s} = 1$, has exactly $2^{\gcd(s,4k)} = 2^t$ solutions.

So we only have to show that $P(a) = c(ab + 1)^{2^s+1} + a^{2^s}(bc + 1)^{2^s+1}$ does not vanish.

Assume $P(a) = 0$, i.e.

$$\frac{c}{a^{2^s}} = \left(\frac{bc + 1}{ab + 1}\right)^{2^s+1}.$$

We have that $\frac{c}{a^{2^s}}$ is the third power of an element of the field since $3 \mid 2^s+1, 2^n-1$ (since s is odd and n is even). On the other hand,

$$\frac{c}{a^{2^s}} = a^{2^{2k}-2^s} = a^{2^s(2^{2k-s}-1)} = \left(w^{2^k-1}v^{2^{3k}+2^{k+s}-(2^s+1)}\right)^{2^s(2^{2k-s}-1)}$$
$$= w^{(2^k-1)2^s(2^{2k-s}-1)}v^{(2^{3k}+2^{k+s}-(2^s+1))2^s(2^{2k-s}-1)}$$

and $2^{3k} + 2^{k+s} - (2^s + 1) = 2^s(2^{3k-s} - 1) + (2^{k+s} - 1)$ is divisible by 3 because $3 \mid 2^{3k-s} - 1$ and $3 \mid 2^{k+s} - 1$ due to k and s being odd. But since k and $2k - s$ are odd, we have $3 \nmid 2^k - 1$ and $3 \nmid 2^{2k-s} - 1$, which means that $w^{(2^k-1)2^s(2^{2k-s}-1)}$ is not a third power, therefore $\frac{c}{a^{2^s}}$ is not a third power either, and we get a contradiction.

As the following proposition illustrates, the binomials from (3) also behave in the same way as the Gold functions from the point of view of bijectivity.

Proposition 1. *A function of the form* (3) *is a permutation if and only if it is EA-equivalent to a 2^t-differentially uniform permutation of the form x^{2^s+1} for some positive integer s.*

Proof. Recall that the power function x^{2^s+1} over $\mathbb{F}_{2^n}$ is 2^t-uniform for some positive integer t if and only if $\gcd(s, n) = t$, and it is a permutation if and only if n/t is odd.

Let $F(x) = wx^{2^s+1} + w^{2^k}x^{2^{ik}+2^{mk+s}}$ be a function satisfying the conditions of Theorem 1. If F is a permutation, then $4k/\gcd(s, 4k)$ is odd. Indeed, assume that F is a permutation and $4k/\gcd(s, 4k)$ is even. Since k is odd, we have that $\gcd(s, 4k)$ should be odd or $\gcd(s, 4k) = 2 \mod 4$. If $\gcd(s, 4k)$ is odd, then so is s, and therefore $3 \mid 2^s + 1$. Since $i = (sk \mod 4)$ and s, k are odd, then i is an odd number, and hence $(m - i)k + s$ is also odd; hence $3 \mid 2^{ik}(1 + 2^{(m-i)k+s}) = 2^{ik} + 2^{mk+s}$. Thus, for any $\gamma \in \mathbb{F}_{2^2}$, we have $F(\gamma x) = F(x)$. On the other hand, if $\gcd(s, 4k) = 2 \mod 4$, then s is even, and therefore i is also even due to $i = sk \mod 4$. Hence, as we discussed in the proof of Theorem 1, F is EA-equivalent to x^{2^s+1} which is not a permutation since $4k/\gcd(s, 4k)$ is even. Therefore $4k/\gcd(s, 4k)$ is necessarily odd if F is a permutation. However, when $4k/\gcd(4k, s)$ is odd, $\gcd(4k, s)$ is divisible by 4, and therefore s is also divisible by 4 since k is odd. This means that F is EA-equivalent to a 2^t-differentially uniform permutation of the form x^{2^l+1} for some positive integer l.

4   Magnitude of the Walsh Coefficients

In following theorem, we compute an upper bound on the absolute values of the Walsh coefficients of the functions from (3). In the proof we make use of the following result.

Lemma 1 ([14]). *Let n, l, d be positive integers such that $\gcd(n, s) = 1$ and let $G(x) = \sum_{i=0}^{d} a_i x^{li} \in \mathbb{F}_{2^n}[x]$. Then the equation $G(x) = 0$ has at most 2^d solutions.*

We are now ready to present the main result of this section.

Theorem 2. *Let s, k, t be positive integers and let $n = 4k$. Let $\gcd(s, 2k) = t$, $2 \nmid k$, $i = sk \mod 4$, $m = 4 - i$ and let w be a primitive element of $\mathbb{F}_{2^n}$. Then the Walsh coefficients of the function F from* (3) *satisfy*

$$|W_F(a, b)| \leq 2^{2k+t}$$

for any $a \in \mathbb{F}_{2^n}^*$ and $b \in \mathbb{F}_{2^n}$.

Proof. For simplicity, instead of $F(x) = wx^{2^s+1} + w^{2^k}x^{2^{ik}+2^{mk+s}}$, we consider the EA-equivalent function $F'(x) = x^{2^s+1} + \alpha x^{2^{ik}+2^{mk+s}}$, where $\alpha = w^{2^k-1}$.

We are going to prove the theorem for $i = 3$, since as we already observed in the proof of Theorem 1, if i is even, the function $F(x)$ is EA-equivalent to a Gold-like differentially 2^t-uniform function; and if i is odd, the functions that we obtain for $i = 1$ and for $i = 3$ are EA-equivalent.

We have

$$W^2_{F'}(a, b) = \sum_x \sum_y (-1)^{\mathrm{Tr}(ax+ay+bF'(x)+bF'(y))}.$$

Substituting $x + y$ for y, we get

$$W^2_{F'}(a, b) = \sum_x \sum_y (-1)^{\mathrm{Tr}(ax+a(x+y)+bF'(x)+bF'(x+y))}.$$

By straightforward calculations, the exponent from the previous expression becomes

$$\mathrm{Tr}\left(ax + a(x + y) + bF'(x) + bF'(x + y)\right)$$

$$= \mathrm{Tr}\left(ay + b\left(x^{2^s+1} + \alpha x^{2^{3k}+2^{k+s}} + (x + y)^{2^s+1} + \alpha(x + y)^{2^{3k}+2^{k+s}}\right)\right)$$

$$= \mathrm{Tr}\left(ay + by^{2^s+1} + b\alpha y^{2^{k+s}+2^{3k}}\right) + \mathrm{Tr}\left(bx^{2^s}y + bxy^{2^s} + b\alpha x^{2^{3k}}y^{2^{k+s}} + b\alpha y^{2^{3k}}x^{2^{k+s}}\right)$$

$$= \mathrm{Tr}\left(ay + by^{2^s+1} + b\alpha y^{2^{k+s}+2^{3k}}\right) + \mathrm{Tr}(x\mathcal{L}(y)),$$

where $\mathcal{L}(y) = (by)^{2^{-s}} + by^{2^s} + (b\alpha)^{2^{-3k}}y^{2^{s-2k}} + (b\alpha)^{2^{3k-s}}y^{2^{2k-s}} = (by)^{2^{-s}} + by^{2^s} + (b\alpha)^{2^{2k}}y^{2^{s+2k}} + (b\alpha)^{2^{3k-s}}y^{2^{2k-s}}$ is a linear function.

Thus

$$W^2_{F'}(a, b) = 2^n \sum_{\{y|\mathcal{L}(y)=0\}} (-1)^{\mathrm{Tr}(ay+by^{2^s+1}+b\alpha y^{2^{k+s}+2^{3k}})}.$$

The next step is to show that the cardinality of the kernel of $\mathcal{L}(y)$ is at most 2^{2t}, where $t = \gcd(2k, s)$. Following the computations of [2], we have

$$b^{2^{-s+2k}}\mathcal{L}(y) + (b\alpha)^{2^{3k-s}}\mathcal{L}^{2^{2k}}(y) = 0 \text{ and } b^{2^{2k}}\mathcal{L}(y) + (b\alpha)^{2^k}\mathcal{L}^{2^{2k}}(y) = 0,$$

from where we get

$$Ay^{2^s} + By^{2^{-s}} + Cy^{2^{s+2k}} = 0, \tag{5}$$

$$B^{2^s}y^{2^s} + A^{2^{2k}}y^{2^{-s}} + Cy^{2^{-s+2k}} = 0, \tag{6}$$

where

$$A = b^{2^{-s+2k}+1} + (b\alpha)^{2^{-k}+2^{3k-s}} \neq 0,$$

$$B = b^{2^{-s}+2^{-s+2k}} + (b\alpha)^{2^{k-s}+2^{3k-s}}, \text{ and}$$

$$C = b^{2^{-s+2k}+2^k}\alpha^{2^k} + b^{2^{2k}+2^{3k-s}}\alpha^{2^{3k-s}} \neq 0,$$

with $B = 0$ if and only if B^{2^s-1} is a cube.

Assume that $B \neq 0$, i.e. B^{2^s-1} is not a cube. Then from (5) and (6) we get

$$B^{2^{2s}} C^{2^{-s}} y^{2^{2s}} + C^{2^{-s}} A^{2^{2k+s}} y + B^{2^{-s}} C^{2^s} y^{2^{-2s}} + A^{2^{-s}} C^{2^s} y = 0.$$

Denote the last expression by $G(y)$. For some $v \neq 0$ in the kernel of $G(y)$, consider the expression $G_v(y) = yG(y) + vG(v) + (y+v)G(y+v)$, i.e.

$$C^{2^s} B^{2^{-s}} \left(y^{2^{-2s}} v + v^{2^{-2s}} y \right) + C^{2^{-s}} B^{2^{2s}} \left(y^{2^{2s}} v + v^{2^{2s}} y \right).$$

Note that the kernel of $\mathcal{L}(y)$ is contained in that of $G_v(y)$. Then from $G_v(y) = 0$ we get

$$C^{2^{-s}-2^s} B^{2^{2s}-1} \left(y^{2^{-2s}} v + v^{2^{-2s}} y \right)^{2^{2s}-1} = B^{2^s-1}.$$

If $y^{2^{-2s}} v + v^{2^{-2s}} y = 0$, i.e. $yv^{-1} = (yv^{-1})^{2^{2s}}$, then $yv^{-1} \in \mathbb{F}_{\gcd(2s,4k)} = \mathbb{F}_{2^{2t}}$ and therefore $\mathcal{L}(y) = 0$ has exactly 2^{2t} solutions. Otherwise, if $y^{2^{-2s}} v + v^{2^{-2s}} y$ does not vanish, then the right-hand side of the previous equation is not a cube by our assumption, while the left-hand side is. Hence, $\mathcal{L}(y) = 0$ has exactly 2^{2t} solutions, where $t = \gcd(2k, s)$.

Suppose now that $B = 0$. Following the computations of [2], the equation $\mathcal{L}(y) = 0$ becomes

$$\left(b + (bw)^{2^k} v^{2^{2k+s}-2^s} \right) y^{2^s} + \left(b^{2^{-s}} + (bw)^{2^{3k-s}} v^{2^{2k-s}-2^{-s}} \right) y^{2^{-s}} = 0.$$

If both coefficients (in front of y^{2^s} and in front of $y^{2^{-s}}$) in the above equation are nonzero, then raising both sides to the power 2^s, we get

$$\left(b + (bw)^{2^k} v^{2^{2k+s}-2^s} \right)^{2^s} y^{2^{2s}} + \left(b^{2^{-s}} + (bw)^{2^{3k-s}} v^{2^{2k-s}-2^{-s}} \right)^{2^s} y = 0.$$

Note that $2s = 2t\frac{s}{t}$ and $\gcd(\frac{s}{t}, 4k) = 1$. Then, applying Lemma 1, we get that $\mathcal{L}(y) = 0$ has at most 2^{2t} solutions. If exactly one of the coefficients is not zero, then the equation will have exactly one solution, namely $y = 0$. If both coefficients are equal to zero, then raising them to the power of 2^s and of 2^{-s}, and adding these powers together, we get $v^{2^{2k}-1} = b^{2^{3k}-2^{k-s}} w^{-2^{k-s}} = b^{1-2^{3k}} w^{-2^{3k}}$ which implies $C = 0$, a contradiction.

Thus, the kernel of $\mathcal{L}(y)$ consists of at most 2^{2t} elements, where $t = \gcd(2k, s)$, and therefore $|W_F^2(a,b)| \leq 2^n 2^{2t}$ and $|W_F(a,b)| \leq 2^{2k+t}$.

The next corollary immediately follows from Theorem 2.

Corollary 1. *Let s, k, t be positive integers and let $n = 4k$. Let $\gcd(s, 2k) = t$, $2 \nmid k$, $i = sk \mod 4$, $m = 4 - i$ and let w be a primitive element of $\mathbb{F}_{2^n}$. Then the nonlinearity of the function F from (3) satisfies*

$$\mathcal{NL}(F) \leq 2^{n-1} - 2^{2k+t-1}.$$

5 A Counterexample: Generalizing a Family of APN Quadrinomials to 2^t-uniform Functions

As discussed above, both families of APN binomials from [5] can be generalized to functions all of whose derivatives are 2^t-to-1 by relaxing conditions; furthermore, the two families are obtained as generalizations of a previously unclassified sporadic APN instance over $\mathbb{F}_{2^{12}}$. Another sporadic APN instance, this time over $\mathbb{F}_{2^{10}}$, was recently also generalized into an infinite family [7]. This immediately raises the question of whether the same approach, i.e. relaxing conditions in order to obtain functions with 2^t-to-1 derivatives, could be applied to the latter family. In this section, we summarize our experimental results, which suggest that this is impossible.

The functions in the infinite family from [7] are defined over $\mathbb{F}_{2^n}$ with $n = 2m$ with m odd such that $3 \nmid m$, and have the form

$$F(x) = x^3 + \beta(x^{2^i+1})^{2^k} + \beta^2(x^3)^{2^m} + (x^{2^i+1})^{2^{m+k}}, \tag{7}$$

where k is a non-negative integer, and β is a primitive element of $\mathbb{F}_{2^2}$. It is shown that the function in (7) is APN for $i = m - 2$ and $i = (m-2)^{-1} \mod n$, as well as for $i = m$ and $i = m - 1$ (however, the last two values yield functions that are trivially EA-equivalent to known ones).

We computationally go through all functions of the form

$$F(x) = x^{2^j+1} + \beta(x^{2^i+1})^{2^k} + \beta^2(x^{2^i+1})^{2^m} + (x^{2^i+1})^{2^{m+k}} \tag{8}$$

with $0 \le i, j \le n - 1$ for all values of $n = 2m$ with $6 \le n \le 14$, disregarding the conditions of $3 \nmid m$ and of m being odd. For each such function, we test whether all of its derivatives are 2^t-to-1 functions for some positive integer t. We restrict ourselves to the cases $k = 0$ and $k = 1$, as the APN functions constructed for $k \in \{0, 1\}$ appear to exhaust all CCZ-equivalence classes [7].

Besides the already known APN functions, for $k = 0$, we only encounter functions with 2^t-to-1 derivatives when $j = i$, i.e. when all exponents are in the same cyclotomic coset. In the case of $k = 1$, the only exceptions are for $n = 12$ where each pair (j, i) with $2 \le j, i \le 12$ and i, j even yields a 2^2-to-1, i.e. 4-to-1 function. However, since we do not observe other such non-trivial functions for other dimensions n, this does not suggest that (7) can be generalized to 2^t-functions in general.

These computational results constitute convincing evidence that the quadrinomials of the form (7) cannot be generalized to 2^t-to-1 functions in the same way as the binomials from [5].

6 Conclusion

The APN binomial $x^3 + \alpha x^{258}$ over $\mathbb{F}_{2^{12}}$ was generalized in 2008 to two infinite APN families over $\mathbb{F}_{2^n}$, one for $3 \mid n$, and one for $4 \mid n$. The family for $3 \mid n$ was generalized to a family of functions with 2^t-to-1 derivatives in 2012 [3] by relaxing conditions. We have shown that the same approach can be applied to the family for $4 \mid n$, and have computed the differential uniformity of the resulting functions. We have also given an upper bound on their nonlinearity, and have shown that this construction cannot be applied to any infinite family of quadratic APN functions by computationally verifying that the quadrinomial family from [7] constitutes a counterexample.

Acknowledgment. This research was supported by the Trond Mohn foundation (TMS).

References

1. Biham, E., Shamir, A.: Differential cryptanalysis of DES-like cryptosystems. J. Cryptol. **4**(1), 3–72 (1991). https://doi.org/10.1007/BF00630563
2. Bracken, C., Byrne, C., Markin, N., Mcguire, G.: Fourier spectra of binomial APN functions. SIAM J. Discrete Math. **23**(2), 596–608 (2009)
3. Bracken, C., Tan, C., Tan, Y.: Binomial differentially 4 uniform permutations with hight nonlinearity. Finite Fields Appl. **18**, 537–546 (2012)
4. Budaghyan, L.: Construction and Analysis of Cryptographic Functions. Springer, Cham (2015). https://doi.org/10.1007/978-3-319-12991-4
5. Budaghyan, L., Carlet, C., Leander, G.: Two classes of quadratic APN binomials inequivalent to power functions. IEEE Trans. Inf. Theory **54**(9), 4218–4229 (2008)
6. Budaghyan, L., Carlet, C., Pott, A.: New classes of almost bent and almost perfect nonlinear functions. IEEE Trans. Inf. Theory **52**(3), 1141–1152 (2006)
7. Budaghyan, L., Helleseth, T., Kaleyski, N.: A new family of APN quadrinomials. IEEE Trans. Inf. Theory (2020, early access article)
8. Carlet, C.: Vectorial (multi-output) Boolean functions for cryptography, chapter of the monography boolean methods and models, In: Crama, Y., Hammer, P. (eds.) Cambridge University Press, to appear soon. Preliminary version available at http://www-rocq.inria.fr/codes/Claude.Carlet/pubs.html
9. Carlet, C., Charpin, P., Zinoviev, V.: Codes, bent functions and permutations suitable for DES-like cryptography. Des. Codes Crypt. **15**(2), 125–156 (1998). https://doi.org/10.1023/A:1008344232130
10. Chabaud, F., Vaudenay, S.: Links between differential and linear cryptanalysis. In: De Santis, A. (ed.) EUROCRYPT 1994. LNCS, vol. 950, pp. 356–365. Springer, Heidelberg (1995). https://doi.org/10.1007/BFb0053450
11. Edel, Y., Kyureghyan, G., Pott, A.: A new APN function which is not equivalent to a power mappings. IEEE Trans. Inf. Theory **52**(2), 744–747 (2006)
12. Matsui, M.: Linear cryptanalysis method for DES cipher. In: Helleseth, T. (ed.) EUROCRYPT 1993. LNCS, vol. 765, pp. 386–397. Springer, Heidelberg (1994). https://doi.org/10.1007/3-540-48285-7_33
13. Nyberg, K.: Differentially uniform mappings for cryptography. In: Helleseth, T. (ed.) EUROCRYPT 1993. LNCS, vol. 765, pp. 55–64. Springer, Heidelberg (1994). https://doi.org/10.1007/3-540-48285-7_6
14. Trachtenberg, H.M.: On the Cross-Correlation Functions of Maximal Linear Sequences, Ph.D. dissertation, University of Southern California, Los Angeles (1970)

On Subspaces of Kloosterman Zeros and Permutations of the Form $L_1(x^{-1}) + L_2(x)$

Faruk Göloğlu[1], Lukas Kölsch[2(✉)], Gohar Kyureghyan[2], and Léo Perrin[3]

[1] Department of Mathematics, Faculty of Mathematics and Physics,
Charles University, Prague, Czech Republic
farukgologlu@gmail.com
[2] Department of Mathematics, University of Rostock, Rostock, Germany
{lukas.koelsch,gohar.kyureghyan}@uni-rostock.de
[3] Inria, Paris, France
leo.perrin@inria.fr

Abstract. Permutations of the form $F(x) = L_1(x^{-1}) + L_2(x)$ with linear functions L_1, L_2 are closely related to several interesting questions regarding CCZ-equivalence and EA-equivalence of the inverse function. In this paper, we show that F cannot be a permutation on binary fields if the kernel of L_1 or L_2 is large. A key step of our proof is an observation on the maximal size of a subspace V of $\mathbb{F}_{2^n}$ that consists of Kloosterman zeros, i.e. a subspace V such that $K_n(v) = 0$ for every $v \in V$ where $K_n(v)$ denotes the Kloosterman sum of v.

Keywords: Inverse function · Permutation polynomials · Kloosterman sums · EA-equivalence · CCZ-equivalence

1 Introduction

Vectorial Boolean functions play an important role in the design of symmetric cryptosystems as design choices for S-boxes. The linear and differential properties of vectorial Boolean functions are a measure of resistance against linear [31] and differential [1] attacks.

Definition 1. *A function* $F\colon \mathbb{F}_{2^n} \to \mathbb{F}_{2^n}$ *has differential uniformity* d, *if*

$$d = \max_{a \in \mathbb{F}_{2^n}^*, b \in \mathbb{F}_{2^n}} |\{x\colon F(x) + F(x + a) = b\}|.$$

A function with differential uniformity 2 is called almost perfect nonlinear (APN) on $\mathbb{F}_{2^n}$.

To resist differential attacks, a vectorial Boolean function should have low differential uniformity. As the differential uniformity is always even, the APN functions yield the best resistance against differential attacks.

Faruk Göloğlu was supported by the GAČR Grant 18-19087S -301-13/201843.

© Springer Nature Switzerland AG 2021
J. C. Bajard and A. Topuzoğlu (Eds.): WAIFI 2020, LNCS 12542, pp. 207–221, 2021.
https://doi.org/10.1007/978-3-030-68869-1_12

Definition 2. *The Walsh transform* $W_F \colon \mathbb{F}_{2^n} \times \mathbb{F}_{2^n} \to \mathbb{Z}$ *of a function* $F \colon \mathbb{F}_{2^n} \to \mathbb{F}_{2^n}$ *is defined as follows:*

$$W_F(a,b) = \sum_{x \in \mathbb{F}_{2^n}} (-1)^{\mathrm{Tr}(aF(x)+bx)}.$$

The nonlinearity of F *is defined as*

$$nl(F) = 2^{n-1} - \frac{1}{2} \max_{a \in \mathbb{F}_{2^n}^*, b \in \mathbb{F}_{2^n}} |W_F(a,b)|. \tag{1}$$

The higher the nonlinearity of a vectorial Boolean function, the better is its resistance to linear attacks.

There are several operations on the set of Boolean functions under which linear and differential properties are invariant. They lead to several equivalence concepts for vectorial Boolean functions. We denote by

$$G_F = \{(x, F(x)) \colon x \in \mathbb{F}_{2^n}\} \subset \mathbb{F}_{2^n} \times \mathbb{F}_{2^n}$$

the graph of the function $F \colon \mathbb{F}_{2^n} \to \mathbb{F}_{2^n}$. In the next definition and in the remainder of the paper we use the term linear function to refer to an $\mathbb{F}_2$-linear one. Similarly, we will call a function affine if it is sum of a linear function and a constant.

Definition 3. *Two functions* $F_1, F_2 \colon \mathbb{F}_{2^n} \to \mathbb{F}_{2^n}$ *are called extended affine equivalent (EA-equivalent) if there are affine permutations* A_1, A_2 *and an affine mapping* A_3 *mapping from* $\mathbb{F}_{2^n}$ *to itself such that*

$$A_1(F_1(A_2(x))) + A_3(x) = F_2(x). \tag{2}$$

F_1 *and* F_2 *are called* affine equivalent *if they are EA-equivalent and it is possible to choose* $A_3 = 0$ *in Eq. (2).*

Moreover, F_1 *and* F_2 *are called* CCZ-equivalent *if there are linear functions* $\alpha, \beta, \gamma, \delta \colon \mathbb{F}_{2^n} \to \mathbb{F}_{2^n}$ *and* $a, b \in \mathbb{F}_{2^n}$ *such that* $\mathcal{L} \colon \mathbb{F}_{2^n}^2 \to \mathbb{F}_{2^n}^2$ *defined by*

$$\mathcal{L}(x,y) = (\alpha(x) + \beta(y), \gamma(x) + \delta(y))$$

is bijective and

$$\mathcal{L}(G_{F_1}) + (a,b) = G_{F_2}.$$

F_1 *and* F_2 *are EA-equivalent if and only if a mapping* $\mathcal{L}$ *defined as above can be found with* $\beta = 0$, *and affine equivalent if and only if a mapping* $\mathcal{L}$ *can be found with* $\beta = \gamma = 0$.

The concept of CCZ-equivalence was introduced in [8] in 1998. It has been extensively studied, since it is a powerful tool for constructing and studying cryptological functions [5–7,14]. Clearly, affine equivalence implies EA-equivalence, which in turn implies CCZ-equivalence. The size of the image set is invariant under affine equivalence but in general it is changed under EA-equivalence. Non-linearity and differential uniformity are invariant under CCZ-equivalence.

Outline. In this paper, we consider EA- and CCZ-equivalence to the inverse function. This is a particularly interesting case because of the good crypto-graphic properties of the inverse function. In the second section, we show that some questions about CCZ- and EA-equivalence to a function F are related to the existence of permutations of the form $L_1(F(x)) + L_2(x)$. Accordingly, we investigate the existence of permutations of the form $L_1(x^{-1}) + L_2(x)$. This problem is related to Kloosterman zeros, i.e. elements whose Kloosterman sum is zero. In Sect. 3 we give an upper bound on the maximal size of a subspace of $\mathbb{F}_{2^n}$ that contains only Kloosterman zeros. Using this result, we show in Sect. 4 that there are no permutations of the form $L_1(x^{-1}) + L_2(x)$ if $\ker(L_1)$ or $\ker(L_2)$ is large.

2   EA- and CCZ-equivalence and Specific Permutations

The only known examples of APN permutations on $\mathbb{F}_{2^n}$ with n even are con-structed for $n = 6$ via study of the set of CCZ-equivalent functions to a known non-bijective function in [4]. The question about existence of APN permuta-tions for an even $n \geq 8$ is considered as the biggest challenge in the research on APN functions. As the examples in [4] suggest, a better understanding of CCZ-equivalence for permutations could be essential for progressing on this topic. Proposition 1 shows that this is closely related to study of permutations of form $L_1(F(x)) + L_2(x)$ with linear L_1, L_2. We would like to note that similar results are mentioned in various papers, for instance in [5–7].

Proposition 1. – **(a)** *Let* $F: \mathbb{F}_{2^n} \longrightarrow \mathbb{F}_{2^n}$ *and no permutation of the form* $F(x) + L(x)$ *exist with non-zero linear* $L(x)$. *Then every permutation that is EA-equivalent to* F *is already affine equivalent to it. In particular, if such an* F *is not bijective, then there are no EA-equivalent permutations to* F.
- **(b)** *Let* $F: \mathbb{F}_{2^n} \to \mathbb{F}_{2^n}$ *and no permutation of the form* $L_1(F(x)) + L_2(x)$ *exist with non-zero linear* L_1, L_2. *Then every function that is CCZ-equivalent to* F *is EA-equivalent to* F *or* F^{-1} *(if it exists). Moreover, all permutations that are CCZ-equivalent to* F *are affine equivalent to* F *or* F^{-1}.

Proof. (a) Let F_2 be a permutation EA-equivalent to F. By the definition of EA-equivalence, there exist $(a, b) \in \mathbb{F}_{2^n}^2$ and a bijective mapping $\mathcal{L}: \mathbb{F}_{2^n}^2 \to \mathbb{F}_{2^n}^2$ defined by $\mathcal{L}(x, y) = (\alpha(x), \gamma(x) + \delta(y))$ with linear functions $\alpha, \gamma, \delta: \mathbb{F}_{2^n} \to \mathbb{F}_{2^n}$ such that

$$\mathcal{L}(x, F(x)) + (a, b) = (\alpha(x) + a, \gamma(x) + \delta(F(x)) + b) = (\pi(x), F_2(\pi(x)))$$

where $\pi: \mathbb{F}_{2^n} \to \mathbb{F}_{2^n}$ is the permutation given by $\pi(x) = \alpha(x) + a$. Note that the function δ is bijective on $\mathbb{F}_{2^n}$, since $\mathcal{L}$ is bijective on $\mathbb{F}_{2^n}^2$. Also the composition

$F_2(\pi(x))$ is bijective on $\mathbb{F}_{2^n}$, implying that $\gamma(x) + \delta(F(x))$ is bijective, and hence also $\delta^{-1}(\gamma(x)) + F(x)$ is a permutation. Since $\delta^{-1}(\gamma(x))$ is linear, our assumption on F yields that $\gamma = 0$, completing the proof.

(b) Let now F_2 be a function CCZ-equivalent to F. By the definition of CCZ-equivalence, there exist $(a, b) \in \mathbb{F}_{2^n}^2$ and a bijective mapping $\mathcal{L} \colon \mathbb{F}_{2^n}^2 \to \mathbb{F}_{2^n}^2$ given by $\mathcal{L}(x, y) = (\alpha(x) + \beta(y), \gamma(x) + \delta(y))$ with linear $\alpha, \beta, \gamma, \delta \colon \mathbb{F}_{2^n} \to \mathbb{F}_{2^n}$ such that

$$\mathcal{L}(x, F(x)) + (a, b) = (\alpha(x) + \beta(F(x)) + a, \gamma(x) + \delta(F(x)) + b)$$
$$= (\pi(x), F_2(\pi(x)))$$

where $\pi \colon \mathbb{F}_{2^n} \to \mathbb{F}_{2^n}$ is the permutation on $\mathbb{F}_{2^n}$ given by $\pi(x) = \alpha(x) + \beta(F(x)) + a$. By our assumption on F, either $\alpha = 0$ or $\beta = 0$. Assume first that $\alpha = 0$. Then $\pi(x) = \beta(F(x)) + a$ and in particular both F and β are bijective. Further, γ is bijective since $\mathcal{L}$ is bijective. We then have

$$\gamma(x) + \delta(F(x)) + b = F_2(\pi(x)) = F_2(\beta(F(x)) + a).$$

The composition with the inverse $F^{-1}(x)$ yields

$$\gamma(F^{-1}(x)) + \delta(x) + b = F_2(\beta(x) + a),$$

and hence F_2 is EA-equivalent to F^{-1}. In the case $\beta = 0$ we get similarly $\pi(x) = \alpha(x) + a$ and

$$\gamma(x) + \delta(F(x)) + b = F_2(\pi(x)) = F_2(\alpha(x) + a),$$

where the mappings α and δ are bijective. Hence F_2 is EA-equivalent to F.

Now assume that F_2 is additionally a permutation. If F_2 is EA-equivalent to F then F_2 is affine equivalent to F using the statement in (a). Let us now consider the case that F_2 is EA-equivalent to F^{-1}. Observe that $F^{-1}(x) + L(x)$ is a permutation if and only if $L(F(x)) + x$ is a permutation, so there are no permutations of the form $F^{-1}(x) + L(x)$ by the assumption stated in the proposition. Again using (a), we conclude that F_2 is affine equivalent to F^{-1}.   □

The following proposition gives a criterion when a function $L_1(F(x)) + L_2(x)$ is bijective. For a linear mapping L, we denote by L^* its adjoint mapping with respect to the bilinear form

$$\langle x, y \rangle = \mathrm{Tr}(xy)$$

where Tr is the absolute trace mapping, i.e. we have

$$\mathrm{Tr}(L(x)y) = \mathrm{Tr}(xL^*(y))$$

for all $x, y \in \mathbb{F}_{2^n}$. Further, for a subset $A \subseteq \mathbb{F}_{2^n}$ we denote by $A^{\perp}$ its orthogonal complement, that is

$$A^{\perp} = \{x \in \mathbb{F}_{2^n} : \mathrm{Tr}(ax) = 0 \text{ for all } a \in A\}.$$

Proposition 2. *Let* $F: \mathbb{F}_{2^n} \to \mathbb{F}_{2^n}$ *and* L_1, L_2 *be linear mappings. The function* $L_1(F(x)) + L_2(x)$ *is a permutation if and only if*

$$W_F(L_1^*(b), L_2^*(b)) = 0$$

for all $b \in \mathbb{F}_{2^n}^*$.

Proof. It is well-known that a function is a permutation if and only if all of its component functions are balanced (for a proof, see [29, Theorem 7.7]). Consequently, $L_1(F(x)) + L_2(x)$ is a permutation if and only if

$$0 = \sum_{x \in \mathbb{F}_{2^n}} (-1)^{\text{Tr}(b(L_1(F(x))+L_2(x)))}$$

$$= \sum_{x \in \mathbb{F}_{2^n}} (-1)^{\text{Tr}(L_1^*(b)F(x)+L_2^*(b)x)} = W_F(L_1^*(b), L_2^*(b))$$

for all $b \in \mathbb{F}_{2^n}^*$.   □

Permutations of form $L_1(F(x)) + L_2(x)$ are characterized for some special choices of F and L_1, L_2. It was shown in [12] that no permutation of the form $x^d + L(x)$ exists when there is an $a \in \mathbb{F}_{2^n}$ such that $\text{Tr}(ax^d)$ is bent. Corollary 2.3 from [16] implies that $x^d + L(K(x))$ is not bijective on $\mathbb{F}_q$ for an arbitrary function K whenever $\gcd(d, q-1) \neq 1$ and L is a non-bijective linear function. In [27] a characterization of all permutations of the form $x^{2^i+1} + L(x)$ over $\mathbb{F}_{2^n}$ with $\gcd(i, n) = 1$ was given, as well as some results for the more general case $x^d + L(x)$. Permutations of the form $x^{2^i+1} + L(x)$ over $\mathbb{F}_{2^n}$ with $\gcd(i, n) > 1$ were recently considered in [3]. A particularly interesting case are the functions of shape $L_1(x^{-1}) + L_2(x)$ because of their good cryptographic properties. (Here we use as usual the convention $0^{-1} = 0$.) It was shown in [17] that such functions are never permutations in characteristic ≥ 5 (except for the trivial cases $L_1 = 0$ or $L_2 = 0$). In characteristic 3, no permutations of the type $x^{-1} + L(x)$ with $L \neq 0$ exist, except for sporadic cases in the small fields $\mathbb{F}_3$ and $\mathbb{F}_9$. In this paper we are interested in the case of characteristic 2. If L_1 or L_2 is bijective, then $L_1(x^{-1}) + L_2(x)$ cannot be bijective on $\mathbb{F}_{2^n}$ for $n \geq 5$ as shown in [28].

Theorem 1. ([28]). *Let* $F: \mathbb{F}_{2^n} \to \mathbb{F}_{2^n}$ *be defined by* $F(x) = x^{-1} + L(x)$ *with some linear mapping* $L(x) \neq 0$. *If* $n \geq 5$ *then* F *is not a permutation.*

The following result is an immediate consequence of Theorem 1.

Corollary 1. *Let* $n \geq 5$ *and* $F: \mathbb{F}_{2^n} \to \mathbb{F}_{2^n}$ *be defined by* $F(x) = L_1(x^{-1}) + L_2(x)$, *where* L_1, L_2 *are non-zero linear functions of* $\mathbb{F}_{2^n}$. *If* L_1 *or* L_2 *is bijective, then* F *is not a permutation on* $\mathbb{F}_{2^n}$.

Proof. Note that $F(x)$ is bijective if and only if $F(x^{-1}) = L_1(x) + L_2(x^{-1})$ is so. Hence without loss of generality suppose L_1 is bijective. Then the composition $L_1^{-1}(F(x)) = x^{-1} + L_1^{-1}(L_2(x))$ is bijective if and only if F is so, and Theorem 1 completes the proof.   □

In this paper, we continue the study of functions $L_1(x^{-1})+L_2(x)$ where L_1, L_2 are linear polynomials over $\mathbb{F}_{2^n}$. In the case of the inverse function $x \mapsto x^{-1}$, the Walsh transform is closely connected to Kloosterman sums.

Definition 4. *For $a \in \mathbb{F}_{2^n}$, the Kloosterman sum of a over $\mathbb{F}_{2^n}$ is defined as*

$$K_n(a) = \sum_{x \in \mathbb{F}_{2^n}} (-1)^{\mathrm{Tr}(x^{-1}+ax)}.$$

An element $a \in \mathbb{F}_{2^n}$ with $K_n(a) = 0$ is called a Kloosterman zero.

Note $K_n(a) = W_F(1,a)$ for $F(x) = x^{-1}$. More precisely, for $a \neq 0$ we have

$$W_F(a,b) = \sum_{x \in \mathbb{F}_{2^n}} (-1)^{\mathrm{Tr}(ax^{-1}+bx)} = \sum_{x \in \mathbb{F}_{2^n}} (-1)^{\mathrm{Tr}(x^{-1}+abx)} = K_n(ab)$$

using the substitution $x \mapsto ax$. For $a = 0$ and $b \neq 0$, we have $K_n(ab) = W_F(a,b) = 0$.

Proposition 2 can thus be stated using Kloosterman sums:

Corollary 2. *Let L_1, L_2 be linear functions of $\mathbb{F}_{2^n}$. Then $L_1(x^{-1}) + L_2(x)$ is a permutation on $\mathbb{F}_{2^n}$ if and only if $\ker(L_1^*) \cap \ker(L_2^*) = \{0\}$ and*

$$K_n(L_1^*(b)L_2^*(b)) = 0$$

for all $b \in \mathbb{F}_{2^n}$.

Proof. By Proposition 2, $L_1(x^{-1}) + L_2(x)$ is a permutation if and only if $W_F(L_1^*(b), L_2^*(b)) = 0$ for all $b \neq 0$. If $b \in \ker(L_1^*) \cap \ker(L_2^*)$, then $W_F(L_1^*(b), L_2^*(b)) = 2^n \neq 0$. In the other cases $W_F(L_1^*(b), L_2^*(b)) = K(L_1^*(b)L_2^*(b))$ by the considerations above. □

Corollary 2 shows that a function $L_1(x^{-1}) + L_2(x)$ is bijective on $\mathbb{F}_{2^n}$ only if the set $\{L_1^*(x)L_2^*(x) | x \in \mathbb{F}_{2^n}\}$ is a subset of the set of Kloosterman zeroes. Conversely, in [19] specific functions of shape $L_1(x^{-1})+L_2(x)$ are used to obtain identities for Kloosterman sums.

3   Vector Spaces of Kloosterman Zeros

The Kloosterman sums provide a powerful tool for studying additive properties of the inversion on finite fields. Kloosterman zeros are used for the construction of bent and hyperbent functions (see for example [10,13,26]). Vector spaces of Kloosterman zeros of dimension d in $\mathbb{F}_{2^n}$ can be used to construct vectorial bent functions from $\mathbb{F}_{2^{2n}}$ to $\mathbb{F}_{2^d}$ by modifying Dillon's construction, as shown in [26, Proposition 5].

Few results about the distribution of Kloosterman zeros are known. There is a way to compute the number of Kloosterman zeros [24], which relies on determining the class number of binary quadratic forms. However, it is difficult to use this method to derive a theoretical result on the number and distribution of Kloosterman sums. It was shown that for all n, Kloosterman zeros exist [23] (note that this is not true in characteristic ≥ 5 [22]). Moreover, it is known that for $n > 4$, Kloosterman zeros are never contained in proper subfields of $\mathbb{F}_{2^n}$ [30]. In [32], it is noted that $|\{a \in \mathbb{F}_{2^n} : K_n(a) = 0\}| = O(2^{3n/4})$. In this section, we give an upper bound for the size of vector spaces that contain exclusively Kloosterman zeros.

Let B be a bilinear form from $\mathbb{F}_{2^n}$ to $\mathbb{F}_2$. We denote by $\mathrm{rad}(B) = \{y \in \mathbb{F}_{2^n} : B(x,y) = 0$ for all $x \in \mathbb{F}_{2^n}\}$ the radical of B. Given a quadratic form $f : \mathbb{F}_{2^n} \to \mathbb{F}_2$, let $B_f(x,y) = f(x)+f(y)+f(x+y)$ be the bilinear form associated to it. The radical of the quadratic form f is defined as $\mathrm{rad}(B_f) \cap f^{-1}(\{0\})$. A quadratic form is called non-degenerate if $\mathrm{rad}(f) = \{0\}$.

Let $Q \colon \mathbb{F}_{2^n} \to \mathbb{F}_2$ be the quadratic form defined by

$$Q(x) = \sum_{0 \leq i < j < n} x^{2^i + 2^j}$$

for all $x \in \mathbb{F}_{2^n}$. Note that if m_a is the minimal polynomial of $a \in \mathbb{F}_{2^n}$ over $\mathbb{F}_2$ of degree d, then $Q(a)$ is the third coefficient of $m_a^{n/d}$. Indeed, recall that $m_a^{n/d} = \chi_a$ is the characteristic polynomial of a over $\mathbb{F}_2$ and $\chi_a(x) = \sum_{i=0}^{n-1}(x + a^{2^i})$. By expanding the product, we see that $Q(a)$ is the coefficient of x^{n-2} as claimed. This in particular shows that $Q(a) \in \mathbb{F}_2$.

The dyadic approximation of Kloosterman sums are often used to study Kloosterman zeroes. A nice survey on this topic is given in [34]. The main tool for results in this section is the following characterization of Kloosterman sums divisible by 2^4.

Theorem 2. ([18]). *Let* $n \geq 4$ *and* $a \in \mathbb{F}_{2^n}$. *Then* $K_n(a) \equiv 0 \pmod{16}$ *if and only if* $\mathrm{Tr}(a) = 0$ *and* $Q(a) = 0$.

Theorem 2 implies that the Kloosterman zeroes are contained in the intersection of the quadric $\{x \in \mathbb{F}_{2^n} : Q(x) = 0\}$ and the hyperplane

$$H = \{x \in \mathbb{F}_{2^n} : \mathrm{Tr}(x) = 0\}.$$

Therefore we consider the quadratic form $Q|_H$ which is induced by Q on H. We first determine its radical.

Lemma 1. *We have*

$$\mathrm{rad}(Q|_H) = \begin{cases} \{0,1\}, & n \equiv 0 \pmod 4 \\ \{0\}, & else. \end{cases}$$

Proof. First we compute the bilinear form associated to Q:

$$B_Q(x,y) = \sum_{0 \le i < j < n} x^{2^i + 2^j} + \sum_{0 \le i < j < n} y^{2^i + 2^j} + \sum_{0 \le i < j < n} (x+y)^{2^i + 2^j}$$

$$= \sum_{i \ne j} x^{2^i} y^{2^j} = \sum_{i=0}^{n-1} x^{2^i} \sum_{j \ne i} y^{2^j}$$

$$= \sum_{i=0}^{n-1} x^{2^i} (\mathrm{Tr}(y) + y^{2^i}) = \sum_{i=0}^{n-1} (xy)^{2^i} + \mathrm{Tr}(y) \sum_{i=0}^{n-1} x^{2^i}$$

$$= \mathrm{Tr}(xy) + \mathrm{Tr}(x)\,\mathrm{Tr}(y) = \mathrm{Tr}((y + \mathrm{Tr}(y))x).$$

Since $\mathrm{Tr}(y) = 0$ for all $y \in H$, we have

$$B_{Q|_H}(x,y) = \mathrm{Tr}(xy).$$

Then $y \in \mathrm{rad}(B_{Q|_H})$, if $B_{Q|_H}(x,y) = \mathrm{Tr}(xy) = 0$ for all $x \in H$. Hence $\mathrm{rad}(B_{Q|_H}) = \mathbb{F}_2 \cap H$. Observe that $1 \in H$ if and only if n is even, so $\mathrm{rad}(B_{Q|_H}) = \{0\}$ if n is odd and $\mathrm{rad}(B_{Q|_H}) = \mathbb{F}_2$ if n is even. One can easily verify that

$$Q(1) = \frac{n(n-1)}{2} = \begin{cases} 0 & n \equiv 0,1 \pmod 4 \\ 1 & n \equiv 2,3 \pmod 4 \end{cases}$$

and the result follows. □

Let $N(Q|_H(x) = u)$ denote the number of solutions of $Q|_H(x) = u$ for $u \in \mathbb{F}_2$. Observe that $N(Q|_H(x) = 0)$ is precisely the number of elements $x \in \mathbb{F}_{2^n}$ whose second and third coefficients of the characteristic polynomial χ_x are zero. The value $N(Q|_H(x) = a)$ was investigated in [9, 15, 33], where irreducible polynomials with prescribed coefficients were studied. In particular, the value $N(Q|_H(x) = 0)$ was determined. We summarize some of their results in the following theorem.

Theorem 3. *Let $N(Q|_H(x) = 0)$ be the number of $x \in H$ with $Q|_H(x) = 0$. Then $N(Q|_H(x) = 0) = 2^{n-2} + e$ where*

$$e = \begin{cases} -2^{\frac{n-2}{2}}, & n \equiv 0 \pmod 8 \\ 2^{\frac{n-3}{2}}, & n \equiv 1,7 \pmod 8 \\ 0, & n \equiv 2,6 \pmod 8 \\ -2^{\frac{n-3}{2}}, & n \equiv 3,5 \pmod 8 \\ 2^{\frac{n-2}{2}}, & n \equiv 4 \pmod 8. \end{cases}$$

Two quadratic forms f and g on a vector space V are called equivalent if f can be transformed into g with a non-singular linear transformation of V. The following result is well known (see e.g. [20, 29]).

Theorem 4 (Classification of quadratic forms). *Let $f: V \to \mathbb{F}_2$ with* $\dim(V) = n$ *be a quadratic form with* $\dim(\mathrm{rad}(f)) = w$. *Then f is equivalent to one of three forms:*

$$f \simeq \sum_{i=1}^{v} x_i y_i \qquad \text{(hyperbolic case)}$$

$$f \simeq z + \sum_{i=1}^{v} x_i y_i \qquad \text{(parabolic case)}$$

$$f \simeq x_1^2 + x_1 y_1 + y_1^2 + \sum_{i=2}^{v} x_i y_i \qquad \text{(elliptic case)},$$

where $v = \lfloor (n-w)/2 \rfloor$.

The value of $N(f(x) = 0)$ depends only on n, w and the type of the quadratic form. More precisely,

$$N(f(x) = 0) = 2^{n-1} + \Lambda(f) 2^{\frac{n+w-2}{2}},$$

with

$$\Lambda(f) = \begin{cases} 1, & \text{if } f \text{ is hyperbolic} \\ 0, & \text{if } f \text{ is parabolic} \\ -1, & \text{if } f \text{ is elliptic.} \end{cases}$$

The *Witt index* of a quadratic form is the number of pairs $x_i y_i$ that appear in the decomposition described above. In particular, the Witt index of f is v in the hyperbolic and parabolic case, and $v - 1$ in the elliptic case.

Remark 1. Just using the classification of quadratic forms in Theorem 4 and the determination of the radical in Lemma 1 we can give a simple alternative proof of the cases $n \equiv 2, 6 \pmod 8$ in Theorem 3. Indeed, in these cases $Q|_H$ is necessarily parabolic which immediately gives the value for $N(Q|_H(x) = 0)$.

We are now interested in the maximal dimension of a subspace contained in a quadric. Let f be a quadratic form on V. A subspace W of V is called totally isotropic if $f(w) = 0$ for all $w \in W$. And a subspace W is called maximal totally isotropic if there is no subspace W_2 with $f(w) = 0$ for all $w \in W_2$ and $W \subsetneq W_2 \subseteq V$. Any two maximal totally isotropic subspaces have the same dimension, which is the sum of the Witt index and the dimension of the radical of the quadratic form, as the following result implies.

Proposition 3 ([25, Corollary 4.4.]). *Let $f: V \to \mathbb{F}_2$ be a non-degenerate quadratic form on a vector space V over $\mathbb{F}_2$ with $\dim(V) = n$. Let W be a maximal totally isotropic subspace of V. Then, the dimension of W is equal to the Witt index of f. In particular, we have*

$$\dim(W) = \begin{cases} \frac{n}{2}, & \text{if } f \text{ is hyperbolic} \\ \frac{n-1}{2}, & \text{if } f \text{ is parabolic} \\ \frac{n-2}{2}, & \text{if } f \text{ is elliptic.} \end{cases}$$

We collect the above observations to give a sharp upper bound on the size of vector spaces that consist of elements with Kloosterman sum divisible by 16.

Proposition 4. *Let W be a subspace of $\mathbb{F}_{2^n}$ with $K_n(w) \equiv 0 \pmod{16}$ for all $w \in W$ and $n \geq 5$. Then $\dim W \leq d$ where*

$$
d = \begin{cases}
\frac{n-2}{2}, & n \equiv 0, 2, 6 \pmod 8 \\
\frac{n-1}{2}, & n \equiv 1, 7 \pmod 8 \\
\frac{n-3}{2}, & n \equiv 3, 5 \pmod 8 \\
\frac{n}{2}, & n \equiv 4 \pmod 8.
\end{cases}
$$

The bounds are sharp.

Proof. From the Theorems 3 and 4 we deduce that $Q|_H$ is elliptic if $n \equiv 0, 3, 5$ (mod 8), hyperbolic if $n \equiv 1, 4, 7 \pmod 8$ and parabolic if $n \equiv 2, 6 \pmod 8$. In the cases $n \not\equiv 0, 4 \pmod 8$ the quadratic form $Q|_H$ is non-degenerate by Lemma 1 and we immediately get bounds on $\dim(W)$ from Proposition 3 (recall that $Q|_H$ is a quadratic form on an $(n - 1)$ dimensional space). If $n \equiv 0, 4$ (mod 8) then $\dim(\mathrm{rad}(Q|_H)) = 1$, so $\dim V \leq 1 + \frac{n-4}{2} = \frac{n-2}{2}$ if $n \equiv 0 \pmod 8$ and $\dim V \leq 1 + \frac{n-2}{2} = \frac{n}{2}$ if $n \equiv 4 \pmod 8$. □

Remark 2. Every vector space W that contains exclusively Kloosterman zeros is of course also a vector space that contains only Kloosterman sums divisible by 16. In particular, by Propositions 3 and 4, all vector spaces of Kloosterman zeros are necessarily contained in a maximal totally isotropic vector space of $Q|_H$. However, these vector spaces are generally not unique.

Using Proposition 4, we get the following result.

Theorem 5. *Let W be a subspace of $\mathbb{F}_{2^n}$ such that $K_n(v) = 0$ for all $v \in W$ and $n \geq 5$. Then $\dim W \leq d$ where*

$$
d = \begin{cases}
\frac{n-2}{2}, & n \equiv 0, 2, 4, 6 \pmod 8 \\
\frac{n-1}{2}, & n \equiv 1, 7 \pmod 8 \\
\frac{n-3}{2}, & n \equiv 3, 5 \pmod 8.
\end{cases}
$$

Proof. The bound follows from Proposition 4 for all cases except $n \equiv 4 \pmod 8$. In the latter case the bound of Proposition 4 can be improved by one using the following observation for even n. [1] Let $n = 2k$ be even. As noted in [30], there are no non-zero Kloosterman zeros in the subfield $\mathbb{F}_{2^k}$. We have $\mathbb{F}_{2^k} \subset H$, $W \subset H$ and $W \cap \mathbb{F}_{2^k} = \{0\}$, implying $\dim(V) \leq \frac{n-2}{2}$. □

We would like to mention that the following approach yields a slightly weaker bound than the one given in Theorem 5. The following identity for sums of

---

[1] This is due to an anonymous referee.

Kloosterman sums over a vector space was given in [11, Proposition 3]: For any subspace V of $\mathbb{F}_{2^n}$ with $\dim(V) = k$ we have

$$\sum_{a \in V}(K_n^2(a) - K_n(a)) = 2^{n+k} - 2^{n+1} + 2^k \sum_{u \in V^\perp} K_n(u^{-1}).$$

If V contains exclusively Kloosterman zeros, we get

$$0 = 2^{n+k} - 2^{n+1} + 2^k \sum_{u \in V^\perp} K_n(u^{-1}),$$

recall we set $0^{-1} = 0$. Bounding the Kloosterman sum in the right hand side of the equation using the Weil bound $|K_n(a)| \leq 2^{\frac{n}{2}+1}$, we get

$$0 \geq 2^{n+k} - 2^{n+1} - 2^k 2^{n-k} 2^{\frac{n}{2}+1} = 2^{n+k} - 2^{n+1} - 2^{\frac{3n}{2}+1}.$$

This shows that $k = \dim(V) \leq \frac{n}{2} + 1$ for $n \geq 3$.

Remark 3. Theorem 5 provides to our knowledge the first general upper bound on the maximal size of subspaces of Kloosterman zeros. However, experimental results indicate that our bound is weak, see Table 1. Our bound is sharp for very small n (see right table in Table 1), which is not surprising since the approximation modulo 16 is strong for small n. Numerics in Table 1 were computed using [21, 24] for the left table and [2] for the right table. The left table shows that the total number of Kloosterman zeros in the field $\mathbb{F}_{2^n}$ is close to $2^{n/2}$ for $n \leq 60$. It is of course not to expect that the set of Kloosterman zeros has a strong additive structure, so we believe that the bound of Theorem 5 can be significantly improved.

Problem 1. Find a better bound on the maximal size of a subspace containing exclusively Kloosterman zeros.

4   Permutations of the Form $L_1(x^{-1}) + L_2(x)$

We now apply the results from the previous section. The following lemma is well-known. We include a simple proof of it for the convenience of the reader.

Lemma 2. *Let* $L\colon \mathbb{F}_{2^n} \to \mathbb{F}_{2^n}$ *be linear and* L^* *be its adjoint mapping. Then* $\dim(\mathrm{im}(L^*)) = \dim(\mathrm{im}(L))$ *and* $\dim(\ker(L^*)) = \dim(\ker(L))$.

Proof. Let $v \in \mathrm{im}(L^*)$ and $w \in \ker(L)$. We can write $v = L^*(x)$ for some $x \in \mathbb{F}_{2^n}$. Then $\langle v, w \rangle = \langle L^*(x), w \rangle = \langle x, L(w) \rangle = \langle x, 0 \rangle = 0$, so $\mathrm{im}(L^*) \subseteq \ker(L)^\perp$, in particular $\dim(\mathrm{im}(L^*)) \leq \dim(\mathrm{im}(L))$. The other inequality holds with $L^{**} = L$.
    The statement on the kernel follows from $\dim(\mathrm{im}(L)) + \dim(\ker(L)) = n$. $\square$

Table 1. Left Table: Comparison of the number of Kloosterman zeros over $\mathbb{F}_{2^n}$ to the value $2^{n/2}$. Here, $\mathcal{Z}(n)$ denotes the number of Kloosterman zeros over $\mathbb{F}_{2^n}$. Right table: the maximal dimension of a subspace W of $\mathbb{F}_{2^n}$ that contains exclusively Kloosterman zeros.

n	$2^{\frac{-n}{2}} \mathcal{Z}(n)$
5	0.88
10	1.87
15	1.57
20	0.86
25	0.67
30	1.29
35	1.15
40	1.15
45	1.14
50	0.91
55	1.32
60	1.25

n	$\dim(V)$
5	1
6	2
7	3
8	1
9	1
10	2
11	2
12	2
13	1
14	3
15	4
16	2

Corollary 1 shows that a function $L_1(x^{-1}) + L_2(x)$ cannot be bijective on $\mathbb{F}_{2^n}$ if at least one of L_1 or L_2 is bijective, equivalently has a trivial kernel. The next result shows that such a function is not bijective also in the case when the kernel of L_1 or L_2 is large.

Theorem 6. *Let $n \geq 5$ and $F(x) = L_1(x^{-1}) + L_2(x)$ where L_1 and L_2 are non-bijective non-zero linear functions of $\mathbb{F}_{2^n}$. Further, let d be defined as in Theorem 5. If $\max(\dim(\ker(L_1)), \dim(\ker(L_2))) > d$, then F does not permute $\mathbb{F}_{2^n}$.*

Proof. Observe that $F(x)$ is a permutation if and only if $F(x^{-1}) = L_1(x) + L_2(x^{-1})$ is so. Hence we may assume without loss of generality that $\dim(\ker(L_1)) \geq \dim(\ker(L_2)) \geq 1$. Suppose F is a permutation. Then by Corollary 2 we have $\ker(L_1^*) \cap \ker(L_2^*) = \{0\}$ and $K_n(L_1^*(b)L_2^*(b)) = 0$ for all $b \in \mathbb{F}_{2^n}$. Set $e = \dim \ker L_1 = \dim \ker L_1^*$. Choose $0 \neq c \in \ker(L_2^*)$. The set

$$V = L_1^*(c + \ker(L_1^*)) \cdot L_2^*(c + \ker(L_1^*)) = L_1^*(c) \cdot L_2^*(\ker(L_1^*))$$

is a vector space that is contained in the image set of $L_1^*(b)L_2^*(b)$. In particular $K_n(v) = 0$ for all $v \in V$. Since $\ker(L_1^*) \cap \ker(L_2^*) = \{0\}$ we have $\dim(V) = e$. Theorem 5 then implies that $e \leq d$. □

We conjecture that the following statements hold: [2]

Conjecture 1. Let $F = L_1(x^{-1}) + L_2(x)$ where $L_1 \neq 0$ and $L_2 \neq 0$ are linearized polynomials over $\mathbb{F}_{2^n}$ with $n \geq 5$. Then F does not permute $\mathbb{F}_{2^n}$.

---

[2] After the acceptance of this submission, Lukas Kölsch found a proof for Conjecture 1.

With Proposition 1, Conjecture 1 implies the following (recall that the inverse mapping is an involution):

Conjecture 2. Let $n \geq 5$. Every function $F \colon \mathbb{F}_{2^n} \to \mathbb{F}_{2^n}$ that is CCZ equivalent to the inverse function is already EA equivalent to it. Moreover, if F is additionally a permutation then F is affine equivalent to the inverse function.

Acknowledgements. We would like to thank the anonymous referees for their careful reading of our paper and their comments, which helped us to improve its presentation. We especially thank a referee who suggested an improvement in Theorem 5 for the case n even and provided background information on Kloosterman sums that helped us to improve the tutorial value of our paper. Remark 3 is based on comments from her/his report. We thank Petr Lisonek for interesting discussions on Kloosterman zeroes and sending us the reference [21], which we used to compute Table 1. This work was supported by the GAČR Grant 18-19087S -301-13/201843

References

1. Biham, E., Shamir, A.: Differential cryptanalysis of DES-like cryptosystems. J. Cryptol. **4**(1), 3–72 (1991). https://doi.org/10.1007/BF00630563
2. Bonnetain, X., Perrin, L., Tian, S. :Anomalies and vector space search: Tools for S-box analysis (full version). Cryptology ePrint Archive, Report 2019/528 (2019). https://eprint.iacr.org/2019/528
3. Boura, C., Perrin, L., Tian, S.: Boomerang uniformity of popular S-box constructions. In: WCC 2019 - The Eleventh International Workshop on Coding and Cryptography, Saint-Jacut-de-la-Mer, France, March 2019
4. Browning, K.A., Dillon, J.F., McQuistan, M.T., Wolfe, A.J.: An APN permutation in dimension six. In: McGuire, G., Mullen, G.L., Panario, D., Shparlinski, I.E. (eds.) Finite Fields: Theory Applications, vol. 518. Comtemporary Mathematics, pp. 33–42 (2010)
5. Budaghyan, L., Calderini, M., Villa, I.: On relations between CCZ- and EA-equivalences. Crypt.Commun. **12**, 95–100 (2020). https://doi.org/10.1007/s12095-019-00367-5
6. Budaghyan, L., Carlet, C., Pott, A.: New classes of almost bent and almost perfect nonlinear polynomials. IEEE Trans. Inf. Theory **52**(3), 1141–1152 (2006)
7. Canteaut, A., Perrin, L.: On CCZ-equivalence, extended-affine equivalence, and function twisting. Finite Fields Appl. **56**, 209–246 (2019)
8. Carlet, C., Charpin, P., Zinoviev, V.: Codes, bent functions and permutations suitable for des-like cryptosystems. Des. Codes Crypt. **15**(2), 125–156 (1998). https://doi.org/10.1023/A:1008344232130
9. Cattell, K., Miers, C.R., Ruskey, F., Sawada, J., Serra, M.: The number of irreducible polynomials over GF(2) with given trace and subtrace. J. Comb. Math. Comb. Comput. **47**, 31–64 (2003)
10. Charpin, P., Gong, G.: Hyperbent functions, Kloosterman sums and Dickson polynomials. In: 2008 IEEE International Symposium on Information Theory, pp. 1758–1762, July 2008
11. Charpin, P., Helleseth, T., Zinoviev, V.: Propagation characteristics of $x \mapsto x^{-1}$ and Kloosterman sums. Finite Fields Appl. **13**(2), 366–381 (2007)

12. Charpin, P., Pasalic, E.: Some results concerning cryptographically significant mappings over GF(2^n). Des. Codes Crypt. **57**, 257–269 (2010)
13. Dillon, J.: Elementary Hadamard Difference Sets. Ph.D. thesis, University of Maryland (1974)
14. Edel, Y., Pott, A.: On the equivalence of nonlinear functions. In: Preneel, B., Dodunekov, S., Rijmen, V., Nikova, S., (eds.) Enhancing Cryptographic Primitives with Techniques from Error Correcting Codes. Nato Science for Peace and Security, vol. 23, pp. 87–103. IOS Press (2009)
15. Fitzgerald, R.W., Yucas, J.L.: Irreducible polynomials over GF(2) with three prescribed coefficients. Finite Fields Appl. **9**(3), 286–299 (2003)
16. Gerike, D., Kyureghyan, G.: Results on permutation polynomials of shape $x^t + \gamma tr(x^d)$. In: Schmidt, K.-U., Winterhof, A., (eds.) Combinatorics and Finite Fields. Radon Series on Computational and Applied Mathematics, vol. 23, pp. 67–78. De Gruyter, Berlin (2019)
17. Göloğlu, F., McGuire, G.: On theorems of Carlitz and Payne on permutation polynomials over finite fields with an application to $x^{-1} + l(x)$. Finite Fields Appl. **27**, 130–142 (2014)
18. Göloğlu, F., McGuire, G., Moloney, R.: Binary Kloosterman sums using Stickelberger's theorem and the Gross-Koblitz formula. Acta Arith. **148**(3), 269–279 (2011)
19. Hollmann, H.D., Xiang, Q.: Kloosterman sum identities over $\mathbb{F}_{2^m}$. Discrete Math. **279**, 277–286 (2004)
20. Hou, X.-D.: Lectures on finite fields Graduate Studies in Mathematics. Graduate Studies in Mathematics, vol. 190. American Mathematical Society, Providence (2018)
21. Kim, Y.-J.: Algorithms for Kloosterman zeroes. Master Thesis at Simon Fraser University. Supervised by Petr Lisonek (2011)
22. Kononen, K.P., Rinta-aho, M.J., Väänänen, K.O.: On integer values of Kloosterman sums. IEEE Trans. Inf. Theory **56**(8), 4011–4013 (2010)
23. Lachaud, G., Wolfmann, J.: Sommes de Kloosterman, courbes elliptiques et codes cycliques en caractéristique 2. CR Acad. Sci. Paris (I) **305**, 881–883 (1987)
24. Lachaud, G., Wolfmann, J.: The weights of the orthogonals of the extended quadratic binary Goppa codes. IEEE Trans. Inf. Theory **36**(3), 686–692 (1990)
25. Lam, T.: Introduction to Quadratic Forms over Fields. Graduate Studies in Mathematics, vol. 67. American Mathematical Society (2005)
26. Lapierre, L., Lisonek, P.: On vectorial bent functions with Dillon-type exponents. In 2016 IEEE International Symposium on Information Theory (ISIT), pp. 490–494 (2016)
27. Li, Y., Wang, M.: On EA-equivalence of certain permutations to power mappings. Des. Codes Crypt. **58**, 1259–269 (2011). https://doi.org/10.1007/s10623-010-9406-8
28. Li, Y., Wang, M.: Permutation polynomials EA-equivalent to the inverse function over GF(2^n). Crypt. Commun. **3**, 175–186 (2011). https://doi.org/10.1007/s12095-011-0045-3
29. Lidl, R., Niederreiter, H.: Finite Fields, vol. 20, 2nd. Encyclopedia of Mathematics and Its Applications. Cambridge University Press (1997)
30. Lisonek, P., Moisio, M.: On zeros of Kloosterman sums. Des. Codes Crypt. **59**(1), 223–230 (2011). https://doi.org/10.1007/s10623-010-9457-x
31. Matsui, M.: Linear cryptanalysis method for DES cipher. In: Helleseth, T. (ed.) EUROCRYPT 1993. LNCS, vol. 765, pp. 386–397. Springer, Heidelberg (1994). https://doi.org/10.1007/3-540-48285-7_33

32. Shparlinski, I.E.: On the values of Kloosterman sums. IEEE Trans. Inf. Theory **55**(6), 2599–2601 (2009)
33. Yucas, J.L., Mullen, G.L.: Irreducible polynomials over GF(2) with prescribed coefficients. Discrete Math. **274**(1), 265–279 (2004)
34. Zinoviev, V.: On classical Kloosterman sums. Crypt. Commun. **11**, 461–496 (2019). https://doi.org/10.1007/s12095-019-00357-7

Explicit Factorization of Some Period Polynomials

Gerardo Vega$^{(\boxtimes)}$ [iD]

Dirección General de Cómputo y de Tecnologías de Información y Comunicación, Universidad Nacional Autónoma de México, 04510 Ciudad de México, Mexico
gerardov@unam.mx

Abstract. Let p, t, q, n, m and r be positive integers, such that p is a prime number, $q = p^t$, $\gcd(q, n) = 1$, $m = \mathrm{ord}_n(q)$, and suppose that the prime factors of r divide n but not $(q^m - 1)/n$, and that $q^m \equiv 1 \pmod 4$, if $4|r$. Also let u such that $u = \gcd(\frac{q^m-1}{q-1}, \frac{q^m-1}{n})$. Assume that $u = 1$ or p is semiprimitive modulo u. Under these conditions, we are going to obtain the explicit factorization of the period polynomial of degree $\gcd(\frac{q^{mr}-1}{q-1}, \frac{q^{mr}-1}{nr})$ for the finite field $\mathbb{F}_{q^{mr}}$. In fact, we will see that such polynomial has always integer roots, meaning that the corresponding Gaussian periods are also integer numbers. As an application, we also determine the number of solutions of certain diagonal equations with constant exponent.

Keywords: Period polynomials · Gaussian periods · Irreducible cyclic codes · Weight distribution

1 Introduction

Let $\mathbb{F}_q$ be a finite field of characteristic p with $q = p^t$, and let γ be a primitive element of $\mathbb{F}_q$. Let e and f be positive integers such that $q - 1 = ef$. For $i = 0, 1, \cdots, e - 1$, define $\mathcal{D}_i^{(e,q)} := \gamma^i \langle \gamma^e \rangle$, where $\langle \gamma^e \rangle$ denotes the subgroup of $\mathbb{F}_q^*$ generated by γ^e. The coset $\mathcal{D}_i^{(e,q)}$ is called the i-th *cyclotomic class* of order e in $\mathbb{F}_q$. Let χ be the canonical additive character of $\mathbb{F}_q$ (see for example [10, Chap. 5]).

For $i = 0, 1, \cdots, e - 1$, the i-th *Gaussian period*, $\eta_i^{(e,q)}$, of order e for $\mathbb{F}_q$ is defined to be

$$\eta_i^{(e,q)} := \sum_{z \in \mathcal{D}_i^{(e,q)}} \chi(z) \,.$$

The *period polynomial*, $\psi_{(e,q)}(X)$, of degree e for $\mathbb{F}_q$ is given by

$$\psi_{(e,q)}(X) := \prod_{i=0}^{e-1} (X - \eta_i^{(e,q)}) \,,$$

Partially supported by PAPIIT-UNAM IN109818.

J. C. Bajard and A. Topuzoğlu (Eds.): WAIFI 2020, LNCS 12542, pp. 222–233, 2021.
https://doi.org/10.1007/978-3-030-68869-1_13

while the *reduced period polynomial*, $\psi^*_{(e,q)}(X)$, of degree e for $\mathbb{F}_q$ is

$$\psi^*_{(e,q)}(X) := \prod_{i=0}^{e-1}(X - \eta_i^{*(e,q)}) \,,$$

where

$$\eta_i^{*(e,q)} := \sum_{z\in\mathbb{F}_q} \chi(\gamma^i z^e) = 1 + e\eta_i^{(e,q)} \,,$$

is the i-th *reduced Gaussian period* of order e for $\mathbb{F}_q$.

The polynomial $\psi_{(e,q)}(X)$ has integer coefficients and is independent of the choice of the primitive element γ (see [11, Theorem 3]). To determine the coefficients of the period polynomial $\psi_{(e,q)}(X)$ (or equivalently, the reduced period polynomial $\psi^*_{(e,q)}(X) = e^e\psi_{(e,q)}((X-1)/e)$) is a classical problem dating back to Gauss (see [2]). It is well known (see [11, Theorem 4]) that the period polynomial $\psi_{(e,q)}(X)$ splits over $\mathbb{Q}$ into $\delta = \gcd(e, (q-1)/(p-1))$ factors of degree e/δ (not necessarily distinct), and each of these factors are irreducible or a power of an irreducible polynomial. The explicit factorization of $\psi_{(e,q)}(X)$, if reducible, is in general very hard to determine, and it has been done only in certain special cases (see for example [1,7–9,11]).

Let p, t, q, n, m and r be positive integers, such that p is a prime number, $q = p^t$, $\gcd(q,n) = 1$, $m = \mathrm{ord}_n(q)$, and suppose that the prime factors of r divide n but not $(q^m - 1)/n$, and that $q^m \equiv 1 \pmod 4$, if $4|r$. Also let u such that $u = \gcd(\frac{q^m-1}{q-1}, \frac{q^m-1}{n})$. Assume that $u = 1$ or p is semiprimitive modulo u (see definition below). Fix $e = \gcd(\frac{q^{mr}-1}{q-1}, \frac{q^{mr}-1}{nr})$. The aim of this work is to obtain the Gaussian periods, $\eta_i^{(e,q^{mr})}$, of order e for $\mathbb{F}_{q^{mr}}$. That is, for $\mathbb{F}_{q^{mr}}$, we are going to give the explicit factorization of the period polynomials of the form $\psi_{(e,q^{mr})}(X)$. In fact, since $\frac{q^{mr}-1}{q-1} | \frac{q^{mr}-1}{p-1}$, $\delta = \gcd(e, (q^{mr} - 1)/(p - 1)) = e$, therefore $\psi_{(e,q^{mr})}(X)$ (or the reduced period polynomial $\psi^*_{(e,q^{mr})}(X)$) will split over $\mathbb{Q}$ into e factors of degree 1. In fact, we will see that such linear factors have always integer roots, meaning that the corresponding Gaussian periods $\eta_i^{(e,q^{mr})}$ are all integer numbers.

The explicit factorization of the period polynomials has several applications and one of them is to determine the weight distributions for some irreducible cyclic codes (see for example [1,5,6,15]). Now, the traditional approach to obtain the explicit factorization of one of these polynomials is to express its corresponding Gaussian periods in terms of Gauss sums and to apply known results about these sums. Instead, to achieve our goal, we are going to proceed in the opposite direction. In other words, in some cases it may be possible to determine the weight distributions of some irreducible cyclic codes without needing to obtain the Gaussian periods (see for example [13,14]). Particularly, an infinite family of irreducible cyclic codes whose weight distributions are explicitly determined through the well-known weight distributions of either a one-weight, or a semiprimitive two-weight irreducible cyclic code, was recently presented in [14].

Thus, our strategy for this work, will be to use these already known weight distributions in this infinite family, in order to determine the explicit factorization for the corresponding period polynomials. As an application of this explicit factorization, we also determine the number of solutions of certain diagonal equations with constant exponent.

This work is organized as follows: Sect. 2 consists of background material and already known results needed later, in particular, it is recalled the result that determines the weight distributions for an infinite family of irreducible cyclic codes. In Sect. 3, some preliminary results are presented. Among them an identity that gives us the values of Gaussian periods in terms of the Hamming weight of the codewords in an irreducible cyclic code. Results in Sects. 2 and 3, are used in Sect. 4 in order to present an explicit factorization of the period polynomials of the form $\psi_{(e,q^{mr})}(X)$. Some examples of such factorization are also presented in Sect. 4. As an application of our results, we determine the number of solutions of certain diagonal equations with constant exponent in Sect. 5. Finally, Sect. 6 is devoted to conclusions.

2   Background Material and Already Known Results

Let v and w be integers, such that $\gcd(v, w) = 1$. Then, the smallest positive integer i, such that $w^i \equiv 1 \pmod{v}$, is called the *multiplicative order* of w modulo v, and is denoted by $\mathrm{ord}_v(w)$. In addition, we are going to say that w is *semiprimitive* modulo v, if there exists a positive integer j, such that $w^j \equiv -1 \pmod{v}$.

By identifying the vector $(c_0, c_1, \ldots, c_{n-1}) \in \mathbb{F}_q^n$ with the polynomial $c_0 + c_1 x + \ldots + c_{n-1} x^{n-1} \in \mathbb{F}_q[x]$, it follows that any linear code $\mathcal{C}$ of length n over $\mathbb{F}_q$ corresponds to a subset of the residue class ring $\mathbb{F}_q[x]/\langle x^n - 1 \rangle$. Moreover, it is well known that the linear code $\mathcal{C}$ is *cyclic* if and only if the corresponding subset is an ideal of $\mathbb{F}_q[x]/\langle x^n - 1 \rangle$ (see for example [10, Theorem 9.36]).

Now, note that every ideal of $\mathbb{F}_q[x]/\langle x^n - 1 \rangle$ is principal. In consequence, if $\mathcal{C}$ is a cyclic code of length n over $\mathbb{F}_q$, then $\mathcal{C} = \langle g(x) \rangle$, where $g(x)$ is a monic polynomial, such that $g(x) \mid (x^n - 1)$. This polynomial is unique, and it is called the *generator polynomial* of $\mathcal{C}$. On the other hand, the polynomial $h(x) = (x^n - 1)/g(x)$ is referred to as the *parity check polynomial* of $\mathcal{C}$.

A cyclic code over $\mathbb{F}_q$ is called *irreducible* (*reducible*) if its parity check polynomial is irreducible (reducible) over $\mathbb{F}_q$.

A precise description of an irreducible cyclic code over $\mathbb{F}_q$ is achieved by means of the following:

Definition 1 [12, Definition 2.2]. *Let n be a positive divisor of $q^m - 1$, write $D = (q^m - 1)/n$, and let ω be a primitive n-th root of unity in $\mathbb{F}_{q^m}$. For each $\beta \in \mathbb{F}_{q^m}$ define $c(q^m, D, \beta)$ as the vector of length n over $\mathbb{F}_q$, given by*

$$c(q^m, D, \beta) = (\mathrm{Tr}_{\mathbb{F}_{q^m}/\mathbb{F}_q}(\beta \omega^j))_{j=0}^{n-1}.$$

where "$\mathrm{Tr}_{\mathbb{F}_{q^m}/\mathbb{F}_q}$" denotes the trace mapping from $\mathbb{F}_{q^m}$ to $\mathbb{F}_q$. Then, an irreducible cyclic code $\mathcal{C}_{(D)}$ of length n over $\mathbb{F}_q$, is the set

$$\mathcal{C}_{(D)} = \{c(q^m, D, \beta) \mid \beta \in \mathbb{F}_{q^m}\}.$$

Remark 1. The dimension of $\mathcal{C}_{(D)}$ is $\mathrm{ord}_n(q)$, which is a divisor of m. Also note that, if γ is a primitive element of $\mathbb{F}_{q^m}$, then, thanks to Delsarte's Theorem (see for example [3]), the parity-check polynomial of the irreducible cyclic code under the previous definition is the *minimal polynomial* of γ^{-D} (see [10, Definition 1.81]), if $\omega = \gamma^D$.

According to [14, Definition 3 and Theorem 4], an irreducible cyclic code $\mathcal{C}$ of length n and dimension m over $\mathbb{F}_q$ is called *semiprimitive* if $u \geq 2$ and p is semiprimitive modulo u, where $u = \gcd(\frac{q^m-1}{q-1}, \frac{q^m-1}{n})$ (recall $q = p^t$).

To determine the Gaussian periods in terms of the weight distribution of an irreducible cyclic code, we need to recall the following:

Lemma 1 [6, Lemma 5]. *Let D be a positive divisor of $q^m - 1$. Fix $e = \gcd(\frac{q^m-1}{q-1}, D)$, and let i be any integer with $0 \leq i < D$. We have the following multiset equality:*

$$\{xy \mid x \in \mathcal{D}_i^{(D,q^m)}, y \in \mathbb{F}_q^*\} = \frac{e(q-1)}{D} * \mathcal{D}_i^{(e,q^m)},$$

where $\frac{e(q-1)}{D} * \mathcal{D}_i^{(e,q^m)}$ denotes the multiset in which each element in the set $\mathcal{D}_i^{(e,q^m)}$ appears in the multiset with multiplicity $\frac{e(q-1)}{D}$.

The following result determines the weight distribution of an infinite family of irreducible cyclic codes, that includes as a particular instance $(r = 1)$ all the one-weight and semiprimitive two-weight irreducible cyclic codes.

Theorem 1 [14, Theorems 4 and 10]. *Let n, m and r be three positive integers, such that $\gcd(n, q) = 1$, $m = \mathrm{ord}_n(q)$, and $r \geq 1$. If $r \geq 2$, suppose that the prime factors of r divide n but not $(q^m - 1)/n$, and that $q^m \equiv 1 \pmod 4$, if $4 | r$. Fix $d = \frac{q^m-1}{n}$, and $u = \gcd(\frac{q^m-1}{q-1}, d)$. Assume also that $u = 1$ or p is semiprimitive modulo u. Fix $D = \frac{q^{mr}-1}{nr}$. Then, any irreducible cyclic code of the form $\mathcal{C}_{(D)}$ is an $[nr, mr]$ code over $\mathbb{F}_q$, whose weight enumerator polynomial is*

$$\left(1 + \frac{(q^m-1)}{u}[(u-1)z^{\frac{(q-1)q^{m/2}}{dq}(q^{m/2}-(-1)^s)} + z^{\frac{(q-1)q^{m/2}}{dq}(q^{m/2}+(-1)^s(u-1))}]\right)^r,$$

where $s = (mt)/\mathrm{ord}_u(p)$.

Note that if $r = 1$, then all the weight distributions for the one-weight and semiprimitive two-weight irreducible cyclic codes, are determined through the previous theorem (see [14, Theorem 4]). On the other hand, if $r = 2$ and $u = 1$, then the weight distribution of the corresponding irreducible cyclic code in Theorem 1, is the weight distribution of a semiprimitive two-weight irreducible cyclic code, which can also be described through the same theorem when $u > 1$ and $r = 1$. However, if $r \geq 2$ and $u > 1$, or if $r \geq 3$ and $u \geq 1$, then the weight distribution of the corresponding irreducible cyclic code in Theorem 1, cannot be the weight distribution of either a one-weight, or a semiprimitive two-weight irreducible cyclic code.

3   Some Preliminary Results

It is important to note that the weight distribution in Theorem 1, is not given explicitly. Thus, the purpose of the following result is to get rid of this inconvenience.

Lemma 2. *Let u, A_1, A_2, w_1, w_2, and r be non-negative integers such that $u \geq 1$, $u|(q-1)$, $A_1 = \frac{(q^m-1)}{u}(u-1)$, $A_2 = \frac{(q^m-1)}{u}$, and $r \geq 1$. Then the polynomial $(1 + A_1 z^{w_1} + A_2 z^{w_2})^r$ is equal to*

$$
\begin{cases}
1 + \displaystyle\sum_{j=1}^{r} \binom{r}{j} A_2^j z^{jw_2} & \text{if } u = 1, \\[2ex]
1 + \displaystyle\sum_{\substack{0 \leq k \leq j \leq r \\ (j,k) \neq (0,0)}} \binom{r}{j}\binom{j}{k} A_1^{j-k} A_2^k z^{jw_1 + k(w_2 - w_1)} & \text{otherwise.}
\end{cases} \tag{1}
$$

Proof. It is a consequence of the binomial serie: $(a+b)^r = \displaystyle\sum_{j=0}^{r} \binom{r}{j} a^{r-j} b^j$.   □

Table 1. Weight distribution of $\mathcal{C}_{(D)}$, when $u = 1$. The integer j runs from 1 to r.

Weight	Frequency
0	1
$\frac{(q-1)q^m j}{dq}$	$\binom{r}{j}(q^m - 1)^j$

An explicit description of the weight distribution of an irreducible cyclic code in Theorem 1 is as follows.

Corollary 1. *Assume the same notation and conditions as in Theorem 1 and Lemma 2. Then the weight distribution of any irreducible cyclic code of the form $\mathcal{C}_{(D)}$ is given by Tables 1 and 2.*

Proof. If $w_1 = \frac{(q-1)q^{m/2}}{dq}(q^{m/2}-(-1)^s)$, and $w_2 = \frac{(q-1)q^{m/2}}{dq}(q^{m/2}+(-1)^s(u-1))$, then the result follows directly from Theorem 1 and Lemma 2.   □

Example 1. With the notation of Theorem 1, let $q = p = 5$, $n = 3$, and $r = 3$. Thus, $m = 2$, $d = 8$, $u = 2 > 1$, and $s = 1$. Clearly r divides n but not $(q^m-1)/n$, and p is semiprimitive modulo u. Fix $D = \frac{q^{mr}-1}{nr} = 1736$. Then, Table 2 tells us that the explicit weight enumerator polynomial of $\mathcal{C}_{(1736)}$ is $1 + 36z^2 + 36z^3 + 432z^4 + 864z^5 + 2160z^6 + 5184z^7 + 5184z^8 + 1728z^9$.

Table 2. Weight distribution of $\mathcal{C}_{(D)}$, when $u > 1$. Here $s = (mt)/\mathrm{ord}_u(p)$, and the integers j and k are such that $0 \le k \le j \le r$, and $(j,k) \ne (0,0)$.

Weight	Frequency
0	1
$\frac{(q-1)q^{m/2}}{dq}(jq^{m/2} + (-1)^s(ku - j))$	$\binom{r}{j}\binom{j}{k}\frac{(q^m-1)^j(u-1)^{j-k}}{u^j}$

As already stated, the aim of this work is to use irreducible cyclic codes, whose weight distributions are already known, in order to determine the explicit factorization for the corresponding period polynomials. To achieve this objective, we need to find a way to express the value of a Gaussian period in terms of the Hamming weight of a codeword in an irreducible cyclic code. Such way is described below.

Lemma 3. *Let n be a positive divisor of $q^m - 1$, write $D = (q^m - 1)/n$, and let $\mathcal{C}_{(D)}$ be the associated irreducible cyclic code under the Definition 1. Fix $e = \gcd(\frac{q^m-1}{q-1}, D)$, and for $i = 0, 1, \cdots, e-1$, let $\eta_i^{(e,q^m)}$ and $\mathcal{D}_i^{(e,q^m)}$ be, respectively, the i-th Gaussian period of order e for $\mathbb{F}_{q^m}$, and the i-th cyclotomic class of order e in $\mathbb{F}_{q^m}$. Let $c(q^m, D, \beta) \in \mathcal{C}_{(D)}$, and suppose that $\beta \in \mathcal{D}_i^{(e,q^m)}$. Then $\eta_i^{(e,q^m)}$ is an integer number, whose value is given by*

$$\eta_i^{(e,q^m)} = \frac{D}{e}\left(n - \frac{qw_H(c(q^m, D, \beta))}{q - 1}\right), \tag{2}$$

where "w_H" stands for the Hamming weight of a codeword.

Proof. Let ω be a primitive n-th root of unity in $\mathbb{F}_{q^m}$, then the Hamming weight of the codeword $c(q^m, D, \beta) \in \mathcal{C}_{(D)}$ is equal to $n - Z(\beta)$, where

$$Z(\beta) = \#\{\, j \mid 0 \le j < n, \mathrm{Tr}_{\mathbb{F}_{q^m}/\mathbb{F}_q}(\beta\omega^j) = 0\}.$$

If χ' and χ are, respectively, the canonical additive characters of $\mathbb{F}_q$ and $\mathbb{F}_{q^m}$ (see for example [10, Chap. 5]), then χ' and χ are related by $\chi'(\mathrm{Tr}_{\mathbb{F}_{q^m}/\mathbb{F}_q}(\varepsilon)) = \chi(\varepsilon)$ for all $\varepsilon \in \mathbb{F}_{q^m}$. Therefore,

$$Z(\beta) = \frac{1}{q}\sum_{j=0}^{n-1}\sum_{y\in\mathbb{F}_q}\chi'(\mathrm{Tr}_{\mathbb{F}_{q^m}/\mathbb{F}_q}(y(\beta\omega^j)))$$

$$= \frac{n}{q} + \frac{1}{q}\sum_{j=0}^{n-1}\sum_{y\in\mathbb{F}_q^*}\chi(y\beta\omega^j)$$

$$= \frac{n}{q} + \frac{1}{q}\sum_{x\in\mathcal{D}_0^{(D,q^m)}}\sum_{y\in\mathbb{F}_q^*}\chi(y\beta x).$$

because $\mathcal{D}_0^{(D,q^m)} := \langle \omega \rangle$. But, owing to Lemma 1 we have,

$$Z(\beta) = \frac{n}{q} + \frac{e(q-1)}{Dq} \sum_{z \in \mathcal{D}_0^{(e,q^m)}} \chi(\beta z).$$

Now, since $\beta \in \mathcal{D}_i^{(e,q^m)}$,

$$w_H(c(q^m, D, \beta)) = n - Z(\beta) = n - \frac{n}{q} - \frac{e(q-1)}{Dq} \eta_i^{(e,q^m)},$$

which implies (2). Finally, since $e = \gcd(\frac{q^m - 1}{q-1}, D)$, we have, owing to [6, Theorem 13], that $\eta_i^{(e,q^m)} \in \mathbb{Z}$.   □

To determine the number of solutions of certain diagonal equations with constant exponent, we need the following:

Lemma 4. *With our current notation, let e and v be positive integers such that $v \geq 2$, and $e | (q-1)$. Let $b \in \mathbb{F}_q$, and denote by N the number of solutions to the equation $x_1^e + \cdots + x_v^e = b$ in $\mathbb{F}_q^v$. Suppose that $-b \in \mathcal{D}_0^{(e,q)}$. Then*

$$N = q^{v-1} + \frac{1}{q} \sum_{i=1}^{e-1} (\eta_i^{*(e,q)})^v \eta_i^{(e,q)}.$$

Proof. From [16, Proposition 1], we have

$$N = q^{-1} \sum_{a \in \mathbb{F}_q} \chi(-ab) \prod_{j=0}^{v} \sum_{z \in \mathbb{F}_q} \chi(az^e).$$

If γ is a primitive element of $\mathbb{F}_q$, then

$$N = q^{v-1} + \frac{1}{q} \sum_{i=0}^{e-1} \sum_{y \in \langle \gamma^e \rangle} \chi(-\gamma^i y b) \prod_{j=0}^{v} \sum_{z \in \mathbb{F}_q} \chi(\gamma^i y z^e)$$

$$= q^{v-1} + \frac{1}{q} \sum_{i=0}^{e-1} \left[\prod_{j=0}^{v} \sum_{z \in \mathbb{F}_q} \chi(\gamma^i z^e) \right] \sum_{y \in \langle \gamma^e \rangle} \chi(-\gamma^i y b)$$

$$= q^{v-1} + \frac{1}{q} \sum_{i=1}^{e-1} (\eta_i^{*(e,q)})^v \sum_{y \in \langle \gamma^e \rangle} \chi(-\gamma^i y b).$$

But since $-b \in \mathcal{D}_0^{(e,q)}$, $\sum_{y \in \langle \gamma^e \rangle} \chi(-\gamma^i y b) = \eta_i^{(e,q)}$. Thus, the result is proved.   □

4   Explicit Factorization of the Period Polynomials of the Form $\psi_{(e,q^{mr})}(X)$

Prior to seeking for the explicit factorization of a period polynomial of the form $\psi_{(e,q^{mr})}(X)$, we first need to find the Gaussian periods associated with such polynomial.

Table 3. The Gaussian periods of order e for $\mathbb{F}_{q^{mr}}$, when $u = 1$. The integer j runs from 1 to r.

Value	Multiplicity	
$\frac{q^{mr}-1}{e}\left(1 - \frac{q^{m}j}{(q^{m}-1)r}\right)$	$\binom{r}{j}$	$\frac{(q^{m}-1)^{j}e}{q^{mr}-1}$

Table 4. The Gaussian periods of order e for $\mathbb{F}_{q^{mr}}$, when $u > 1$. Here $s = (mt)/\mathrm{ord}_u(p)$, and the integers j and k are such that $0 \leq k \leq j \leq r$, and $(j,k) \neq (0,0)$.

Value	Multiplicity		
$\frac{q^{mr}-1}{e}\left(1 - \frac{q^{m/2}}{(q^{m}-1)r}[jq^{m/2} + (ku - j)(-1)^{s}]\right)$	$\binom{r}{j}$	$\binom{j}{k}$	$\frac{(q^{m}-1)^{j}(u-1)^{j-k}e}{(q^{mr}-1)u^{j}}$

Theorem 2. *Let n, m and r be three positive integers, such that $\gcd(n,q) = 1$, $m = \mathrm{ord}_n(q)$, and $r \geq 1$. If $r \geq 2$, suppose that the prime factors of r divide n but not $(q^m - 1)/n$, and that $q^m \equiv 1 \pmod 4$, if $4|r$. Fix $d = \frac{q^m-1}{n}$, $u = \gcd(\frac{q^m-1}{q-1}, d)$, and $e = \gcd(\frac{q^{mr}-1}{q-1}, \frac{q^{mr}-1}{nr})$. Assume also that $u = 1$ or p is semiprimitive modulo u. Then, the Gaussian periods of order e for $\mathbb{F}_{q^{mr}}$ are integer numbers, whose values and multiplicities are given by Tables 3 and 4.*

Proof. Let $D = \frac{q^{mr}-1}{nr}$, and $\mathcal{C}_{(D)}$ the associated irreducible cyclic code under Definition 1. Also let $\beta \in \mathbb{F}_{q^{mr}}$, and suppose that $\beta \in \mathcal{D}_i^{(e,q^{mr})}$, for some $i = 0, 1, \cdots, e-1$. Thus, owing to Lemma 3, the i-th Gaussian period, $\eta_i^{(e,q^{mr})}$, of order e for $\mathbb{F}_{q^{mr}}$ is an integer number, whose value is,

$$\eta_i^{(e,q^{mr})} = \frac{D}{e}\left(nr - \frac{q w_H(c(q^{mr}, D, \beta))}{q-1}\right).$$

Now, if F is the Frequency of $w_H(c(q^{mr}, D, \beta))$ in the weight distribution of $\mathcal{C}_{(D)}$, then Corollary 1 tells us that the pair $(w_H(c(q^{mr}, D, \beta)), F)$ should be listed either in Table 1 or Table 2. Finally since $|\mathcal{D}_i^{(e,q^{mr})}| = \frac{q^{mr}-1}{e}$, the result follows from the fact that the multiplicity of $\eta_i^{(e,q^{mr})}$ is $\frac{Fe}{q^{mr}-1}$. $\square$

Table 5. The Gaussian periods $\eta_i^{(7,64)}$ $(0 \leq i < 7)$.

Value	1	5	-3
Multiplicity	3	1	3

Example 2. With the notation of Theorem 2, let $q = 4$, $n = 3$, and $r = 3$. Thus, $m = 1$ and clearly r divides n but not $(q^m - 1)/n$. In addition, $d = 1$, $u = 1$, and $e = 7$. Then, through Table 3 we see that the values and their corresponding multiplicities of the Gaussian periods, $\eta_i^{(7,64)}$ $(0 \leq i < 7)$ are given by Table 5.

Remark 2. Note that our previous example coincides with Lemma 3 in [17].

We are now in a position to give an explicit factorization for the period polynomials of the form $\psi_{(e,q^{mr})}(X)$.

Corollary 2. *Assume the same notation and conditions as in Theorem 2. Fix* $h_1 = \frac{q^{mr}-1}{e}$, *and* $h_2 = \frac{q^{m/2}}{(q^m-1)r}$. *Then the explicit factorization for the period polynomial,* $\psi_{(e,q^{mr})}(X)$, *of degree e for* $\mathbb{F}_q^{mr}$ *is*

$$\prod_{j=1}^{r} (X - h_1(1 - h_2 q^{m/2}j))^{\binom{r}{j}\frac{(q^m-1)j}{h_1}},$$

if $u = 1$, and if $u > 1$ we have

$$\prod_{\substack{0 \leq k \leq j \leq r \\ (j,k) \neq (0,0)}} (X - h_1(1 - h_2[jq^{m/2} + (ku - j)(-1)^s]))^{\binom{r}{j}\binom{j}{k}\frac{(q^m-1)j(u-1)^{j-k}}{h_1 u^j}}.$$

Proof. It is a direct consequence of the definition of $\psi_{(e,q^{mr})}(X)$, and Theorem 2. □

Example 3. With the notation of Theorem 2, let $q = p = 5$, $n = 3$, $m = 2$, and $r = 3$ (see Example 1). Thus, $d = 8$, $u = 2 > 1$, $e = 434$, and $s = 1$. Clearly r divides n but not $(q^m - 1)/n$, and p is semiprimitive modulo u. Fix $h_1 = \frac{q^{mr}-1}{e} = 36$, and $h_2 = \frac{q^{m/2}}{(q^m-1)r} = \frac{5}{72}$. Then, through Corollary 2, we see that

$$\psi_{(434,5^6)}(X) = (X - 21)(X - 26)(X - 6)^{12}(X - 11)^{24}(X - 16)^{12}$$
$$(X + 9)^{48}(X + 4)^{144}(X - 1)^{144}(X - 6)^{48}.$$

Take $q = p$ (that is $t = 1$), $m \geq 1$, $r = 1$. Let n and e such that $n|(p^m - 1)$, and $e = \gcd(\frac{p^m-1}{p-1}, \frac{p^m-1}{n})$, and suppose that $e > 2$ and $e|(p^v + 1)$, with v chosen minimal. Thus, in accordance with Theorem 2, $u = e$. Therefore $s = (mt)/\mathrm{ord}_u(p) = m/(2v)$, and p is semiprimitive modulo u. Now, by using Corollary 2 we have:

$$\psi_{(e,p^m)}(X) = (X - h_1(1 - h_2[p^{m/2} - (-1)^s]))^{e-1}$$
$$(X - h_1(1 - h_2[p^{m/2} + (e-1)(-1)^s]))$$
$$= (X - \frac{1}{e}(p^{m/2}(-1)^s - 1))^{e-1}(X - \frac{1}{e}(p^{m/2}(e-1)(-1)^{s+1} - 1)).$$

But $\psi^*_{(e,p^m)}(X) = e^e \psi_{(e,p^m)}((X-1)/e))$, therefore

$$\psi^*_{(e,p^m)}(X) = (X + (-1)^{m/(2v)}(e-1)p^{m/2})(X - (-1)^{m/(2v)}p^{m/2})^{e-1}.$$

The previous reduced period polynomial is the same to that reported in [1, p. 320]. In this way, without considering the problem of locating the Gaussian periods (see [11, Sect. 7]), we end this section by noting that Theorem 2 extends the scope of Proposition 20 in [11].

5   Diagonal Equations with Constant Exponent

For integers $e \geq 1$ and $v \geq 2$, and for finite field elements $a_1, \cdots, a_v, b \in \mathbb{F}_q$, a *diagonal equation* with constant exponent e, over $\mathbb{F}_q$, is an equation of the form:

$$a_1 x_1^e + \cdots + a_v x_v^e = b.$$

Finding the number of solutions $(x_1, \cdots, x_v) \in \mathbb{F}_q^v$, in the general case, is a very difficult problem. The following result gives a solution to this problem for certain diagonal equations.

Theorem 3. *Assume the same notation and conditions as in Theorem 2. Also, let $v \geq 2$ be an integer, and $b \in \mathbb{F}_{q^{rm}}$. Denote by N the number of solutions to the diagonal equation $x_1^e + \cdots + x_v^e = b$ in $\mathbb{F}_{q^{rm}}^v$. Fix $h_1 = \frac{q^{mr}-1}{e}$, and $h_2 = \frac{q^{m/2}}{(q^m-1)r}$. For integers j and k, such that $0 \leq k \leq j \leq r$, define the functions:*

$$\mathfrak{N}_1(j) = h_1(1 - h_2 q^{m/2} j),$$
$$\mathfrak{F}_1(j) = \binom{r}{j} \frac{(q^m-1)^j}{h_1},$$
$$\mathfrak{N}_2(j,k) = h_1(1 - h_2[jq^{m/2} + (ku - j)(-1)^s]),$$
$$\mathfrak{F}_2(j,k) = \binom{r}{j}\binom{j}{k} \frac{(q^m-1)^j(u-1)^{j-k}}{h_1 u^j}.$$

Suppose that $b \in \mathcal{D}_0^{(e,q^{rm})}$. Then $N - q^{rm(v-1)}$ is equal to

$$
\begin{cases}
\dfrac{1}{q^{rm}} \displaystyle\sum_{j=1}^{r} \mathfrak{F}_1(j)(e\mathfrak{N}_1(j)+1)^v \mathfrak{N}_1(j) & \text{if } u = 1, \\[3mm]
\dfrac{1}{q^{rm}} \displaystyle\sum_{\substack{0 \le k \le j \le r \\ (j,k) \neq (0,0)}} \mathfrak{F}_2(j,k)(e\mathfrak{N}_2(j,k)+1)^v \mathfrak{N}_2(j,k) & \text{otherwise.}
\end{cases}
$$

Proof. Since $-1 \in \mathcal{D}_0^{(e,q^{rm})}$, $-b \in \mathcal{D}_0^{(e,q^{rm})}$. Thus, the result is a direct consequence of Theorem 2, and Lemma 4. □

Example 4. Let $q = 4$, $n = 3$, and $r = 3$. Thus, by Example 2, we have $((\mathfrak{N}_1(j), \mathfrak{F}_1(j))_{j=1}^{3} = ((1,3),(5,1),(-3,3))$. Therefore, the number of solutions $(x_1, \cdots, x_v) \in \mathbb{F}_{4^3}^v$ of the diagonal equation $x_1^7 + \cdots + x_v^7 = 1$ is

$$
N = 4^{3(v-1)} + \frac{1}{64}[3(8)^v + (36)^v(5) + 3(-20)^v(-3)].
$$

Example 5. With the notation of Theorem 2, let $q = p = 5$, $n = 8$, and $r = 2$. Thus, $m = 2$, $d = 3$, $u = 3 > 1$, $e = 39$, and $s = 1$. Clearly r divides n but not $(q^m - 1)/n$, and p is semiprimitive modulo u. Fix $h_1 = \frac{q^{mr}-1}{e} = 16$, and $h_2 = \frac{q^{m/2}}{(q^m-1)r} = \frac{5}{48}$. Thus, by Theorem 3, we have that

$$
((\mathfrak{N}_2(j,k), \mathfrak{F}_2(j,k)))_{\substack{0 \le k \le j \le 3 \\ (j,k) \neq (0,0)}} = ((6,2),(11,1),(-4,16),(1,16),(6,4)).
$$

Therefore, the number of solutions $(x_1, \cdots, x_v) \in \mathbb{F}_{5^4}^v$ of the diagonal equation $x_1^{39} + \cdots + x_v^{39} = 4$ is

$$
N = 5^{4(v-1)} + \frac{1}{625}[6(235)^v(6) + (430)^v(11) + 16(-155)^v(-4) + 16(40)^v].
$$

Remark 3. For, $v = 2, 3$, the numerical results in the two previous examples were corroborated with the help of a computer.

6 Conclusion

Let n, m, u, e, and r be positive integers, such that $\gcd(q,n) = 1$, $m = \mathrm{ord}_n(q)$, $u = \gcd(\frac{q^m-1}{q-1}, \frac{q^m-1}{n})$, $e = \gcd(\frac{q^{mr}-1}{q-1}, \frac{q^{mr}-1}{nr})$, and suppose that $u = 1$ or p is semiprimitive modulo u, and suppose also that the prime factors of r divide n but not $(q^m - 1)/n$, and that $q^m \equiv 1 \pmod 4$, if $4|r$. Then, under these conditions, we obtained the values and their multiplicities of all the Gaussian periods of the form $\eta_i^{(e,q^{mr})}$, showing at the same time that such Gaussian periods are always integer numbers (Theorem 2). Then, we used such values and multiplicities to give the explicit factorization of the period polynomials of the form $\psi_{(e,q^{mr})}(X)$ (Corollary 2). As an application of our results, we determined the number of solutions of certain diagonal equations with constant exponent (Theorem 3). Finally, notice that such solutions are a generalisation of those in [4].

Acknowledgments. The author want to express his gratitude to the anonymous referees for their valuable suggestions.

References

1. Baoulina, I.N.: On period polynomials of degree 2^m and weight distributions of certain irreducible cyclic codes. Finite Fields Their Appl. **50**, 319–337 (2018)
2. Berndt, B.C., Evans, R.J., Williams, K.S.: Gauss and Jacobi Sums. Canadian Mathematical Society Series of Monographs and Advanced Texts, vol. 21. Wiley, Hoboken (1998)
3. Delsarte, P.: On subfield subcodes of Reed-Solomon codes. IEEE Trans. Inform. Theory **21**(5), 575–576 (1975)
4. Ding, C., Kohel, D., Ling, S.: Counting the number of points on affine diagonal curves. In: Lam, K.Y., Shparlinski, I., Wang, H., Xing, C. (eds.) Cryptography and Computational Number Theory. Progress in Computer Science and Applied Logic, vol. 20, pp. 15–24. Basel, Birkhäuser (2001). https://doi.org/10.1007/978-3-0348-8295-8_3
5. Ding, C.: The weight distributions of some irreducible cyclic codes. IEEE Trans. Inform. Theory **55**(3), 955–960 (2009)
6. Ding, C., Yang, J.: Hamming weights in irreducible cyclic codes. Discrete Math. **313**, 434–446 (2013)
7. Gurak, S.: Period polynomials for $\mathbb{F}_{p^2}$ of fixed small degree. In: Jungnickel, D., Niederreiter, H. (eds.) Finite Fields and Applications, pp. 196–207. Springer, Heidelberg (2000). https://doi.org/10.1007/978-3-642-56755-1_16
8. Gurak, S. J.: Period polynomials for $\mathbb{F}_q$ of fixed small degree. In: Number Theory. CRM Proceedings and Lecture Notes, vol. 36, pp. 127–145. American Mathematical Society (2004)
9. Hoshi, A.: Explicit lifts of quintic Jacobi sums and period polynomials for $\mathbb{F}_q$. Proc. Jpn. Acad. Ser. A **82**, 87–92 (2006)
10. Lidl, R., Niederreiter, H.: Finite Fields. Cambridge University Press, Cambridge (1983)
11. Myerson, G.: Period polynomials and Gauss sums for finite fields. Acta Arith. **39**, 251–264 (1981)
12. Schmidt, B., White, C.: All two-weight irreducible cyclic codes? Finite Fields Their Appl. **8**, 1–17 (2002)
13. Sharma, A., Bakshi, G.K.: The weight distribution of some irreducible cyclic codes. Finite Fields Appl. **18**(1), 144–159 (2012)
14. Vega, G.: A characterization of all semiprimitive irreducible cyclic codes in terms of their lengths. Appl. Algebra Eng. Commun. Comput. **30**(5), 441–452 (2019). https://doi.org/10.1007/s00200-019-00385-z
15. Ward, R.L.: Weight enumerators of more irreducible cyclic binary codes. IEEE Trans. Inform. Theory **39**(5), 1701–1709 (1993)
16. Wolfmann, J.: The number of solutions of certain diagonal equations over finite fields. J. Number Theory **42**, 247–257 (1992)
17. Zeng, X., Fan, C., Zeng, Q., Qi, Y.: Two classes of binary cyclic codes and their weight distributions. Appl. Algebra Eng. Commun. Comput. (2019). https://doi.org/10.1007/s00200-019-00400-3

Improved Lower Bounds for Permutation Arrays Using Permutation Rational Functions

Sergey Bereg[ID], Brian Malouf[ID], Linda Morales[ID], Thomas Stanley[ID], and I. Hal Sudborough[✉][ID]

Department of Computer Science, University of Texas at Dallas,
Box 830688, Richardson, TX 75083, USA
hal@utdallas.edu

Abstract. We consider rational functions of the form $V(x)/U(x)$, where both $V(x)$ and $U(x)$ are relatively prime polynomials over the finite field $\mathbb{F}_q$. Polynomials that permute the elements of a field, called *permutation polynomials (PPs)*, have been the subject of research for decades. Let $\mathcal{P}^1(\mathbb{F}_q)$ denote $\mathbb{F}_q \cup \{\infty\}$. If the rational function, $V(x)/U(x)$, permutes the elements of $\mathcal{P}^1(\mathbb{F}_q)$, it is called a *permutation rational function (PRF)*. Let $N_d(q)$ denote the number of PPs of degree d over $\mathbb{F}_q$, and let $N_{v,u}(q)$ denote the number of *PRFs* with a numerator of degree v and a denominator of degree u. It follows that $N_{d,0}(q) = N_d(q)$, so *PRFs* are a generalization of PPs. The number of monic degree 3 *PRFs* is known [11]. We develop efficient computational techniques for $N_{v,u}(q)$, and use them to show $N_{4,3}(q) = (q+1)q^2(q-1)^2/3$, for all prime powers $q \leq 307$, $N_{5,4}(q) > (q+1)q^3(q-1)^2/2$, for all prime powers $q \leq 97$, and give a formula for $N_{4,4}(q)$. We conjecture that these are true for all prime powers q. Let $M(n, D)$ denote the maximum number of permutations on n symbols with pairwise Hamming distance D. Computing improved lower bounds for $M(n, D)$ is the subject of much current research with applications in error correcting codes. Using *PRFs*, we obtain significantly improved lower bounds on $M(q, q - d)$ and $M(q + 1, q - d)$, for $d \in \{5, 7, 9\}$.

Keywords: Hamming distance · Permutation array · Rational functions · Permutation polynomials

1 Introduction

Permutation arrays (PAs) with large Hamming distance have been the subject of many recent papers with applications in the design of error correcting codes. New lower bounds for the size of such permutation arrays are given, for example [1–7, 12, 14, 15, 19, 20, 22].

Let X be a set of n symbols, and let π and σ be permutations over X. The Hamming distance between π and σ, denoted by $hd(\pi, \sigma)$, is the number

S. Bereg—Research of the first author is supported in part by NSF award CCF-1718994.

© Springer Nature Switzerland AG 2021
J. C. Bajard and A. Topuzoğlu (Eds.): WAIFI 2020, LNCS 12542, pp. 234–252, 2021.
https://doi.org/10.1007/978-3-030-68869-1_14

of positions $x \in X$ such that $\pi(x) \neq \sigma(x)$. Define the Hamming distance of a PA A, by $hd(A) = \min\{hd(\pi, \sigma) \mid \pi, \sigma \in A,\ \pi \neq \sigma\}$. Let $M(n, D)$ denote the maximum number of permutations in any PA A on n symbols with Hamming distance D.

Let $\mathbb{F}_q$ denote the finite field with $q = p^m$ elements, where p is prime and $m \geq 1$. The prime p is called the *characteristic* of the field. A polynomial $V(x)$ over $\mathbb{F}_q$ is a *permutation polynomial* (*PP*) if it permutes the elements of $\mathbb{F}_q$. Permutation polynomials have been studied for many decades, for example [2, 8–10, 13, 16, 17, 21].

In this paper, we focus on permutation rational functions (*PRFs*), defined as follows:

Definition 1. *Let $V(x)$ and $U(x)$ be polynomials over $\mathbb{F}_q$, such that $\gcd(V(x), U(x)) = 1$. Let $\mathcal{P}^1(\mathbb{F}_q)$ denote $\mathbb{F}_q \cup \{\infty\}$. If the rational function $V(x)/U(x)$ permutes the elements of $\mathcal{P}^1(\mathbb{F}_q)$, then it is called a* **permutation rational function (*PRF*)**.

Yang *et al.* [23] used *PRFs* to compute, for example, an improved lower bound for $M(19, 14)$. Ferraguti and Micheli [11] enumerated all *PRFs* of degree 3.

Let $a \in \mathbb{F}_q$ and $a' \in \mathbb{F}_q \setminus \{0\}$. We use these conventions to evaluate expressions involving ∞:

$$a/\infty = 0, \quad a'/0 = \infty. \tag{1}$$

Let $\mathcal{W}(x) = \frac{V(x)}{U(x)}$ be a *PRF*, where $V(x)$ has degree v, $U(x)$ has degree u, and their high order coefficients are a_v and b_u, respectively. We use Eq. 1 to evaluate $\mathcal{W}(x)$ at ∞:

$$\mathcal{W}(\infty) = \mathcal{W}(1/x) \text{ when } x = 0. \tag{2}$$

Specifically, Eq. 2 implies that

$$\mathcal{W}(\infty) = \begin{cases} \infty, & \text{when } v > u \\ 0, & \text{when } v < u \\ a_v/b_v, & \text{when } v = u. \end{cases} \tag{3}$$

Observe that when $v > u$, *PRFs* over $\mathcal{P}^1(\mathbb{F}_q)$ can be viewed as permutations of $\mathbb{F}_q$ by eliminating ∞ from the domain.

Example. Let $V(x) = x^3 + x$ and $U(x) = x^2 + 5$ be polynomials over $\mathbb{F}_7$, where our computations are based on the primitive polynomial $x + 4$. Observe that $V(0) = 0$, $V(1) = 3$, $V(2) = 3$, $V(3) = 2$, $V(4) = 6$, $V(5) = 6$, $V(6) = 5$ and $U(0) = 5$, $U(1) = 6$, $U(2) = 4$, $U(3) = 1$, $U(4) = 6$, $U(5) = 4$, $U(6) = 1$. Let $\mathcal{W}(x)$ be the rational function defined by $\mathcal{W}(x) = V(x)/U(x) = (x^3 + x)/(x^2 + 5)$. Then

$$\mathcal{W}(x) = \begin{pmatrix} 0 & 1 & 2 & 3 & 4 & 5 & 6 & \infty \\ \frac{0}{5} & \frac{3}{6} & \frac{3}{4} & \frac{2}{1} & \frac{6}{6} & \frac{6}{4} & \frac{5}{1} & \frac{1}{0} \end{pmatrix} = \begin{pmatrix} 0 & 1 & 2 & 3 & 4 & 5 & 6 & \infty \\ 0 & 4 & 6 & 2 & 1 & 3 & 5 & \infty \end{pmatrix}.$$

Clearly $\mathcal{W}(x)$ is a permutation of the elements of $\mathcal{P}^1(\mathbb{F}_7)$. Hence $\mathcal{W}(x)$ is a *PRF*. Observe also that when $\mathcal{W}(x)$ is restricted to $\mathbb{F}_7$, the result is a permutation of the elements of $\mathbb{F}_7$. Also observe that $\mathcal{W}(1/x)$ is a *PRF*:

$$W(1/x) = \begin{pmatrix} 0 & 1\ 2\ 3\ 4\ 5\ 6\ \infty \\ \infty & 4\ 5\ 3\ 1\ 2\ 6\ 0 \end{pmatrix}$$

In general, many of the same concepts and techniques discussed for polynomials over finite fields apply to *PRFs*. Let $N_d(q)$ be the number of PPs of degree d over $\mathbb{F}_q$ [17]. We generalize this notion by defining $N_{v,u}(q)$ for *PRFs*.

Definition 2. $N_{v,u}(q)$ *is the number of PRFs* $V(x)/U(x)$, *where* $V(x)$ *has degree* v, *and* $U(x)$ *has degree* u.

Note that $N_d(q)$ is the same as $N_{d,0}(q)$. Note also that $N_{u,v}(q) = N_{v,u}(q)$, because $\frac{V(x)}{U(x)}$ is a *PRF* if and only if $\frac{U(x)}{V(x)}$ is also a *PRF*. That is, if $(a_0, a_1, \ldots, a_q)$ is a permutation of $P^1(\mathbb{F}_q)$, then $(a_0^{-1}, a_1^{-1}, \ldots, a_q^{-1})$ is also a permutation of $P^1(\mathbb{F}_q)$.

We compute values of $N_{v,u}(q)$, for many values of v, u and q, and use the computed values to give significantly improved lower bounds for $M(q, D)$ and $M(q+1, D)$. We show that the Hamming distance between permutations defined by *PRFs*, $\frac{V(x)}{U(x)}$ and $\frac{R(x)}{S(x)}$, where $V(x)$ is of degree v, $U(x)$ is of degree u, $R(x)$ is of degree r, and $S(x)$ is of degree s, is at least $q - \max\{v + s, u + r\}$. In this paper we focus on *PRFs* with numerators of degree v and denominators of degree either v or $v - 1$; however, $N_{v,u}(q)$ is computed also for other pairs of v, u for the sake of computing $M(q, D)$.

Definition 3. *Define* $T_d(q) = \sum_{v,u} N_{v,u}(q)$, *for all* $v, u \le (d+1)/2$.

We obtain improved lower bounds for $M(q, q-d)$ and $M(q+1, q-d)$ by showing that $M(q+1, q-d) \ge T_d(q)$. In addition, by computation, we show that:

$$N_{4,3}(q) = (q+1)q^2(q-1)^2/3, \ q \le 307,$$
$$N_{5,4}(q) > (q+1)q^3(q-1)^2/2, \ q \le 97, \text{ and}$$
$$N_{4,4}(q) = (q+1)q^2(q-1)^3/3, \text{ for odd } q \le 307.$$

Based on our experimental evidence, we conjecture that these formulas are valid for all prime powers q. We have also computed $N_{3,2}(q)$ and $N_{3,3}(q)$, not included in the above list as Ferraguti et al. [11], described all *PRFs* of degree 3. However, we do use the results for degree 3 *PRFs* to give improved lower bounds for $M(q, q-d)$ and $M(q+1, q-d)$ for $d \in \{5, 7, 9\}$.

Our paper is organized as follows. In Sect. 2 we discuss Hamming distance properties of *PRFs*, and give proofs of our new lower bounds for $M(q, q-d)$ and $M(q+1, q-d)$. In Sect. 3 we consider various forms of normalization that are useful for speeding up the search for *PRFs*. In Sect. 4 we discuss functions that map *PRFs* into *PRFs* that are also useful for speeding up our computations. In Sect. 5 we give formulas and compute values for $N_{4,3}(q)$, $N_{4,4}(q)$, and $N_{5,4}(q)$. The formulas are verified computationally and conjectured to be valid for all prime powers q. In Tables 3 and 5, we list new results, derived from *PRFs*, for $M(q, D)$, for various q and D. Table 3 shows new results for $M(q, q-5)$ and $M(q, q-7)$, for $16 \le q \le 149$. Table 4 shows new results for $N_{5,4}(q)$ and values obtained by our formula given in Conjecture 2 (our 5/4 conjecture). Table 5 shows new results for $M(q, q-9)$, for $13 \leq q \leq 97$. Tables 6 and 7 give new results for $M(q+1, q-5)$ and $M(q+1, q-7)$, respectively. We have improved lower bounds for $M(q, D)$ for several other values of q and D, but they are not included here due to space restrictions.

Notation. We use the following notation throughout this paper. $\mathbb{F}_q$ is a finite field where $q = p^m$ for some $m \geq 1$. We use the convention that t denotes a generator of the group of non-zero elements of $\mathbb{F}_q$. Using this notation, the elements of $\mathbb{F}_q$ are 0, $t^0 = 1$, $t^1 = 2, \ldots, t^{q-2} = q - 1$. Lidl and Niederreiter [18] give this as one way to represent the elements of a finite field. Another representation lists the elements of $\mathbb{F}_{p^m}$ by degree m polynomials with coefficients from $\mathbb{F}_p$. *PRFs* can easily be converted from one notation to the other. As a primitive polynomial is needed to do the appropriate arithmetic, we give explicit primitive polynomials for our computations and results. For notational clarity, we let V, U, R and S denote polynomials of degree v, u, r and s, with coefficients a_i, b_i, c_i and d_i respectively, That is, $V(x) = \sum_{i=0}^{v} a_i x^i$, $U(x) = \sum_{i=0}^{u} b_i x^i$, $R(x) = \sum_{i=0}^{r} c_i x^i$, and $S(x) = \sum_{i=0}^{s} d_i x^i$, Lastly, we let $\mathcal{W}$, $\mathcal{Y}$, and $\mathcal{Z}$ denote *PRFs*. So if $\mathcal{W}(x) = \frac{V(x)}{U(x)}$, then $\mathcal{W}(x) = \sum_{i=0}^{v} a_i x^i / \sum_{i=0}^{u} b_i x^i$.

2   Hamming Distance of *PRFs*

Recall that by Definition 1, $\gcd(V(x), U(x)) = 1$ for any *PRF*. This property is implicit in our counting arguments for *PRFs*. For example, see Corollary 5 and Corollary 7.

We now discuss properties of *PRFs* that are useful for improving lower bounds for $M(q, D)$ and $M(q+1, D)$. Some similar ideas were given in [23]. For the proofs in this section, we consider the *PRFs* $\mathcal{W}(x) = \frac{V(x)}{U(x)}$ and $\mathcal{Y}(x) = \frac{R(x)}{S(x)}$ that permute the elements of $\mathcal{P}^1(\mathbb{F}_q)$ such that $V(x)S(x) - U(x)R(x)$ is not a constant. For this discussion, the degrees of the *PRFs* need be not be the same.

Theorem 4. *Let* $v + s \leq d$ *and* $u + r \leq d$, *for some* d. *Let* π *and* σ *be the permutations of* $\mathcal{P}^1(\mathbb{F}_q)$ *generated by* $\mathcal{W}(x)$ *and* $\mathcal{Y}(x)$ *respectively. Then* $hd(\pi, \sigma) \geq q - d$.

Proof. We consider the values of the *PRFs* for elements of $\mathbb{F}_q$, and simultaneously note that $\frac{V(\infty)}{U(\infty)}$ and $\frac{R(\infty)}{S(\infty)}$ may be the same. Assume that for some $a \in \mathbb{F}_q$, $\frac{V(a)}{U(a)} = \frac{R(a)}{S(a)}$. Then $V(a)S(a) = U(a)R(a)$, so $V(a)S(a) - U(a)R(a) = 0$. Observe that $V(x)S(x)$ and $U(x)R(x)$ are polynomials of degree $v + s \leq d$ and $u + r \leq d$, respectively. Hence, $V(x)S(x) - U(x)R(x)$ is a polynomial of degree at most d and has at most d roots. That is, there are at most d values $a \in \mathbb{F}_q$ such that $V(a)S(a) - U(a)R(a) = 0$. Note also that if $\frac{V(a)}{U(a)} = \frac{R(a)}{S(a)} = \infty$, then $U(a) = S(a) = 0$. So, $V(a)S(a) - U(a)R(a) = 0$, and a is a root. This means that $\frac{V(a)}{U(a)} = \frac{R(a)}{S(a)}$ for at most d values $a \in \mathbb{F}_q$. By including the values of the *PRFs* at ∞, there may be $d+1$ agreements. Thus, there are at least $q+1-(d+1) = q-d$ disagreements. Hence, $hd(\pi, \sigma) \geq q - d$.   □

It follows also that the permutations corresponding to different *PRFs* are different, because the permutations have non-trivial Hamming distance.

Corollary 5. $M(q+1, q-d) \geq T_d(q)$.

Proof. Let $v, u, r, s \leq (d+1)/2$, and consider any pair of distinct *PRFs* $\mathcal{W}(x) = \frac{V(x)}{U(x)}$ and $\mathcal{Y}(x) = \frac{R(x)}{S(x)}$. Observe that, by Eq. 3, $\mathcal{W}(\infty) = \frac{a_v}{b_u} \in \mathbb{F}_q \setminus \{0\}$ if and only if $v = u$, and $\mathcal{Y}(\infty) = \frac{c_r}{d_s} \in \mathbb{F}_q \setminus \{0\}$ if and only if $r = s$.

Case 1. $v = u = r = s$.
For Case 1, observe that $\mathcal{W}(\infty) = \mathcal{Y}(\infty)$ if and only if the ratios of the high order coefficients in the numerator and denominator are the same in $\mathcal{W}(x)$ and $\mathcal{Y}(x)$. That is, $\mathcal{W}(\infty) = \mathcal{Y}(\infty)$ if and only if $a_v/b_u = c_r/d_s$. Call this property (=). Observe that the coefficients of the high order terms in the polynomials $V(x)S(x)$ and $U(x)R(x)$ are $a_v d_s x^{v+s}$ and $b_u c_r x^{u+r}$, respectively, where $v + s = u + r$. So the high order term of the polynomial $V(x)S(x) - U(x)R(x)$ is $(a_v d_s - b_u c_r)x^{v+s}$. If (=) is true, then the polynomial $V(x)S(x) - U(x)R(x)$ is of degree at most d, not $d + 1$, since the high order terms, if they are of the same degree, disappear through subtraction. It follows that, if $\mathcal{W}(x)$ and $\mathcal{Y}(x)$ have the same value at ∞, then there are at most d agreements when $x \in \mathbb{F}_q$, hence, at most $d + 1$ agreements counting the agreement at infinity. On the other hand, if (=) is not true, then the polynomial $V(x)S(x) - U(x)R(x)$ is of degree at most at $d + 1$, so it too has at most $d + 1$ agreements. Either way, the permutations defined by these *PRFs* have Hamming distance at least $q + 1 - (d + 1) = q - d$.

Case 2. $v = u$ and $r > s$.
It follows that $\mathcal{W}(\infty) \in \mathbb{F}_q \setminus \{0\}$ and $\mathcal{Y}(\infty) = \infty$, so $\mathcal{W}(\infty) \neq \mathcal{Y}(\infty)$. Furthermore, $V(x)S(x) - U(x)R(x)$ is of degree at most $d + 1$, so it has at most $d + 1$ roots. That is, there are at most $d+1$ values $a \in \mathbb{F}_q$ such that $\mathcal{W}(a) = \mathcal{Y}(a)$. Consequently, there are at least $q+1-(d+1) = q-d$ positions b where $\mathcal{W}(b) \neq \mathcal{Y}(b)$. So, the permutations defined by these *PRFs* have Hamming distance at least $q - d$.

Case 3. $v = u$ and $r < s$.
This is similar to Case 2. The difference is that $\mathcal{Y}(\infty) = 0$.

Case 4. $v < u$ and $r > s$.
This is similar to Case 2. The difference is that $\mathcal{W}(\infty) = 0$.

Case 5. $v < u$ and $r < s$.
It follows that $\mathcal{W}(\infty) = \mathcal{Y}(\infty) = 0$. Since, $V(x)S(x) - U(x)R(x)$ is of degree at most d, it has at most d roots. Hence, there are at most d values $a \in \mathbb{F}_q$ such that $\mathcal{W}(a) = \mathcal{Y}(a)$, and counting the agreement at infinity, the result is a total of $d + 1$ agreements. That is, at least $q + 1 - (d + 1) = q - d$ positions b where $\mathcal{W}(b) \neq \mathcal{Y}(b)$. So, the permutations defined by these *PRFs* have Hamming distance at least $q - d$. $\square$

Definition 6. *Define* $\mathcal{S}_d(q) = N_{t,t}(q)/(q-1) + \sum_{v,u} N_{v,u}(q)$, *where* $t = (d-3)/2$, *and in the sum,* v *and* u *are evaluated as*

$$v, u = \begin{cases} v \le (d+1)/2, \ u \le (d-1)/2 & \text{when } v > u, \\ v \le (d-1)/2, \ u \le (d-3)/2 & \text{when } v < u, \\ u, v \le (d-5)/2 & \text{when } v = u. \end{cases}$$

Corollary 7. $M(q, q-d) \ge \mathcal{S}_d(q)$.

Proof. Consider the *PRFs* $\mathcal{W}(x) = \frac{V(x)}{U(x)}$ and $\mathcal{Y}(x) = \frac{R(x)}{S(x)}$. Observe that by the definition of $\mathcal{S}_d(q)$, the largest values for v and s are $(d+1)/2$ and $(d-1)/2$, respectively, and similarly for u and r. Let $v, r \le (d+1)/2$ and $u, s \le (d-1)/2$. It follows that $V(x)S(x) - U(x)R(x)$ is of degree $\le d$. As seen in the proof of Theorem 4, this means that permutations defined by $\mathcal{W}(x)$ and $\mathcal{Y}(x)$ have at most d agreements.

As we want to consider permutations on $\mathbb{F}_q$ (not $\mathcal{P}^1(\mathbb{F}_q)$) we need to eliminate occurrences of the symbol ∞ in the permutations corresponding to $\mathcal{W}(x)$ and $Y(x)$ using an operation called *contraction* [1]. If $\mathcal{W}(\infty) = \infty$, then we can simply eliminate the symbol ∞ in the corresponding permutation, which of course makes no new agreements. If $\mathcal{W}(\infty) = a$, with $a \in \mathbb{F}_q$, then we exchange the symbol ∞ wherever it occurs in the permutation with a. This moves the symbol ∞ to the last position in the permutation, so it can be eliminated. One, or at most two, new agreements could be created, the latter situation arising when $v = u$. Consequently, if $\mathcal{W}(\infty) \ne \infty$ (so an exchange with ∞ is required), stronger conditions are needed to ensure that there are a total of at most d agreements. The terms in the sum $\mathcal{S}_d(q)$ are calculated to ensure that the Hamming distance between permutations (after all needed contractions are performed) is at least $q - d$.

We do a proof by cases based on the values of v, u, r, s and t.

Case 1. $v = u = t = (d-3)/2$.
Suppose that $U(x)$ and $V(x)$ are monic polynomials. Then the related permutations always end with the same symbol, namely 1. Contraction applied to the permutation associated with $\mathcal{W}(x)$ creates at most one new agreement with any other permutation that already has the symbol 1 in the exchanged position. The number of such permutations produced by *PRFs* with $U(x)$ and $V(x)$ both monic and both of degree t is $N_{t,t}(q)/(q-1)$.

Case 2. v and u have their maximum values, and $r = s \le (d-5)/2$.
It follows that the polynomial $V(x)S(x) - U(x)R(x)$ has degree at most $d-2$. Then permutations defined by $\mathcal{W}(x)$ and $\mathcal{Y}(x)$ have at most $d-2$ agreements. Since contraction creates at most 2 new agreements, it follows that there are at most d agreements. Therefore, the permutations have Hamming distance at least $q - d$.

Case 3. $r \le v \le (d+1)/2$ and $s \le u \le (d-1)/2$, $v > u$ and $r > s$.
It follows that the polynomial $V(x)S(x) - U(x)R(x)$ has degree at most d, so permutations defined by $\mathcal{W}(x)$ and $\mathcal{Y}(x)$ have at most d agreements. Note that

$\mathcal{W}(\infty) = \infty$ and $\mathcal{Y}(\infty) = \infty$. Hence the permutations defined by $\mathcal{W}(x)$ and $\mathcal{Y}(x)$ make no new agreements through contraction, *i.e.*, the ∞ simply disappears in each permutation. Therefore, the permutations have Hamming distance at least $q - d$.

Case 4. $u < v \le (d+1)/2$, and $r < s \le (d-3)/2$.
It follows that the polynomial $V(x)S(x) - U(x)R(x)$ has degree $\le d - 1$, so permutations defined by $\mathcal{W}(x)$ and $\mathcal{Y}(x)$ have at most $d - 1$ agreements. Since $\mathcal{W}(\infty) = \infty$ and $\mathcal{Y}(\infty) = 0$, at most one new agreement is created through contraction. Therefore, the total number of agreements is d, and the permutations have Hamming distance at least $q - d$.

Case 5. $v < u \le (d-3)/2$ and $r < s \le (d-3)/2$.
    It follows that the polynomial $V(x)S(x) - U(x)R(x)$ has degree at most $d - 4$, so permutations defined by $\mathcal{W}(x)$ and $\mathcal{Y}(x)$ have at most $d-4$ agreements. Contraction of these permutations makes at most 2 new agreements. Therefore, the permutations have Hamming distance at least $q - d$.

    Hence the Hamming distance between permutations defined by $\mathcal{W}(x)$ and $\mathcal{Y}(x)$, with the stated numerator/denominator degree bounds given in the sum in the definition of $\mathcal{S}_d(q)$, is at most $q - d$. It follows that total number of permutations on q symbols with pairwise Hamming distance $q - d$ is at least as large as $\mathcal{S}_d(q)$.   □

Examples. (Note: Some of the terms in the sums are not shown because they are zero. Also, some terms are written as $2N_{u,v}$ to denote $N_{u,v} + N_{v,u}$ when applicable).
(a) $M(q, q-5) :\ \mathcal{S}_5(q) = N_{3,2}(q) + N_{3,0}(q) + N_{2,0}(q) + N_{1,1}(q)/(q-1) + 2N_{1,0}(q)$.
(b) $M(q, q - 7) :\ \ \mathcal{S}_7(q) = N_{4,3}(q) + N_{3,2}(q) + N_{4,0}(q) + N_{3,0}(q) + 2N_{2,0}(q) + N_{2,2}(q)/(q - 1) + N_{1,1}(q) + 2N_{1,0}(q)$.
(c) $M(q, q - 9) :\ \ \mathcal{S}_9(q) = N_{5,4}(q) + N_{5,3}(q) + N_{5,0}(q) + N_{4,3}(q) + N_{4,0}(q) + N_{3,3}(q)/(q-1) + 2N_{3,2}(q) + 2N_{3,0}(q) + N_{2,2}(q) + 2N_{2,0}(q) + N_{1,1}(q) + 2N_{1,0}(q)$.

3   Normalization of *PRFs*

The goal of normalization is to enable a more efficient search for *PRFs*. That is, normalization indicates that certain coefficients can be fixed at a specified value and a search algorithm need not try all possibilities. Normalization has been discussed previously in the context of PPs [2,17,21]. Equivalence relations based on normalization [2] allow partitioning of PPs of degree d in $\mathbb{F}_q$ into equivalence classes, each represented by a *normalized permutation polynomial (nPP)*.

    We use normalization to map *PRFs* to normalized *PRFs* (*nPRFs*). Normalization operations [18], listed in Table 1, are essentially the same for PPs and *PRFs*. We point out a few subtleties that arise due to the presence of a denominator in *PRFs*. Let $a, b, c, r, y, z \in \mathbb{F}_q$. Multiplying a *PRF* $\mathcal{W}(x) = \frac{V(x)}{U(x)}$ by a nonzero constant a is equivalent to multiplying by $a = y/z$, for $y, z \ne 0$. Addition of a constant b to the variable is accomplished by replacing x by $x + b$

Table 1. Normalization operations for PPs and *PRFs* in $\mathbb{F}_q$.

Normalization operation	For PPs $V(x)$	For *PRFs* $\mathcal{W}(x) = \frac{V(x)}{U(x)}$
Multiplication by a nonzero constant	$aV(x) : a \in \mathbb{F}_q$	$\frac{yV(x)}{zU(x)} : y, z \in \mathbb{F}_q$
Addition to the variable	$V(x+b) : b \in \mathbb{F}_q$	$\frac{V(x+b)}{U(x+b)} : b \in \mathbb{F}_q$
Addition of a constant	$V(x) + c : c \in \mathbb{F}_q$	$\frac{V(x)+cU(x)}{U(x)} : c \in \mathbb{F}_q$
Multiplication of the variable by a constant	$V(rx) : r \in \mathbb{F}_q$	$\frac{V(rx)}{U(rx)} : r \in \mathbb{F}_q$

in both numerator and denominator. Adding a constant c to $\mathcal{W}(x)$ equates to computing $\frac{V(x)}{U(x)} + c = \frac{V(x)+cU(x)}{U(x)}$. Multiplication of the variable by a constant is accomplished by replacing the argument x by rx, for some constant $r \neq 0$. Note that if $\mathcal{W}(x)$ permutes the elements of $\mathcal{P}^1(\mathbb{F}_q)$, then so does $\mathcal{W}(rx)$. That is, if $\mathcal{W}(x)$ is a *PRF*, then $\mathcal{W}(rx)$ is also a *PRF*. In fact, all of the normalization operations in Table 1 map *PRFs* to *PRFs*.

We now discuss the usage of these operations to map *PRFs* to *nPRFs*. In Table 2 we define define three types of normalized *PRFs* and list the restrictions on each. The definitions are modeled after the definitions of normalization of PPs which are described in [2]. Note that normalization of *PRFs* fixes four coefficients: a_v and b_u both have the value 1, a_0 is 0, and an additional coefficient, determined by the type of normalization, is zero. In the sections that follow, we prove that almost all *PRFs* can be normalized. As explained earlier, this is useful for an efficient search for *PRFs*. We use the following in our proofs for normalization. Let $a, b, c \in \mathbb{F}_q$, $a \neq 0$, let $x, y \in \mathbb{F}_q \backslash \{0\}$ such that $y/z = a$. Let $\mathcal{Y}(x) = a\mathcal{W}(x+b) + c$. Then

$$\mathcal{Y}(x) = a\mathcal{W}(x+b) + c = \frac{yV(x+b)}{zU(x+b)} + \frac{czU(x+b)}{zU(x+b)}$$

$$= \frac{yV(x+b) + czU(x+b)}{zU(x+b)} = \frac{V'(x)}{U'(x)}, \text{ where}$$

$$V'(x) = yV(x+b) + cU'(x)$$
$$= (ya_v(x+b)^v + ya_{v-1}(x+b)^{v-1} + \cdots + ya_1(x+b) + ya_0) \tag{4}$$
$$+ (czb_u(x+b)^u + czb_{u-1}(x+b)^{u-1} + \cdots + czb_1(x+b) + czb_0),$$

and

$$U'(x) = zU(x+b) = zb_u(x+b)^u + zb_{u-1}(x+b)^{u-1} + \cdots + zb_0. \tag{5}$$

3.1   C-Normalization

As seen in Table 2, c-normalization applies to *PRFs* when the field characteristic p does not divide the degree of the denominator. We use *nPRFs* to define an equivalence relation on *PRFs* as follows:

Table 2. Types of normalization for *PRFs* $\mathcal{W}(x) = \frac{V(x)}{U(x)}$, where $V(x) = \sum_{i=0}^{v} a_i x^i$ and $U(x) = \sum_{i=0}^{u} b_i x^i$, with field characteristic p. The degrees of $V(x)$ and $U(x)$ are v and u, respectively.

Normalization type	Degree restriction	*nPRF* properties
c-normalization	$p \nmid u$	$V(x)$ and $U(x)$ are monic, $V(0) = 0$, and
	$v > u$	$b_{u-1} = 0$
m-normalization	$p \mid u$ and $p > 2$	$V(x)$ and $U(x)$ are monic, $V(0) = 0$, and
	$v > u$	in $U(x)$, either $b_{u-1} = 0$ or $b_{u-2} = 0$
b-normalization	$p \mid u$ and $p = 2$	$V(x)$ and $U(x)$ are monic, $V(0) = 0$, and
	$v > u$	if $2^i \leq u \leq 2^{i+1} - 3$ for some i, then either
		$b_r = 0$ or $b_{r-1} = 0$, where $r = 2^i - 1$

Definition 8. *Let* $\mathcal{W}(x) = \frac{V(x)}{U(x)}$ *and* $\mathcal{Y}(x) = \frac{R(x)}{S(x)}$ *be PRFs. We say that* $\mathcal{W}(x)$ *and* $\mathcal{Y}(x)$ *are related by* $\mathcal{R}_c$ *if there is a sequence of the first three normalization operations in Table 1 that converts* $\mathcal{W}(x)$ *into* $\mathcal{Y}(x)$.

It is easily seen that $\mathcal{R}_c$ is an equivalence relation on *PRFs*. That is, observe that each of the three operations has an inverse. For example, the inverse of multiplying by a is multiplying by the inverse of a. So, $\mathcal{W}(x)$ is related to itself by the empty sequence of operations. If $\mathcal{W}(x)$ and $\mathcal{Y}(x)$ are $\mathcal{R}_c$ related, then there is some sequence that transforms $\mathcal{W}(x)$ into $\mathcal{Y}(x)$. A sequence formed by taking the inverse of each operation in backwards order transforms $\mathcal{Y}(x)$ into $\mathcal{W}(x)$. So, $\mathcal{Y}(x)$ and $\mathcal{W}(x)$ are also $\mathcal{R}_c$ related. That is, $\mathcal{R}_c$ is symmetric. Finally, if there is a sequence of operations that transforms $\mathcal{W}(x)$ into $\mathcal{Y}(x)$ and a sequence that transforms $\mathcal{Y}(x)$ into $\frac{P(x)}{Q(x)}$, then a concatenation of these sequences transforms $\mathcal{W}(x)$ into $\frac{P(x)}{Q(x)}$. So, $\mathcal{R}_c$ is transitive.

The equivalence class under the relation $\mathcal{R}_c$ containing $\mathcal{W}(x)$, denoted by $[\mathcal{W}]$, is the set

$$[\mathcal{W}] = \{a\mathcal{W}(x+b) + c \mid a, b, c \in \mathbb{F}_q \text{ and } a \neq 0\}$$

$$= \{a\frac{V(x+b)}{U(x+b)} + c \mid a, b, c \in \mathbb{F}_q \text{ and } a \neq 0\}.$$

We show that $[\mathcal{W}]$ contains exactly $q^2(q-1)$ *PRFs*. Theorem 9 below is nearly identical to one proved (for PPs) in [2].

Theorem 9 [2]. *Let* $\mathcal{W}(x) = \frac{V(x)}{U(x)}$ *be a PRF with* $v > u$. *Then there is a unique triple* (a, b, c) *such that* $\mathcal{Y}(x) = a\mathcal{W}(x+b) + c = \frac{V'(x)}{U'(x)}$ *is c-normalized.*

Lemma 10. *All* $q^2(q-1)$ *PRFs in* $[\mathcal{W}]$ *are different.*

Proof. Let $\mathcal{Y}(x) \in [\mathcal{W}]$ be a *PRF* that is not normalized, and let $\mathcal{W}(x)$ be the *nPRF* that represents $[\mathcal{W}]$. That is, let $\mathcal{Y}(x) = a\mathcal{W}(x+b) + c$. We compute the triple (a', b', c') such that $\mathcal{Y}'(x) = a'\mathcal{Y}(x + b') + c'$ is normalized as follows.

$$\mathcal{Y}'(x) = a'\mathcal{Y}(x + b') + c' = a'(aW(x + b) + b') + c) + c'$$
$$= a'aW(x + (b + b')) + a'c + c' = W(x),$$

where the last equality is achieved by letting $a' = a^{-1}$, $b' = -b$, and $c' = -(a'c)$. By Theorem 9, the triple (a', b', c') is unique for normalizing the specific *PRF* $\mathcal{Y}(x)$. By the uniqueness properties of inverses in a field, a, b and c are are unique as well. Thus each triple in the set $\{(a, b, c) \mid a, b, c \in \mathbb{F}_q$ and $a \neq 0\}$ is unique. Since there are $q^2(q - 1)$ such triples, the claim follows. $\qquad\square$

Note that Theorem 9 implies that each equivalence class of $\mathcal{R}_c$ contains one and only one *nPRF*. By Lemma 10, each equivalence class contains exactly $q^2(q - 1)$ members (including the representative *nPRF*). Equivalence classes by definition are disjoint, so, if the number of *nPRFs* is k, there are $kq^2(q-1)$ *PRFs*. Note that c-normalization indicates that we can fix four coefficients, namely the first coefficient of both $V(x)$ and $U(x)$, the second coefficient of $U(x)$, and the last coefficient of $V(x)$. There are, in general, q possible values for each coefficient. Furthermore, $V(x)$ and $U(x)$ are of degrees v and u, respectively, so there are $v + u + 2$ coefficients altogether. This means a naive search program (which exhaustively tries all combinations of coefficients) needs to examine q^{u+v+2} rational functions. Normalization allows the number to be reduced to q^{u+v-2}.

3.2   M-Normalization

As seen in Table 2, m-normalization is used when $p \mid u$ and $p \neq 2$. See Eqs. 4 and 5 for the definitions of $V'(x)$ and $U'(x)$.

Theorem 11. *Let $v, u \in \mathbb{F}_{p^m}$, where $v > u$ and $p \mid u$ and $p \neq 2$. Any PRF $W(x) = \frac{V(x)}{U(x)}$ can be transformed to an m-normalized PRF $\mathcal{Y}(x) = \frac{V'(x)}{U'(x)}$ by the normalization operations.*

Proof. For m-normalization, we need to show that

(A) either the coefficient of x^{u-1}, or the coefficient of x^{u-2} in $U'(x)$ is zero,
(B) $U'(x)$ is monic,
(C) $V'(x)$ is monic, and
(D) $V'(0)$, the constant term of $V'(x)$, is zero.

To show that (A) holds, we must show that either $zb_{u-1} = 0$ or $zb_{u-2} = 0$.

Case 1. $b_{u-1} = 0$. So, $zb_{u-1} = 0$.

Case 2. $b_{u-1} \neq 0$. Consider zb_{u-2} in $U'(x)$. Since u is a multiple of p, the expansion of $(x+b)^u$ will derive nonzero coefficients only for terms whose degrees are multiples of p. Since $p > 2$, this means that $p \nmid (u - 2)$, so $(x + b)^u$ will have a coefficient of 0 for the degree $u - 2$ term. Hence b_{u-2} is calculated solely by the expansion of $(x + b)^{u-1}$ and $(x + b)^{u-2}$.

The expansion of $(x+b)^{u-1}$ will produce a term of degree $u-2$ with coefficient $zb_{u-1}b'$ where $b' = \sum_1^{u-1} b$. The expansion of $(x+b)^{u-2}$ will produce a term of degree $u-2$ with coefficient zb_{u-2}. Therefore the coefficient of x^{u-2} in $U'(x)$ is $zb_{u-1}b' + zb_{u-2} = z(b_{u-1}b' + b_{u-2})$, which is zero if $b_{u-1}b' + b_{u-2} = 0$. Since $b_{u-1} \neq 0$, and $u-1$ is not a multiple of p, we can choose b such that b' is the additive inverse of b_{u-2}/b_{u-1}, making the coefficient of x^{u-2} in $U'(x)$ equal to zero.

It follows that, in $U'(x)$, either $b_{u-1} = 0$ or $b_{u-2} = 0$, so (A) holds.

To show that (B) holds, observe that the degree u term of $U'(x)$ has the coefficient zb_u. If we choose z to be the multiplicative inverse of b_u, then $U'(x)$ will be monic. To show that (C) holds, observe that every term in $U'(x)$ has smaller degree than v. Hence none of them affect the coefficient of degree v term in $V'(x)$. This means that the coefficient of x^v term of $V'(x)$ is ya_v. Since $a_v \neq 0$, we choose $y = a_v^{-1}$, making $V'(x)$ monic. To show that (D) holds, observe that the coefficient of x^0 in $V'(x)$ is $y \sum_{j=0}^v a_j b^j + cz \sum_{j=0}^u b_j b^j$. We choose c to be $(-y \sum_{j=0}^v a_j b^j)/(z \sum_{j=0}^u b_j b^j)$, making the coefficient of x^0 in $V'(x)$ equal to zero.

It follows that $\mathcal{Y}(x)$ is m-normalized. $\qquad\square$

3.3   B-Normalization

In this section, we consider the remaining case, namely, $p \mid u$ and $p = 2$, and show that *b-normalization* can be achieved except when $u = 2^i - 2$, for some $i \geq 2$.

We begin with a brief description of the Gap Lemma for polynomials (Lemma 12 below), and its application for normalization of polynomials (Lemma 13 below). Both were proven in [2]. We use these lemmas in the proof of Theorem 14 which describes b-normalization for *PRFs*.

We say that the integer interval $[r, s]$ has a $[t, w]$ gap, if for all $d \in [r, s]$, the expansion of $(x + b)^d$, does not include any nonzero x^e monomials, where $e \in [t, w]$.

Lemma 12 *[Gap Lemma [2]]. For all $i > 1$, the expansion of $(x + b)^d$, for $d \in [2^i, 2^{i+1} - 3]$, has a $[2^i - 2, 2^i - 1]$ gap.*

Lemma 13 *[2]. Let $i > 1$, $m > 2$ and let $d \in [2^i, 2^{i+1} - 3]$ be even. For any PP $P(x)$ over $\mathbb{F}_{2^m}$, there is a constant b in $\mathbb{F}_{2^m}$ such that in the PP $P(x + b)$, either the $x^{2^i - 1}$ term or the $x^{2^i - 2}$ term is zero.*

For example, let $d = 2^3$, and let $P(x) = a_8 x^8 + a_7 x^7 + a_6 x^6 + \cdots + a_1 x + a_0$. Adding b to the argument gives:

$$\begin{aligned}
P(x + b) &= a_8(x + b)^8 + a_7(x + b)^7 + a_6(x + b)^6 + \cdots + a_1 x + a_0 \\
&= a_8(x^8 + b^8) + a_7(x^7 + bx^6 + b^2x^5 + \ldots) + a_6(x^6 + b^2x^4 + \ldots) + \cdots \\
&= a_8 x^8 + a_8 b^8 + (a_7 x^7 + a_7 bx^6 + \ldots) + (a_6 x^6 + a_6 b^2 x^4 + \ldots) + \cdots
\end{aligned}$$

We want to solve for the value of b that makes the coefficient of the x^6 term of $P(x+b)$ zero. So $a_7bx^6 + a_6x^6 = 0$ is satisfied by $b = -a_7/a_6$.

We now use Lemma 13 in our proof that certain $PRFs$ can be b-normalized.

Theorem 14. *Any PRF $\frac{V(x)}{U(x)}$ in $\mathbb{F}_{2^m}$ with $v > u, m > 2$, and $2 \mid u$, can be transformed to a b-normalized PRF $\frac{V'(x)}{U'(x)}$ by the normalization operations, except when $u = 2^i - 2$, for some $i \geq 2$.*

Proof. See Eqs. 4 and 5 for the definitions of $V'(x)$ and $U'(x)$. Observe first that the degree u term of $U'(x)$ has the coefficient zb_u. Noting that $b_u \neq 0$, we choose $z = b_u^{-1}$, making $U'(x)$ monic. Observe further that every term in $U'(x)$ has smaller degree than v. Hence none of them affect the coefficient of degree v term in $V'(x)$. This means that the coefficient of x^v term of $V'(x)$ is ya_v. Noting that $a_v \neq 0$, we choose $y = a_v^{-1}$, making $V'(x)$ monic. To see that $V'(0) = 0$, observe that the coefficient of x^0 in $V'(x)$ is $y\sum_{j=0}^{v} a_jb^j + cz\sum_{j=0}^{u} b_jb^j$. We choose c to be $(-y\sum_{j=0}^{v} a_jb^j)/(z\sum_{j=0}^{u} b_jb^j)$, making the coefficient of x^0 in $V'(x)$ equal to zero. Finally, by Lemma 13, there is a b such that in $U'(x)$, the coefficient of either the degree $2^i - 1$ term or degree $2^i - 2$ term equal to 0, except when $u = 2^i - 2$, for some $i \geq 2$. Hence $\frac{V'(x)}{U'(x)}$ is b-normalized.   □

4   Mapping *nPRFs* to *nPRFs*

We are interested in methods to optimize the search for *PRFs*. In [2] we described several operations on permutation polynomials that allow certain coefficients of PPs to be fixed, making the search space smaller. These operations include normalization and the *F-map* and the *G-map*. The F-map allows an additional coefficient to be fixed. The G-map partitions *nPRFs* into disjoint cycles, and each cycle can be described by a representative *nPRF*. We show that the F-map and the G-map can be extended to *nPRFs*, allowing again faster searches.

Definition 15. *Define the **F-map** on a polynomial $V(x)$ over $\mathbb{F}_q$ [2] by*

$$F(V(x)) = t^0a_vx^v + t^1a_{v-1}x^{v-1} + \cdots + t^{v-1}a_1x + t^va_0.$$

*Define the **F-map** on a PRF $W(x) = V(x)/U(x)$ over $\mathbb{F}_q$ by*

$$F(W(x)) = \frac{F(V(x))}{F(U(x))} = \frac{t^0a_vx^v + t^1a_{v-1}x^{v-1} + \cdots + t^{v-1}a_1x + t^va_0}{t^0b_ux^u + t^1b_{u-1}x^{u-1} + \cdots + t^{u-1}b_1x + t^ub_0}. \quad (6)$$

It is shown in [2] that $F(V(x)) = t^vV(x/t)$. So for a *PRF* $W(x)$, we have

$$F(W(x)) = W'(x) = \frac{t^vV(x/t)}{t^uU(x/t)} = \frac{t^{v-u}V(x/t)}{U(x/t)} = t^{v-u}W(x/t).$$

Thus, if $W(x)$ is a *PRF*, then so is $W'(x) = t^{v-u}W(x/t)$. That is, if $W(x)$ permutes the elements of $\mathcal{P}^1(\mathbb{F}_q)$, then so does $W'(x)$. Consequently, the F-map maps *PRFs* to *PRFs*. In fact, referring to Eq. 6, we see that the F-map maps *nPRFs* to *nPRFs*, as the first coefficients of both numerator and denominator map to themselves, and any zero coefficient is mapped to itself.

We use the F-map to fix an additional coefficient in a *PRF*, resulting in a total of 5 fixed coefficients for each *nPRF*. For example, consider searching for *nPRFs* of the form $W(x) = \frac{V(x)}{U(x)}$. By the definition of normalization, $V(x)$ and $U(x)$ are monic, the coefficient of x^0 in $V(x)$ is zero, and one other coefficient in $U(x)$ is zero, as determined by the type of normalization. Using the F-map, the coefficient of x^{v-1} in $V(x)$ can also be fixed to either 0 or 1. That is, if the coefficient of x^{v-1} is not zero, then consider the cycle, $V(x), F(V(x)), F^2(V(x)), \ldots, F^i(V(x)), \ldots$. By the definition of the F-map, the coefficient of x^{v-1} in $F^i(V(x))$ is $t^i a_{v-1}$, and for some i, $t^i a_{v-1} = 1$. Thus, if $\frac{V(x)}{U(x)}$ is an *nPRF* and the coefficient of x^{v-1} in $V(x)$ is nonzero and is not equal to 1, then there is also an *nPRF* where the coefficient of x^{v-1} in $V(x)$ is equal to 1. We now discuss the G-map [2] and how it can be applied to *PRFs*. The G-map raises each coefficient in a polynomial to the p-th power, where p is the field characteristic.

Definition 16. *Define the* **G-map on polynomials** *[2] over* $\mathbb{F}_q$ *by*

$$G(V(x)) = a_v^p x^v + a_{v-1}^p x^{v-1} + \cdots + a_1^p x + a_0^p$$

Define the **G-map on PRFs** *over* $\mathbb{F}_q$ *by*

$$G(W(x)) = \frac{G(V(x))}{G(U(x))} = \frac{a_v^p x^v + a_{v-1}^p x^{v-1} + \cdots + a_1^p x + a_0^p}{b_u^p x^u + b_{u-1}^p x^{u-1} + \cdots + b_1^p x + b_0^p}.$$

It is shown in [2] that, if $V(x)$ is a PP (nPP), then $G(V(x))$ is a PP (nPP), and that $G(V(x^p)) = V(x)^p$. Similarly,

$$G(W(x^p)) = \frac{G(V(x^p))}{G(U(x^p))} = \frac{V(x)^p}{U(x)^p} = W(x)^p.$$

Consequently, if $W(x)$ is a *PRF*, then $G(W(x))$ is a *PRF*. That is, $(0^p, 1^p, \ldots, (q-1)^p, \infty^p)$ is a permutation of $P^1(\mathbb{F}_q)$, and, if $W(x)$ is a *PRF*, then $W(x)^p$ is a *PRF*. This follows from the fact that $(x+y)^p = x^p + y^p$, when p is the characteristic of the field.

Iterating the G-map gives a cycle based on orbits of elements in $\mathbb{F}_q$. For example, consider the field $\mathbb{F}_{2^3}$, when defined by the primitive polynomial $x^3 + x^2 + 1$, and the *PRF* $W(x) = \frac{x^3 + x^2 + 2x}{x^2 + 4x + 1}$. We have $G(W(x)) = \frac{x^3 + x^2 + 3x}{x^2 + 7x + 1}$, $G^2(W(x)) = \frac{x^3 + x^2 + 5x}{x^2 + 6x + 1}$, $G^3(W(x)) = W(x)$. Consequently, it is not necessary to search separately for cases when the coefficient of x in the numerator is either 3 or 5. It is sufficient to search with the coefficient 2. In general, cycles partition the elements of $\mathbb{F}_q$ into disjoint sets, so for a chosen coefficient, the search can be limited to one value in each set.

5   Results

A basic computational strategy computes $N_{v,u}$ by considering all possible rational functions, $\frac{V(x)}{U(x)}$, where $V(x)$ and $U(x)$ are polynomials of degree v and u, respectively. This entails evaluating q^{u+v+2} different rational functions. We use a more efficient strategy that implements the normalization theorems in Table 1, and the F-map and the G-map described in Sect. 4. This fixes five coefficients, thus requiring at most q^{u+v-3} different rational functions to be evaluated. This computational strategy yields equivalence class representatives, which in turn yield the total number of *PRFs*, as indicated in Sects. 3 and 4. Our results are presented in Tables 3 through 7.

We have found several interesting classes of *PRFs*. Specifically, there are good classes with degree ratios 3/2, 4/3, and 5/4 for *PRFs* of $\mathbb{F}_q$, and with degree ratios 3/3, 4/4, and 5/5 for *PRFs* of $\mathcal{P}^1(\mathbb{F}_q)$. Note that when the degree of the numerator is larger than the degree of the denominator, the permutations of $\mathcal{P}^1(\mathbb{F}_q)$ end with ∞, and ∞ can just be deleted giving a permutation of $\mathbb{F}_q$.

Theorem 17 below justifies substantial improvements on lower bounds for $M(q, q-5)$ and $M(q+1, q-5)$, for many prime powers q, as shown in Table 3 and Table 6. As mentioned earlier, Ferraguti *et al.* [11] have recently given a complete characterization of monic degree 3 *PRFs*, which subsumes our results for degree ratios 3/2 and 3/3. They gave essentially Theorem 17 based on monic *PRFs*, hence we omit a proof.

Theorem 17. *For all q,*

$$N_{3,2}(q) = q^2(q-1)^2/2,$$
$$N_{3,3}(q) = q^2(q-1)^2(q+1)/2, \qquad \text{if } q \equiv 2 \pmod 3,$$
$$N_{3,3}(q) = q^2(q-1)^3/2, \qquad \text{if } q \equiv 1 \pmod 3,$$
$$N_{3,3}(q) = (q^4 - q^3 + q^2 - q)/2, \qquad \text{if } q \equiv 0 \pmod 3.$$

For degree ratios 4/3, 4/4, and 5/4, the number of *nPRFs* is also predictable. Formulas for the number of *PRFs* for ratios 4/3, 4/4, and 5/4 are given in Conjectures 1, 2 and Theorem 18 below. Experimentally, we have verified that $N_{4,3}(q)$ is exactly $(q+1)q^2(q-1)^2/3$, for all $q \le 307$. Again, this justifies substantial improvements on previous lower bounds for $M(q, q-7)$ for many prime powers q, as shown in Table 3.

Table 3. Lower bounds for $M(q, q-5)$ and $M(q, q-7)$ using $\mathcal{S}_5(q)$ and $\mathcal{S}_7(q)$, respectively. Improved bounds are shown in bold. (* see [23]).

q	$M(q, q-5) \geq$	Previous	$M(q, q-7) \geq$	Previous
16	29,792	40,320	381,120	1,377,360
17	42,466	83,504	490,960	1,240,320
19	59,546	65,322*	845,766	1,221,624
23	141,220	291,456	2,201,100	10,200,960
25	181,850	192,000	**3,316,800**	867,000
27	248,562	522,288	**4,866,966**	1,280,448
29	**355,656**	58,968	6,971,020	42,033,992
31	**435,240**	58,968	**9,687,810**	3,056,919
32	496,000	1,388,800	**11,691,712**	3,420,416
37	891,108	1,824,480	**23,411,232**	3,648,348
41	**1,416,960**	68,880	**39,135,320**	1,720,944
43	1,636,236	3,341,100	**49,547,610**	413,280
47	2,445,232	4,879,634	**77,330,416**	9,655,492
49	**2,773,008**	117,600	**95,081,952**	9,433,872
53	3,952,104	7,887,928	**140,812,308**	15,632,032
59	**6,067,206**	407,218	**240,463,940**	12,319,200
61	**6,708,780**	226,920	**283,767,120**	13,622,520
64	**8,144,640**	5,773,824	**360,991,078**	13,622,032
67	9,790,308	19,854,780	**453,303,642**	39,705,138
71	**12,718,230**	357,840	**605,882,760**	12,355,419
73	13,828,536	28,014,480	**695,631,600**	56,023,704
79	**19,003,608**	492,960	**1,032,017,922**	38,950,002
81	**21,280,320**	571,704	**1,169,529,840**	42,787,440
83	23,746,134	47,858,238	**1,321,303,228**	48,423,136
89	**31,390,656**	1,401,920	**1,872,278,760**	63,439,192
97	43,384,604	87,625,920	**2,876,904,792**	88,529,184
101	**52,045,300**	1,030,200	**3,520,385,300**	104,060,300
103	55,209,030	111,458,154	**3,881,578,278**	222,926,814
107	65,556,760	129,854,358	**4,696,631,464**	260,934,052
109	**69,313,536**	1,294,920	**5,150,579,616**	141,158,052
113	80,112,480	161,604,464	**6,166,737,248**	163,047,248
121	**105,444,240**	1,771,440	**8,679,213,840**	212,601,840
125	122,093,500	123,935,000	**10,212,593,500**	125,472,500
127	128,064,006	258,112,260	**11,053,461,510**	258,112,260
128	**132,161,280**	90,903,592	**11,495,251,584**	95,861,632
131	**145,044,510**	2,247,960	**12,905,964,110**	294,409,790
137	173,612,976	349,704,008	**16,142,578,480**	701,979,232
139	184,012,926	370,634,604	**17,354,972,046**	741,250,026
149	**243,189,456**	6,593,548	**24,557,724,656**	496,170,000

Conjecture 1. (4/3 - conjecture) For all q, $N_{4,3}(q) = (q+1)q^2(q-1)^2/3$.

Conjecture 2. (5/4 - conjecture) $N_{5,4}(q) > (q+1)q^3(q-1)^2/2$.

The 5/4-conjecture is true for all $q \leq 97$. $N_{5,4}(q)$ for $q \leq 97$ is shown in Table 4. This justifies substantial improvements on previous lower bounds on $M(q, q-9)$, which are shown in Table 5. We can show the following.

Table 4. Computed results for $N_{5,4}(q)$ and 5/4-conjectured lower bound.

q	$N_{5,4}(q)$	$(q+1)q^3(q-1)^2/2$	q	$N_{5,4}(q)$	$(q+1)q^3(q-1)^2/2$
17	16,189,440	11,319,552	53	11,074,291,488	10,869,212,016
19	24,503,958	22,223,160	59	21,084,006,242	20,726,848,680
23	74,762,512	70,665,936	61	25,753,041,000	25,331,079,600
25	125,820,000	117,000,000	64	34,351,091,712	33,814,609,920
27	193,179,168	86,279,912	67	45,315,700,848	44,544,203,352
29	297,858,652	286,814,640	71	64,037,083,250	63,135,500,400
31	444,126,150	428,990,400	73	75,638,717,568	74,616,572,736
32	553,107,456	519,585,792	79	121,523,765,922	119,985,971,040
37	1,280,989,728	1,247,279,472	81	141,192,720,000	139,450,118,400
41	2,373,572,000	2,315,745,600	83	163,422,731,808	161,477,223,096
43	3,157,263,648	3,085,507,656	89	248,458,577,312	245,667,597,120
47	5,384,729,088	5,272,547,232	97	416,397,477,888	412,148,524,032
49	6,917,645,952	6,776,582,400			

Table 5. Lower bounds for $M(q, q-9)$ using $S_9(q)$. Improved bounds are shown in bold. Previous values are obtained by permutation polynomials, except where indicated: (a, b) = Mathieu group M_{24} and contraction [3], (a, d) = Mathieu group M_{24} and $M(n+1, d) \geq M(n, d)$, (c) coset search [5].

n	$M(q, q-9) \geq$	Previous	q	$M(q, n-9) \geq$	Previous
13	4,926,480	60,635,520[c]	49	**6,877,311,504**	20,497,680
16	12,629,280	70,804,800[c]	53	**11,025,653,600**	23,373,636
17	12,342,272	75,176,640[c]	59	**20,979,628,398**	35,941,256
19	**23,218,380**	12,421,152	61	**25,628,242,320**	13,622,520
23	73,414,022	244,823,040[a,b]	64	**34,192,366,054**	332,236,800
25	121,108,200	244,823,040[a,d]	67	**45,036,911,436**	39,705,686
27	**191,914,893**	28,928,802	71	**63,766,789,800**	605,529,877
29	**294,515,648**	42,033,992	73	**75,367,839,096**	56,023,418
31	**439,831,410**	22,084,310	79	**121,056,446,004**	38,930,002
32	**533,338,880**	32,759,808	81	**140,641,174,881**	3,100,641,122
37	**1,274,288,436**	3,648,348	83	**162,892,864,290**	94,909,620
41	**2,357,705,000**	22,392,560	89	**247,603,307,248**	125,475,872
43	**3,141,656,196**	10,125,360	97	**415,199,758,776**	88,529,184
47	**5,359,530,978**	42,883,412			

Table 6. Lower bounds for $M(q+1, q-5)$ using $T_5(q)$.

$q+1$	$M(q+1, q-5) \geq$	Previous	$q+1$	$M(q+1, q-5) \geq$	Previous
14	172,536	380,160	50	**138,415,200**	2,768,309
17	**497,730**	187,600	54	**213,115,968**	7,890,428
18	**753,984**	83,504	60	**363,621,720**	821,240
20	**1,176,480**	177,840	62	**415,490,520**	13,622,520
24	**3,363,888**	291,456	65	**528,877,440**	5,515,776
26	**4,695,600**	218,418	68	**665,139,552**	19,854,780
28	**788,346**	522,288	72	**914,996,880**	28,014,480
30	**10,620,960**	170,520	74	**1,022,533,776**	28,014,480
32	**13,868,160**	1,388,810	80	**1,519,302,720**	38,930,002
33	**17,320,320**	1,388,810	82	**64,314,000**	21,001,679
38	**33,760,872**	1,824,800	84	**1,993,531,848**	47,458,238
42	**59,374,560**	1,419,680	90	**2,823,749,280**	125,475,872
44	**71,835,456**	1,632,624	98	**4,249,866,432**	87,625,920
48	**117,163,104**	4,879,634			

Table 7. Lower bounds for $M(q+1, q-7)$ using $T_7(q)$.

$q+1$	$M(q+1, q-7) \geq$	Previous	$q+1$	$M(q+1, q-7) \geq$	Previous
14	1,762,488	10,834,560	30	**208,424,160**	14,326,150
17	6,087,810	6,617,760	32	**309,087,360**	22,887,424
18	8,744,256	12,421,152	33	**375,214,080**	14,076,480
20	**16,771,680**	10,745,640	38	**887,754,024**	6,529,464
24	52,522,800	244,823,040	42	**1,640,859,360**	10,125,360
26	**85,815,600**	9,313,200	44	**2,176,677,888**	3,341,100
28	**129,574,458**	10,511,196	48	**3,706,982,496**	143,116,896

Theorem 18. *For any* $v > 1$, $N_{v,v}(q) = (q-1)\sum_{u<v} N_{v,u}(q)$.

We use the $N_{4,4}(q)$ results to obtain improved lower bounds for $M(q+1, q-7)$ as shown in Table 7.

Examples of overall results:
(a) $\mathcal{S}_5(23) = N_{3,2}(23) + N_{3,0}(23) + N_{1,1}(23)/22 + 2N_{1,0}(23) = 140,688$. So, $M(23,18) \geq 140,688$.
(b) $\mathcal{S}_7(23) = N_{4,3}(23) + N_{3,2}(23) + N_{3,0}(23) + N_{2,2}(23)/22 + 2N_{2,0}(23) + N_{1,1}(23) + 2N_{1,0}(23) = 2,201,100$, since $N_{4,3}(23) = 2,048,288$, $N_{3,2}(23) = 128,018$, $N_{3,0}(23) = 11,638$, $N_{1,1}(23) = 12,144$, and $N_{1,0}(23) = 506$. So, $M(23,16) \geq 2,201,100$.
(c) $N_{5,5}(19) = 446,802,480$. This yields $M(20,10) \geq N_{5,5}(19) + 2N_{5,4}(19) + 2N_{5,0}(19) + N_{4,4}(19) + 2N_{4,3}(19) + N_{3,3}(19) + 2N_{3,2}(19) + N_{1,1}(19) + 2N_{1,0}(19) = 508,177,876$.

(d) $N_{5,5}(23) = 1,650,664,092$. This yields $M(24,14) \geq N_{5,5}(23) + 2N_{5,4}(23) + 2N_{5,0}(23) + N_{4,4}(23) + 2N_{4,3}(23) + N_{3,3}(23) + 2N_{3,2}(23) + 2N_{3,0}(23) + N_{1,1}(23) + 2N_{1,0}(23) = 1,845,054,112$.

6 Conclusions, Acknowledgments, and Future Work

We have substantially improved many lower bounds for $M(n, D)$. We conjecture that many of our bounds represent formulas that are true for all powers of a prime q. We wish to thank Dr. Carlos Arreche in the Department of Mathematical Sciences at the University of Texas at Dallas for many helpful discussions.

References

1. Bereg, S., Levy, A., Sudborough, I.H.: Constructing permutation arrays from groups. Des. Codes Crypt. **86**(5), 1095–1111 (2017). https://doi.org/10.1007/s10623-017-0381-1
2. Bereg, S., Malouf, B., Morales, L., Stanley, T., Sudborough, I.H., Wong, A.: Equivalence relations for computing permutation polynomials. arXiv e-prints arXiv:1911.12823 (2019)
3. Bereg, S., Miller, Z., Mojica, L.G., Morales, L., Sudborough, I.H.: New lower bounds for permutation arrays using contraction. Des. Codes Crypt. **87**(9), 2105–2128 (2019). https://doi.org/10.1007/s10623-019-00607-y
4. Bereg, S., Mojica, L.G., Morales, L., Sudborough, H.: Constructing permutation arrays using partition and extension. Des. Codes Crypt. **88**(2), 311–339 (2019). https://doi.org/10.1007/s10623-019-00684-z
5. Bereg, S., Morales, L., Sudborough, I.H.: Extending permutation arrays: improving MOLS bounds. Des. Codes Crypt. **83**(3), 661–683 (2016). https://doi.org/10.1007/s10623-016-0263-y
6. Chu, W., Colbourn, C.J., Dukes, P.: Constructions for permutation codes in powerline communications. Des. Codes Crypt. **32**(1–3), 51–64 (2004). https://doi.org/10.1023/b:desi.0000029212.52214.71
7. Colbourn, C., Kløve, T., Ling, A.C.: Permutation arrays for powerline communication and mutually orthogonal latin squares. IEEE Trans. Inf. Theory **50**(6), 1289–1291 (2004). https://doi.org/10.1109/tit.2004.828150
8. Fan, X.: A classification of permutation polynomials of degree 7 over finite fields. Finite Fields Appl. **59**, 1–21 (2019). https://doi.org/10.1016/j.ffa.2019.05.001
9. Fan, X.: Permutation polynomials of degree 8 over finite fields of characteristic 2. Finite Fields Their Appl. **64**, 101662 (2020). https://doi.org/10.1016/j.ffa.2020.101662
10. Fan, X.: Permutation polynomials of degree 8 over finite fields of odd characteristic. Bull. Aust. Math. Soc. **101**(1), 40–55 (2020). https://doi.org/10.1017/S0004972719000674
11. Ferraguti, A., Micheli, G.: Full classification of permutation rational functions and complete rational functions of degree three over finite fields. Des. Codes Crypt. **88**(5), 867–886 (2020). https://doi.org/10.1007/s10623-020-00715-0
12. Gao, F., Yang, Y., Ge, G.: An improvement on the Gilbert-Varshamov bound for permutation codes. IEEE Trans. Inf. Theory **59**(5), 3059–3063 (2013). https://doi.org/10.1109/tit.2013.2237945

13. Hou, X.: Permutation polynomials over finite fields - a survey of recent advances. Finite Fields Appl. **32**, 82–119 (2015). https://doi.org/10.1016/j.ffa.2014.10.001
14. Janiszczak, I., Lempken, W., Östergård, P.R.J., Staszewski, R.: Permutation codes invariant under isometries. Des. Codes Crypt. **75**(3), 497–507 (2014). https://doi.org/10.1007/s10623-014-9930-z
15. Janiszczak, I., Staszewski, R.: Isometry invariant permutation codes and mutually orthogonal latin squares. J. Combin. Des. **27**(9), 541–551 (2019). https://doi.org/10.1002/jcd.21661
16. Li, J., Chandler, D.B., Xiang, Q.: Permutation polynomials of degree 6 or 7 over finite fields of characteristic 2. Finite Fields Appl. **16**, 406–419 (2010). https://doi.org/10.1016/j.ffa.2010.07.001
17. Lidl, R., Mullen, G.L.: When does a polynomial over a finite field permute the elements of the fields? II. Am. Math. Monthly **100**(1), 71–74 (1993). https://doi.org/10.1080/00029890.1993.11990369
18. Lidl, R., Niederreiter, H.: Introduction to Finite Fields and Their Applications. Cambridge University Press, Cambridge (1994). https://doi.org/10.1017/cbo9781139172769. Revised edn.
19. Micheli, G., Neri, A.: New lower bounds for permutation codes using linear block codes. IEEE Trans. Inf. Theory **66**(7), 4019–4025 (2020). https://doi.org/10.1109/tit.2019.2957354
20. Pavlidou, N., Vinck, A.H., Yazdani, J., Honary, B.: Power line communications: state of the art and future trends. IEEE Commun. Mag. **41**(4), 34–40 (2003). https://doi.org/10.1109/mcom.2003.1193972
21. Shallue, C.J., Wanless, I.M.: Permutation polynomials and orthomorphism polynomials of degree six. Finite Fields Appl. **20**, 84–92 (2013). https://doi.org/10.1016/j.ffa.2012.12.003
22. Wang, X., Zhang, Y., Yang, Y., Ge, G.: New bounds of permutation codes under Hamming metric and Kendall's τ-metric. Des. Codes Crypt. **85**(3), 533–545 (2016). https://doi.org/10.1007/s10623-016-0321-5
23. Yang, L., Chen, K., Yuan, L.: New constructions of permutation arrays. arXiv e-prints (2006). https://arxiv.org/pdf/0801.3987.pdf

Bases

Existence and Cardinality of k-Normal Elements in Finite Fields

Simran Tinani$^{(\boxtimes)}$ and Joachim Rosenthal

University of Zürich, Winterthurerstrasse, 8057 Zurich, Switzerland
simran.tinani@math.uzh.ch
https://www.math.uzh.ch/aa/

Abstract. Normal bases in finite fields constitute a vast topic of large theoretical and practical interest. Recently, k-normal elements were introduced as a natural extension of normal elements. The existence and the number of k-normal elements in a fixed extension of a finite field are both open problems in full generality, and comprise a promising research avenue. In this paper, we first formulate a general lower bound for the number of k-normal elements, assuming that they exist. We further derive a new existence condition for k-normal elements using the general factorization of the polynomial $x^m - 1$ into cyclotomic polynomials. Finally, we provide an existence condition for normal elements in $\mathbb{F}_{q^m}$ with a non-maximal but high multiplicative order in the group of units of the finite field.

Keywords: Finite fields · Normal bases

1 Introduction

Let q denote a power of a prime p, and $\mathbb{F}_q$ denote the finite field of order q. If $\mathbb{F}$ is an extension field of the field $\mathbb{K}$, we denote by $\mathrm{Gal}(\mathbb{F}/\mathbb{K})$ the Galois group of the extension field $\mathbb{F}$ over $\mathbb{K}$. We are interested in studying elements in a finite extension $\mathbb{F}_{q^m}$ of degree m over $\mathbb{F}_q$. An element $\alpha \in \mathbb{F}_{q^m}$ is called a normal element over $\mathbb{F}_q$ if all its Galois conjugates, i.e. the m elements $\{\alpha, \alpha^q, \ldots, \alpha^{q^{m-1}}\}$, form a basis of $\mathbb{F}_{q^m}$ as a vector space over $\mathbb{F}_q$. A basis of this form is called a normal basis.

We let ϕ denote the usual Euler-phi function for integers. Let $f \in \mathbb{F}_q[x]$ be a polynomial with positive degree m. Then $\Phi_q(f)$ is defined to be the order of the ring $\left(\frac{\mathbb{F}_q[x]}{\langle f \rangle} \right)^\times$, where $\langle f \rangle$ denotes the ideal generated by f in $\mathbb{F}_q[x]$. In other words, $\Phi_q(f)$ is the number of polynomials co-prime to f and with degree less than m. It is well known that normal elements exist in every finite extension $\mathbb{F}_{q^m}$ of $\mathbb{F}_q$ and that there are precisely $\Phi_q(x^m - 1)$ normal elements, and thus $\frac{\Phi_q(x^m-1)}{m}$ normal bases in $\mathbb{F}_{q^m}$ [14, Theorem 2.35, Theorem 3.73], [8,18]).

Normal elements are a topic of major significance and interest because they offer an avenue for efficient arithmetic in a finite field $\mathbb{F}_q$: for instance, raising

© Springer Nature Switzerland AG 2021
J. C. Bajard and A. Topuzoğlu (Eds.): WAIFI 2020, LNCS 12542, pp. 255–271, 2021.
https://doi.org/10.1007/978-3-030-68869-1_15

an element to the power q is simply a cyclic shift in normal base representation. Normal bases and related concepts such as optimal normal bases and self-dual normal bases find several applications, both theoretical and practical. We refer the interested reader to [1,6], and [16] for more on this topic.

In [9], Huczynska et al. introduced the concept of k-normal elements as a natural generalization of normal elements. One of the many equivalent ways to define a k-normal element $\alpha \in \mathbb{F}_{q^m}$ is as an element whose conjugates $\{\alpha, \alpha^q, \alpha^{q^2}, \ldots \alpha^{q^{m-1}}\}$ span a vector space of dimension $m - k$ over $\mathbb{F}_q$. It is then of natural interest to examine the existence and the number of k-normal elements. These problems have been shown to be closely tied to the factorization of the polynomial $x^m - 1$ [9]. In this paper, we denote by n_k the number of k-normal elements in an extension $\mathbb{F}_{q^m}$ of $\mathbb{F}_q$. There are numerous known results on bounds on the number n_0, several of which build on the lower bounds proved in [5] using properties of the function Φ_q (see also the improvements on these results in [7]). For arbitrary k, $0 < k < m - 1$, neither a general rule for the existence of k-normal elements nor a general formula for their number n_k, when they exist, is known.

Huczynska et al. [9] have used the approach of Frandsen [5] to give a lower bound on n_k which holds asymptotically, as well as an upper bound which holds in general. However, both their upper and lower bounds depend directly on the number of divisors of $x^m - 1$ with degree $m - k$, and are thus difficult to calculate. Moreover, when $x^m - 1$ has no divisor of degree $m - k$, the bounds equal zero, which means that the statement about lower bounds does not yield any existence result.

In a recent paper, Saygı et al. [21] give formulas (in terms of q and m) for n_k for cases where m is a power of a prime or of the form $2^v \cdot r$ where $r \neq 2$ is a prime and $v \geq 1$, using known results on the explicit factorization of cyclotomic polynomials. In particular, their formulae guarantee existence for certain cases. A recent result by Reis [20, Theorem 5.5] provides a sufficient condition on m for which k-normal elements exist for every $0 \leq k \leq m$. Some relevant interesting results on the construction of k-normal elements, as well as alternate proofs of existing results, are found in [22].

In 1987, Lenstra and Schoof [13] proved (also see partial proofs by Carlitz [2] and Davenport [4]) the Primitive Normal Basis theorem, which states the existence of an element that is simultaneously normal and primitive (i.e. has multiplicative order $q^m - 1$ in $\mathbb{F}_{q^m}^*$). By extension, elements that have high multiplicative orders and also span large subspaces along with their conjugates are of interest. In particular, problems along this line have found mention in [9,11,12] and [17]. The question of the existence of elements in $\mathbb{F}_{q^m}$ that are both 1-normal over $\mathbb{F}_q$ and primitive has been answered in entirety in [20], after a partial proof and formulation of the problem in [9].

In this paper, we first present a result that guarantees a general lower bound on n_k (for arbitrary $0 \leq k \leq m - 1$), provided that k-normal elements exist. This proves a link between n_0 and n_k for $k > 0$. Since this result does not make any additional assumption about k, q or m, it is not derivable from any of the known formulas for n_k.

We further present an existence condition for k-normal elements (over $\mathbb{F}_q$) in $\mathbb{F}_{q^m}$ based on inequalities involving m and k. It turns out that under certain constraints on m (loosely put, m must have a sufficiently large common divisor with $q^m - 1$), k-normal elements exist for k above a minimum lower bound. This result is independent of the factorization of $x^m - 1$. Moreover, the conditions on m and q required are weaker than the special forms required in [21], and also cannot be derived from the conditions in [19, Theorem 5.5]. In fact, when $p \nmid m$, our theorem is a generalization of this result.

Our final contribution is an existence condition for normal elements of multiplicative order $\frac{q^m-1}{q-1}$ in $\mathbb{F}_{q^m}$ when m and $q - 1$ are co-prime. Using the terminology of [17], this is the same as talking about 0-normal, $(q - 1)$-primitive elements. With this result, we answer a special case of Problem 6.4 posed in [9], which deals with high multiplicative order k-normal elements in $\mathbb{F}_{q^m}$ over $\mathbb{F}_q$. Our proof follows the method used by Lenstra and Schoof in proving the Primitive Normal Basis Theorem [13].

2 Preliminaries

Definition 1. *An element $\alpha \in \mathbb{F}_{q^m}$ is called k-normal if*

$$\dim_{\mathbb{F}_q}\left(\operatorname{span}_{\mathbb{F}_q}\left\{\alpha, \alpha^q, \ldots, \alpha^{q^{m-1}}\right\}\right) = m - k.$$

Remark 1. It is clear from the definition that an element α is 0-normal if and only if it is normal by the usual definition. Also, the only m-normal element in $\mathbb{F}_{q^m}$ is 0.

Given $\alpha \in \mathbb{F}_{q^m}$, we denote by $\operatorname{ord}(\alpha)$ the usual multiplicative order of α in the group $\mathbb{F}_{q^m}^*$. $\mathbb{F}_{q^m}$ may be seen as a module over the ring $\mathbb{F}_q[x]$, under the action

$$\sum_{i=0}^{n} a_i x^i \cdot \alpha = \sum_{i=0}^{n} a_i \alpha^{q^i}, \ \alpha \in \mathbb{F}_{q^m}. \tag{1}$$

In other words, the value of the image of α under the action of a polynomial $f(x) = \sum_{i=0}^{n} a_i x^i$ is the evaluation of α at the q-associate [14, Definition 3.58] of $f(x)$. Note that this is the same as the action of $\mathbb{F}_q$-linear maps on $\mathbb{F}_{q^m}$. This module structure has been explored in more detail, for instance, in [22]. Through this module structure, we also have another concept of order, as defined in [13] as an additive analogue of the multiplicative order.

Definition 2. *Define the function*

$$\operatorname{Ord} : \mathbb{F}_{q^m} \to \mathbb{F}_q[x]$$

as follows. For any $\alpha \in \mathbb{F}_{q^m}$, $\mathrm{Ord}(\alpha)$ *is the unique monic polynomial generating the annihilator of* α *under the action defined by Eq. (1), i.e.*

$$\mathrm{Ann}(\alpha) = \langle \mathrm{Ord}(\alpha) \rangle \text{ in } \mathbb{F}_q[x].$$

We now state an important result which provides several equivalent characterizations of k-normal elements.

Theorem 1 [9, Theorem 3.2]. *Let* α *be an element of* $\mathbb{F}_{q^m}$ *and*

$$g_\alpha(x) := \sum_{i=0}^{m-1} \alpha^{q^i} \cdot x^{m-1-i} \in \mathbb{F}_{q^m}[x].$$

Then the following conditions are equivalent:

1. α *is* k-*normal.*
2. $\gcd(x^m - 1, g_\alpha(x))$ *over* $\mathbb{F}_{q^m}$ *has degree* k.
3. $\deg(\mathrm{Ord}(\alpha)) = m - k$.
4. *The matrix* A_α *defined below has rank* $m - k$.

$$A_\alpha = \begin{bmatrix} \alpha & \alpha^q & \alpha^{q^2} & \cdots & \alpha^{q^{m-1}} \\ \alpha^{q^{m-1}} & \alpha & \alpha^q & \cdots & \alpha^{q^{m-2}} \\ \vdots & \vdots & \cdots & \vdots & \vdots \\ \alpha^q & \alpha^{q^2} & \alpha^{q^3} & \cdots & \alpha \end{bmatrix}.$$

The following result on the number of k-normal elements will also prove useful.

Theorem 2 [9, Theorem 3.5]. *The number of* k-*normal elements of* $\mathbb{F}_{q^m}$ *over* $\mathbb{F}_q$ *equals 0 if there is no* $h \in \mathbb{F}_q[x]$ *of degree* $m - k$ *dividing* $x^m - 1$; *otherwise it is given by*

$$\sum_{\substack{h | x^m - 1 \\ \deg(h) = m-k}} \Phi_q(h), \tag{2}$$

where divisors are monic and polynomial division is over $\mathbb{F}_q$.

It is known that $x^m - 1$ factorizes over $\mathbb{F}_q$ into the product of cyclotomic polynomials of degrees dividing m [14, Theorem 2.45]. Moreover, for $p \nmid d$ (recall that p is defined as $p = \mathrm{char}(\mathbb{F}_q)$), each of the irreducible factors of the cyclotomic polynomial $Q_d(x)$ has degree $\frac{\phi(d)}{r}$, where r is the multiplicative order of d mod q [14, Theorem 2.47]. Since there is no known closed formula for this number, there is also no closed-form complete factorization (i.e. factorization into irreducibles) of $x^m - 1$ over $\mathbb{F}_q$. Thus, the above theorem does not give direct answers about the existence of k-normal elements for $k > 0$. However, it may be used to ascertain the existence of k-normal elements for certain values of k. In the next two sections, we look at some interesting results on k-normal elements which can be derived in certain special cases using Thereom 2.

3   Number of k-Normal Elements

For $k = 0$, the formula in Thereom 2 yields the well-known value $\Phi_q(m)$ for the number of normal elements over $\mathbb{F}_q$ in $\mathbb{F}_{q^m}$ [14, Theorem 3.37]. Since $x^m - 1$ always has the divisor $x - 1$ of degree 1 and hence also a divisor of degree $m - 1$ (and since $\Phi_q(f(x)) \neq 0$ for any nonzero polynomial $f(x)$), we always have 1-normal and $(m - 1)$-normal elements in $\mathbb{F}_{q^m}$. It has been observed in [9] that the only values of k for which k-normal elements are guaranteed to exist for every pair (q, m) are 0, 1 and $m - 1$. In fact, as noted in [20], if q is a primitive root modulo m, $\frac{x^m - 1}{x - 1}$ is irreducible and so for $1 < k < m - 1$, k-normal elements do not exist.

In certain other cases, it is possible to use information about the factorization of $x^m - 1$ along with Theorem 2 to gain insights into the number of k-normal elements for different values of k. In [21], the authors provide explicit formulas for k-normal elements for degrees m that are either prime powers or numbers of the form $2^v \cdot r$, for a prime $r \neq 2$, under certain other constraints on q and m. Below we state one of their noteworthy results.

Proposition 1 ([21, Proposition 1]). *Let* $\mathrm{char}(\mathbb{F}_q) = p$ *and* $m = p^r$ *for some positive integer* r. *Then the number of* k-normal elements of $\mathbb{F}_{q^m}$ *over* $\mathbb{F}_q$ *is given by*

$$(q - 1) \cdot q^{m-k-1},$$

where $k = 0, 1, \ldots, m - 1$.

The following result by Huczynska et al. [9] formulates a lower bound for the number of k-normal elements when the extension degree m is large enough.

Theorem 3 ([9, Theorem 4.6]). *Let* c_{m-k} *denote the number of divisors of* $x^m - 1$ *with degree* $m - k$. *There is a constant* c *such that for all* $q \geq 2$ *and* $m > q^c$, *the number of* k-normal elements of $\mathbb{F}_{q^m}$ *over* $\mathbb{F}_q$ *is at least*

$$0.28477 \cdot q^{m-k} \cdot \frac{c_{m-k}}{\sqrt{\log_q(m)}}.$$

Note that there is no simple rule or formula for the value c_{m-k} in terms of m, k and q, and it may equal zero. So, the above result does not yield an existence condition.

We now proceed to build a general result on the number of k-normal elements, assuming that they exist. For this purpose, we consider the structure of $\mathbb{F}_{q^m}$ as an $\mathbb{F}_q[x]$-module under the action defined by Eq. (1). We follow the approach in [10], which is based on the observation that for $\mathbb{K} = \mathbb{F}_{q^m}$ and $G = \mathrm{Gal}(\mathbb{K}/\mathbb{F}_q)$, the group of invertible elements $\mathbb{K}[G]^\times$ of the group algebra $\mathbb{K}[G]$ acts on the set of normal elements of $\mathbb{F}_{q^m}$. Using this group action, the author of [10] formulates an alternative method to count normal elements. We adapt the same argument to find a lower bound on the number of k-normal elements in $\mathbb{F}_{q^m}$ when they exist.

Theorem 4. Let $k \in \{0, 1, \ldots, m\}$ and let n_k denote the number of k-normal elements in $\mathbb{F}_{q^m}$. If $n_k > 0$, i.e. if k-normal elements exist in $\mathbb{F}_{q^m}$, then

$$n_k \geq \frac{\Phi_q(x^m - 1)}{q^k}.$$

Proof. Denote $G := \mathrm{Gal}(\mathbb{F}_{q^m}/\mathbb{F}_q)$ and $\mathbb{K} := \mathbb{F}_q$. Let S_k be the set of k-normal elements over $\mathbb{F}_q$ in $\mathbb{F}_{q^m}$, and assume that $S_k \neq \emptyset$. Let $\mathbb{K}[G]^\times$ be the group of invertible elements of the group algebra $\mathbb{K}[G]$. The map

$$\mathbb{K}[G]^\times \times S_k \to S_k, \text{ given by}$$

$$\left(\sum_{h \in G} a_h \cdot h \right) \cdot \alpha = \sum_{h \in G} a_h \cdot (h \cdot \alpha) \tag{3}$$

for $\alpha \in \mathbb{F}_{q^m}$ and coefficients $a_h \in \mathbb{K}$ defines a group action. The rest of the axioms are clear, and only thing that needs to be verified is that k-normal elements map to k-normal elements. To see this, note that an element $\psi = \sum_{h \in G} a_h \cdot h$ of $\mathbb{K}[G]^\times$ is a field automorphism of $\mathbb{F}_{q^m}$, and so the images of subspaces of dimension $m - k$ also have dimension $m - k$. So, for a k-normal element α,

$$\dim(\mathrm{span}\{\psi(\alpha), \psi(\alpha^q), \ldots, \psi(\alpha^{q^{m-1}})\}) = \dim(\mathrm{span}\{\psi(\alpha), \psi(\alpha)^q, \ldots, \psi(\alpha)^{q^{m-1}}\})$$

$$= \dim(\mathrm{span}\{\alpha, \alpha^q, \ldots, \alpha^{q^{m-1}}\}) = m - k.$$

Now note that for a generator σ of G we have a ring isomorphism

$$\left(\frac{\mathbb{F}_q[x]}{\langle x^m - 1 \rangle} \right) \mapsto \mathbb{K}[G]$$

$$x \mapsto \sigma. \tag{4}$$

Therefore,

$$\mathbb{K}[G]^\times \cong \left(\frac{\mathbb{F}_q[x]}{\langle x^m - 1 \rangle} \right)^\times \quad \text{(as groups).} \tag{5}$$

We conclude that through the isomorphism (4) the group action (3) induces a group action

$$\left(\frac{\mathbb{F}_q[x]}{\langle x^m - 1 \rangle} \right)^\times \times S_k \mapsto S_k$$

$$\text{given by} \quad \left(\sum_{i=0}^{m-1} f_i \cdot x^i \right) \cdot \alpha = \sum_{i=0}^{m-1} f_i \cdot \sigma^i(\alpha) = \sum_{i=0}^{m-1} f_i \cdot \alpha^{q^i}. \tag{6}$$

Denote $H := \left(\frac{\mathbb{F}_q[x]}{\langle x^m - 1 \rangle} \right)^\times$. For any k-normal element α, we have

$$\mathrm{Stab}(\alpha) = \{p(x) \in H : p(x) \cdot \alpha = \alpha\}$$

$$= \{p(x) \in H : (p(x) - 1) \cdot \alpha = 0\}$$

$$= \{p(x) \in H : \mathrm{Ord}(\alpha) \text{ divides } (p(x) - 1).\} \tag{7}$$

We know from Theorem 1 that $\mathrm{Ord}(\alpha)$ is a polynomial of degree $m - k$. Equation (7) implies that for $p(x) \in H$,

$$p(x) \in \mathrm{Stab}(\alpha) \iff p(x) = \mathrm{Ord}(\alpha) \cdot r(x) + 1, \text{ with } \deg(r(x)) \le k - 1. \quad (8)$$

Hence, the number of possible distinct values for $p(x) \in \mathrm{Stab}(\alpha)$ cannot exceed the number of polynomials with degree less than k. More precisely,

$$|\mathrm{Stab}(\alpha)| \le \min(|H|, q^k) = \min \left(\varPhi_q(x^m - 1), q^k \right) \le q^k. \quad (9)$$

Finally, Eq. (9) and the Orbit-Stabilizer Theorem together give

$$|\mathrm{Orb}(\alpha)| = \left| \frac{H}{\mathrm{Stab}(\alpha)} \right| \ge \frac{\varPhi_q(x^m - 1)}{q^k}.$$

Since the action (6) is on k-normal elements, it is now clear that the number n_k of k-normal elements satisfies $n_k \ge \frac{\varPhi_q(x^m - 1)}{q^k}$, thus completing the proof. $\quad \square$

Remark 2. Note that if a k-normal element α exists, then the lower bound in Theorem 2 is, in fact, for the number of k-normal elements lying in a single orbit, and therefore in $\mathrm{span}_{\mathbb{F}_q}\{\alpha, \alpha^q, \alpha^{q^2}, \dots, \alpha^{q^{m-1}}\}$.

Remark 3. In [10], it is shown that for the case of normal elements (i.e. $k = 0$), the action (6) is both free (i.e. $u \cdot \alpha = \alpha \implies u = 1$) and transitive. This yields an alternate proof of the well-known result that the number of normal elements in $\mathbb{F}_{q^m}$ is equal to $\varPhi_q(x^m - 1)$. For $k > 0$ it is clear that for every k-normal α, there exists $u \in \mathbb{K}[G]$ such that $u \cdot \alpha = \alpha$. However, it is unclear whether such a u can be found in $\mathbb{K}[G]^\times$ or if the action is transitive. So, we cannot directly adapt the argument as in [10] to count the exact number of k-normal elements. However, as shown by the above theorem, the action may nevertheless be used to obtain a lower bound.

4   Existence of k-Normal Elements

From the previous section, it is clear that some results on the number of k-normal elements automatically imply their existence. For instance, the existence of k-normal in $\mathbb{F}_{q^m}$ for m a power of the characteristic p is established as an immediate corollary of Proposition 1. On the other hand, the cardinality formula in Theorem 3 gives the value zero when $x^m - 1$ has no divisor with degree $m - k$, and thus yields no condition for the existence of k-normal elements. Similarly, the statement on cardinalities in Theorem 4 holds only under the assumption that k-normal elements exist in $\mathbb{F}_{q^m}$. We now shift our focus to finding existence conditions for k-normal elements over $\mathbb{F}_q$. We begin by presenting (a slight rewording of) a result by Reis, which is closely related to our existence result.

Theorem 5 ([19]). *Let q be a power of a prime p and let $m \ge 2$ be a positive integer such that every prime divisor of m divides $p \cdot (q - 1)$. Then k-normal elements exist for all $k = 0, 1, 2, \dots, m$.*

Clearly, we get the existence implication of Proposition 1 as a corollary of the above theorem. Although this theorem significantly extends Proposition 1, it still restricts the prime factorization of m to be of a particular form, and thus limits the allowed values of m. It is easy to see that it does not apply to simple examples like $q = 5$, $m = 6$, and $q = 8$, $m = 6$, where k-normal elements are known to exist for every $k = 0, 1, 2, \ldots, m$. We now state the main result of this section, a sufficient condition for the existence of k-normal elements, which does not, unlike Proposition 1 and Theorem 5, require m or its prime factors to be of a fixed type. This result is also independent of the factorization of $x^m - 1$ into irreducibles over $\mathbb{F}_q$, and is derived using only the general factorization into cyclotomic polynomials. Before the main theorem, we prove a number theoretic result which will be used. The proof of the below proposition was inspired by the proof of Theorem 6.3 in [15].

Proposition 2. *Let a and m be arbitrary natural numbers and suppose that $m \nmid a^m - 1$. Then m has a prime factor that does not divide $a^m - 1$.*

Proof. We proceed by contradiction. Suppose that the statement is false and let p be any prime divisor of m. By hypothesis, $p \mid a^m - 1$. Write $m = p^b \cdot s$, with $b \geq 1$ and $p \nmid s$. We have

$$0 = a^m - 1 \mod p$$
$$= (a^{sp^b} - 1) \mod p$$
$$= (a^s - 1)^{p^b} \mod p$$
$$\implies a^s = 1 \mod p. \tag{10}$$

We claim that $a^m - 1 = 0 \mod p^b$, or in other words, $a^{sp^b} - 1 = 0 \mod p^b$. We prove this claim by induction on b.

For $b = 1$, the statement $a^{ps} - 1 = 0 \mod p$ is true by the hypothesis of the proposition. Now assume that $a^{sp^b} - 1 = 0 \mod p^b$ for some $b \geq 1$. Then,

$$a^{sp^{b+1}} - 1 = (a^{sp^b})^p - 1$$
$$= (a^{sp^b} - 1)(1 + a^{sp^b} + a^{2sp^b} + \ldots + a^{(p-1)sp^b}). \tag{11}$$

By the induction hypothesis, $p^b \mid a^{sp^b} - 1$. Also, from (10), we have

$$a^s = 1 \mod p$$
$$\implies a^{isp^b} = 1 \mod p \ \forall \ 0 \leq i \leq p - 1.$$
$$\implies 1 + a^{sp^b} + a^{2sp^b} + \ldots + a^{(p-1)sp^b} = 0 \mod p$$

Combining these results, (11) clearly gives $p^{b+1} \mid a^{sp^{b+1}} - 1$, thus proving the result for $b + 1$. By induction, the result holds for every $b \geq 1$, and therefore for every $m = sp^b$. So, we may now conclude that $a^m - 1 = 0 \mod p^b$ for m, b, p as in the proposition. Since this holds for any prime factor p of m, this implies that $m \mid a^m - 1$, which is a contradiction to the assumption. Hence, we must have $p \nmid a^m - 1$ for some prime divisor p of m. The proof is now complete. $\square$

Remark 4. Note that if $p \nmid m$ and the hypothesis of Theorem 5 by Reis holds, i.e. every prime factor of m divides $p \cdot (q-1)$ then Proposition 2 says that we are in the case $m \mid q^m - 1$. In this case it will become clear that our theorem is a generalization of the result of 5.

Theorem 6. *If $m \mid (q^m - 1)$, then k-normal elements exist in $\mathbb{F}_{q^m}$ for every integer k in the interval $0 \le k \le m - 1$. If $m \nmid q^m - 1$, let $d = \gcd(q^m - 1, m)$. Assume that $\sqrt{m} < d$. Let b denote the largest prime divisor of m that is a non-divisor of $q^m - 1$ (b exists by Proposition 2). Then, for $k \ge m - d - b + 1$, k-normal elements exist in $\mathbb{F}_{q^m}$. In particular, if m is a prime not dividing $q^m - 1$, then we have $b = m$, $d = 1$, and so k-normal elements exist for every k in the interval $0 \le k \le m - 1$.*

Proof. We know from Eq. (2) that the number of k-normal elements in $\mathbb{F}_{q^m}$ is given by

$$\sum_{\substack{h \mid x^m - 1 \\ \deg h = m - k}} \Phi_q(h(x)).$$

Thus, normal elements exist in $\mathbb{F}_{q^m}$ if and only if $x^m - 1$ has a divisor of degree $m - k$. First note that for $d = \gcd(q^m - 1, m)$, we have $d \mid q^m - 1$, the order of $\mathbb{F}_{q^m}^*$, so by the general properties of a finite cyclic group, there are precisely d elements α in the group $\mathbb{F}_{q^m}^*$ satisfying $\alpha^d = 1$, and so d elements must also satisfy $\alpha^m = 1$. Thus, $x^m - 1$ has precisely d linear factors over $\mathbb{F}_{q^m}$. Let its roots in $\mathbb{F}_{q^m}$ be $\alpha_1, \alpha_2, \ldots, \alpha_d$.

If $m \mid q^m - 1$, then $d = m$, and $x^m - 1$ splits into linear factors over $\mathbb{F}_{q^m}$. Thus, in this case, for any $k \in \{0, 1, 2, \ldots, m-1\}$, one may always combine $m - k$ of the m linear factors to obtain a factor of degree $m - k$ of $x^m - 1$. Hence, we are done for this case. Note that the same conclusion could have been drawn by directly applying Theorem 2 and using the fact that the polynomial splits into linear factors.

If $m \nmid q^m - 1$, then $d < m$. Assume that for some $k \in \{0, 1, 2, \ldots, m - 1\}$, no k-normal element exists in $\mathbb{F}_{q^m}$. It is known that $x^m - 1$ has the following factorization over $\mathbb{F}_q$:

$$x^m - 1 = \prod_{t \mid m} Q_t(x).$$

where $Q_t(x)$ denotes the tth cyclotomic polynomial, and is known to have coefficients in $\mathbb{F}_q$ [14, Theorem 2.45]. Write

$$x^m - 1 = \prod_{t \mid d} Q_t(x) \cdot \prod_{\substack{t \mid m \\ t \nmid q^m - 1}} Q_t(x)$$

$$= (x^d - 1) \cdot \prod_{\substack{t \mid m \\ t \nmid q^m - 1}} Q_t(x)$$

$$= (x - \alpha_1) \cdot (x - \alpha_2) \cdot \ldots \cdot (x - \alpha_d) \cdot \prod_{\substack{t \mid m \\ t \nmid q^m - 1}} Q_t(x),$$

where the last step follows from the fact that $d \mid q^m - 1$, so as in the first case, $x^d - 1$ splits in $\mathbb{F}_{q^m}$. Now, let b be the largest prime dividing m but not $q^m - 1$ (such a prime exists by Proposition 2). Then $Q_b(x)$ figures in the latter product of the above equation. Since no k-normal element exists in $\mathbb{F}_{q^m}$, $m - k$ must be greater than the number d of linear factors, and it must be impossible to combine the factors of degree greater than 1, in particular, $Q_b(x)$, with the linear factors to obtain a factor of degree $m - k$. Mathematically, we get, after minor rearrangement,

$$k < m - d, \tag{12}$$

and

$$\text{either } k > m - \phi(b) \text{ or } k < m - d - \phi(b). \tag{13}$$

Now, since b is a prime dividing m but not $q - 1$, b must divide $\frac{m}{d}$. In particular, $b \leq \frac{m}{d}$. From the hypothesis $\sqrt{m} < d$, we get $b \leq \frac{m}{d} < d$, and so

$$m - \phi(b) = m - b + 1$$
$$> m - d + 1 > m - d$$
$$> k, \tag{14}$$

where the last step follows from Eq. (12). We now immediately note that the former condition in Eq. (13) is incompatible with Eq. (14), and so it cannot hold. Therefore, the latter condition of Eq. (13) must be satisfied, i.e. we must have

$$k < m - d - \phi(b) = m - d - b + 1$$

for k such that k-normal elements do not exist. Hence, we conclude that for all $k \geq m - d - b + 1$, k-normal elements exist in $\mathbb{F}_{q^m}$, as required.

Finally, it is clear that if m is a prime, then we have $b = m$, $d = 1$, and so k-normal elements exist for every k in the interval $0 \leq k \leq m - 1$ by the above condition. $\square$

Remark 5. If m is composite and does not divide $q^m - 1$, then we cannot conclude the existence of k-normal elements for every value of k using the above theorem. This follows from the following argument, which was provided by one of the reviewers of this paper. Since b and d are different divisors of m, then $b + d \leq \frac{m}{2} + \frac{m}{3}$, which is incompatible with the condition $m \leq d + b - 1$.

Remark 6. Note that the fact that b is a prime plays a key role in the above proof. If b is, instead, an arbitrary divisor of m that does not divide $q-1$, then it is not guaranteed that b divides $\frac{m}{d}$ (E.g. consider $q = 25, m = 20, b = 10$). So the argument may not hold true even though the inequality $\frac{m}{d} < d$ may hold.

We now reconsider the two examples considered before. For $q = 5, m = 6$, we have $q^m - 1 = 15624$, which is divisible by 6. So, Theorem 6 shows that k-normal elements exist in $\mathbb{F}_{q^m}$ for every $k \in \{0, 1, \ldots, m\}$. For $q = 8, m = 6$, we have $q^m - 1 = 262143$, and so $d = \gcd(q^m - 1, m) = 3 > \sqrt{6}$. The largest prime b that divides 6 and not 262143 is clearly 2. So, Theorem 6 shows that k-normal elements exist in $\mathbb{F}_{q^m}$ for every $k \geq m - d - b + 1$, i.e. for every $k \geq 2$. Since we know that 0- and 1-normal elements always exist in $\mathbb{F}_{q^m}$, we conclude that in this case k-normal elements exist for every $k \in \{0, 1, \ldots, m\}$. The exact numbers for these two examples are listed in Tables 4 and 2, respectively, in Sect. 6.

5   Normal Elements with Large Multiplicative Order

So far, we have studied the "additive" structure of $\mathbb{F}_{q^m}$ as a vector space over $\mathbb{F}_q$. It is also of interest to study the relation between this additive structure and the natural multiplicative structure of $\mathbb{F}_{q^m}^*$. One of the most noteworthy results in this direction is the Primitive Normal Basis Theorem [2,4,13]). We state some of its proposed generalizations of this result in Sect. 6. Below, we state and prove an existence result for normal elements (i.e. $k = 0$) with multiplicative order $\frac{q^m - 1}{q - 1}$ in $\mathbb{F}_{q^m}$. It turns out that such elements always exist if m and $q-1$ are co-prime, and that this may be derived using the same methods as Lenstra and Schoof [13] in the proof of the Primitive Normal Basis Theorem.

Theorem 7. *Suppose that $(m, q-1) = 1$. Then $\mathbb{F}_{q^m}$ has a normal element with multiplicative order $\frac{q^m - 1}{q - 1}$.*

Proof. Let $k := \frac{q^m - 1}{q - 1}$. Define

$$A = \{\alpha \in \mathbb{F}_{q^m} : \operatorname{Ord}(\alpha) = x^m - 1\},$$
$$B = \{\alpha \in \mathbb{F}_{q^m}^* : \operatorname{ord}(\alpha) = k\},$$
$$C = \{\alpha \in \mathbb{F}_{q^m} : \alpha^{(q-1)^2} = 1\},$$

where the sets A and C are defined identically as in the proof of Lenstra and Schoof, and B is defined as the set of elements with order k, rather than primitive elements. Note that C is a subgroup of $\mathbb{F}_{q^m}^*$. Also note that since the definitions of A and C are unchanged, we may use directly the result (1.12) of the original proof in [13]. We state this as follows. For the set CA defined as

$$CA = \{\gamma \cdot \alpha : \gamma \in C, \alpha \in A\},$$

we have

$$CA = A. \tag{15}$$

Let BC denote the set $BC = \{\beta \cdot \gamma : \beta \in B, \gamma \in C\}$. Now, since Eq. (15) holds, the exact same argument as in the original proof also yields the result indexed (1.13) in [13]. Since we have a different B, we prove it below. The proof is identical for B defined as the set of elements of any multiplicative order.

If $\alpha \in A$, $\beta \in B$, $\gamma \in C$ are such that $\alpha = \beta \cdot \gamma \in B \cdot C$, then $\beta = \alpha \cdot \gamma^{-1} \in CA \cap B = A \cap B$, and so we have

$$A \cap B = \emptyset \iff A \cap BC = \emptyset. \tag{16}$$

As in the original paper, we use Eq. (16) and prove that $A \cap B \cdot C \neq \emptyset$ to conclude that $A \cap B \neq \emptyset$.

Let H denote the unique subgroup of order k in $\mathbb{F}_{q^m}^*$. Here,

$$
\begin{aligned}
BC &= \{\beta \cdot \gamma : \beta \in B, \gamma \in C\} \\
&= \{\beta \cdot \gamma : \beta \text{ generates } H, \gamma \in C\} \\
&= \left\{ \beta \cdot \gamma : \beta \cdot C \text{ generates } \frac{H}{C}, \gamma \in C \right\} \\
&= \left\{ \beta \cdot \gamma : \beta \cdot C \cap H \text{ generates } \frac{H}{H \cap C}, \gamma \in C \right\}.
\end{aligned}
$$

Now note that

$$
\begin{aligned}
\gcd(k, (q-1)) &= \gcd\left(\frac{q^m - 1}{q - 1}, q - 1 \right) \\
&= \gcd\left(1 + q + q^2 + \ldots + q^{m-1}, q - 1 \right) \\
&= \gcd(m, q - 1) \\
&= 1,
\end{aligned}
$$

where the second last equality can be checked by direct computation for general values of m and q, and the last equality follows by the hypothesis of the theorem. We now have $|C| = (q - 1) \cdot \gcd(q - 1, m) = (q - 1)$. So, in this case, C is the unique subgroup of $\mathbb{F}_{q^m}^*$ with order $q - 1$. Thus, C and H are subgroups with co-prime orders, and therefore intersect trivially. Now let

$$D = \{\alpha \in \mathbb{F}_{q^m}^* : ord(\alpha) = q^m - 1\}$$

denote the set of generators of $\mathbb{F}_{q^m}^*$. We claim that

$$D \subseteq BC.$$

To see this, pick $\alpha \in D$. Since $\gcd(k, q - 1) = 1$, there exist integers a and b such that

$$a \cdot k + b \cdot (q - 1) = 1.$$

This implies that $(a, q - 1) = 1$ and $(b, k) = 1$. Thus, α^{ka} has order $q - 1$ and $\alpha^{b(q-1)}$ has order k.

Thus, $\alpha = \alpha^{b(q-1)} \cdot \alpha^{ka}$, with $\alpha^{b(q-1)} \in B$ and $\alpha^{ka} \in C$. We have hereby proved that $D \subseteq BC$. We now have $A \cap D \subseteq A \cap BC$. But, by [13, result (1.10)], we have $A \cap D \neq \emptyset$, and so we must also have $A \cap BC \neq \emptyset$. By Eq. (16), we conclude that $A \cap B \neq \emptyset$.

Hence, $\mathbb{F}_{q^m}$ contains a normal element with multiplicative order $k = \frac{q^m-1}{q-1}$, as required. $\square$

6   Examples

We now demonstrate Theorems 4, 6, and 7 by providing concrete examples. The following cardinalities were derived by an exhaustive search using the algebra software package SageMath [23]. Each table below corresponds to the extension $\mathbb{F}_{q^m}$ of $\mathbb{F}_q$, and shows that the number of k-normal elements, whenever nonzero, is greater than or equal to the number $\frac{\Phi_q(x^m-1)}{q^k}$ (which has been rounded off to two decimal places in the table), as stated in Theorem 4. Below each table, we give the number of normal elements with multiplicative order $\frac{q^m-1}{q-1}$. In the terminology of [17], we call these $(q-1)$-primitive normal elements. Clearly, Theorem 7 is validated by the fact that all these numbers are non-zero.

We have already discussed Tables 4 and 6 in the light of Theorems 5 and 6. On the other hand, note that for the example in Table 5, Theorem 5 is applicable, while Theorem 6 is not. As we have noted before, this happens precisely when $p \div m$ and the hypothesis of 5 holds. This shows that neither of these two results is stronger than the other. In the case of Table 8, the assumptions of both theorems hold and both guarantee the existence of k-normal elements for every value of k less than m. For Tables 1, 2, 3, and 7, neither Theorem 5 nor Theorem 6 applies. In fact, Table 3 shows that 3-normal elements and 7-normal elements over $\mathbb{F}_2$ do not exist in $\mathbb{F}_{1024}$.

Table 1. $\mathbb{F}_8/\mathbb{F}_2$ $(q = 2, m = 3)$

k	# of k-normal elements	$\dfrac{\Phi_q(x^m-1)}{q^k}$
0	4	4
1	4	4
2	2	2
3	1	1

of $(q-1)$-primitive normal elements = 4

Table 2. $\mathbb{F}_{59049}/\mathbb{F}_9$ $(q = 9, m = 5)$

k	# of k-normal elements	$\dfrac{\Phi_q(x^m-1)}{q^k}$
0	51200	51200
1	6400	5688.89
2	1280	632.10
3	160	70.23
4	8	7.80

of $(q-1)$-primitive normal elements = 5750

Table 3. $\mathbb{F}_{1024}/\mathbb{F}_2$ ($q = 2$, $m = 10$)

k	# of k-normal elements	$\dfrac{\Phi_q(x^m - 1)}{q^k}$
0	480	480
1	240	240
2	240	120
3	0	60
4	35	30
5	15	15
6	15	7.5
7	0	3.75
8	2	1.875
9	1	0.94

of $(q - 1)$-primitive normal elements = 290

Table 4. $\mathbb{F}_{262144}/\mathbb{F}_8$ ($q = 8$, $m = 6$)

k	# of k-normal elements	$\dfrac{\Phi_q(x^m - 1)}{q^k}$
0	225792	225792
1	28224	28224
2	7560	3528
3	441	441
4	119	55.13
5	7	6.89

of $(q - 1)$-primitive normal elements = 20124

Table 5. $\mathbb{F}_{729}/\mathbb{F}_3$ ($q = 3$, $m = 6$)

k	# of k-normal elements	$\dfrac{\Phi_q(x^m - 1)}{q^k}$
0	324	324
1	216	108
2	108	36
3	60	12
4	16	4
5	4	1.33

of $(q - 1)$-primitive normal elements = 290

Table 6. $\mathbb{F}_{15625}/\mathbb{F}_5$ ($q = 5$, $m = 6$)

k	# of k-normal elements	$\dfrac{\Phi_q(x^m - 1)}{q^k}$
0	9216	9216
1	4608	1843.20
2	1344	368.64
3	384	73.73
4	64	14.75
5	8	2.95

of $(q - 1)$-primitive normal elements = 642

Table 7. $\mathbb{F}_{4913}/\mathbb{F}_{17}$ ($q = 17$, $m = 3$)

k	# of k-normal elements	$\dfrac{\Phi_q(x^m - 1)}{q^k}$
0	4608	4608
1	288	271.06
2	16	15.94

of $(q - 1)$-primitive normal elements = 288

Table 8. $\mathbb{F}_{2401}/\mathbb{F}_7$ ($q = 7$, $m = 4$)

k	# of k-normal elements	$\dfrac{\Phi_q(x^m - 1)}{q^k}$
0	1728	1728
1	576	246.86
2	84	35.26
3	16	5.04

of $(q - 1)$-primitive normal elements = 112

7   Conclusions and Open Problems

In this paper, we dealt with the recently introduced concept of k-normal elements in finite fields [9]. The existence and cardinalities of k-normal elements in $\mathbb{F}_{q^m}$ are both strongly tied to the factorization of the polynomial $x^m - 1$ over $\mathbb{F}_q$, which, in turn, depends on the factorization of cyclotomic polynomials. One does not have an explicit formula for the irreducible factors of cyclotomic polynomials, or of their degrees, and so it is not possible to directly infer the existence or numbers of k-normal elements. However, one may deduce several key results by forcing certain conditions on m, k, and q. In Theorem 6, we used the general factorization of $x^m - 1$ into cyclotomic polynomials to obtain a new existence condition for k-normal elements.

The structure of $\mathbb{F}_{q^m}$ as an additive module over $\mathbb{F}_q[x]$ plays a key role in proofs related to normal and k-normal bases. In Theorem 4, we furnished a lower bound for the number of k-normal elements in $\mathbb{F}_{q^m}$ under the sole assumption that at least one of them exists. The proof is inspired by the observation in [10] that the additive module structure of $\mathbb{F}_{q^m}$ in fact gives rise to a group action on all the normal elements. Our bound does not require a specific form for m or q, and therefore extends beyond the formulas provided in [21]. Two interesting problems arise in this direction.

Problem 1. Given a k-normal element α, which subsets of $\{\alpha, \alpha^q, \alpha^{q^2}, \ldots, \alpha^{q^{m-1}}\}$ with size $m - k$ or smaller, apart from $\{\alpha, \alpha^q, \alpha^{q^2}, \ldots, \alpha^{q^{m-k-1}}\}$ are linearly independent? Computer experiments show that in many cases, there do exist linearly dependent subsets with size smaller than $m - k$.

Problem 2. Given a k-normal element α, does there exist another k-normal element outside $\mathrm{span}_{\mathbb{F}_q}\{\alpha, \alpha^q, \alpha^{q^2}, \ldots, \alpha^{q^{m-1}}\}$? We have noted in Remark 2 that Theorem 2 proves that the number of k-normal elements in this subspace is larger than $\frac{\Phi_q(x^m - 1)}{q^k}$. It would also be interesting to see whether a better bound for the total number can be obtained by bounding above the intersection of the $\mathbb{F}_q$-spans of two distinct k-normal elements.

Problem 3. Under what circumstances is the group action (6) free? Under what circumstances is it transitive?

After the proof of the well-known Primitive Normal Basis Theorem by Lenstra and Schoof [13], several interesting generalizations have been proposed. The existence and numbers of elements with different pairs of additive orders (as in Definition 2) and multiplicative group orders have been investigated by several authors. Some solved and unsolved problems in this domain may be found in [3, 9, 17], and [12]. We state one such relevant open problem below.

Problem 4 ([9, Problem 6.4]). Determine the existence of high-order k-normal elements $\alpha \in \mathbb{F}_{q^m}$ over $\mathbb{F}_q$, where "high order" means $ord(\alpha) = N$, with N a large positive divisor of $q^m - 1$.

With Theorem 7 we answered a special case of Problem 4. Following the method of Lenstra and Schoof [13], we provided an existence condition for elements in $\mathbb{F}_{q^m}$ with maximal additive order (i.e. normal elements) that simultaneously have a non-maximal but high multiplicative order, namely $\frac{q^m - 1}{q - 1}$.

Acknowledgement. This work was partially supported by Swiss National Science Foundation grant no. 188430. The authors are also greatly thankful to Gianira Alfarano for her thorough proofreading and constructive feedback on this manuscript.

References

1. Ash, D.W., Blake, I.F., Vanstone, S.A.: Low complexity normal bases. Discret. Appl. Math. **25**(3), 191–210 (1989). https://doi.org/10.1016/0166-218X(89)90001-2
2. Carlitz, L.: Primitive roots in a finite field. Trans. Am. Math. Soc. **73**, 373–382 (1952). https://doi.org/10.2307/1990797
3. Cohen, S.D., Huczynska, S.: The strong primitive normal basis theorem. Acta Arith. **143**(4), 299–332 (2010). https://doi.org/10.4064/aa143-4-1
4. Davenport, H.: Bases for finite fields. J. Lond. Math. Soc. **43**, 21–39 (1968). https://doi.org/10.1112/jlms/s1-43.1.21
5. Frandsen, G.S.: On the density of normal bases in finite fields. Finite Fields Appl. **6**(1), 23–38 (2000). https://doi.org/10.1006/ffta.1999.0263
6. Gao, S.: Normal bases over finite fields. ProQuest LLC, Ann Arbor, MI, thesis (Ph.D.)-University of Waterloo (Canada) (1993)
7. Gao, S., Panario, D.: Density of normal elements. Finite Fields Appl. **3**(2), 141–150 (1997). https://doi.org/10.1006/ffta.1996.0177
8. Hensel, K.: Ueber die Darstellung der Zahlen eines Gattungsbereiches für einen beliebigen Primdivisor. J. Reine Angew. Math. **103**, 230–237 (1888). https://doi.org/10.1515/crll.1888.103.230
9. Huczynska, S., Mullen, G.L., Panario, D., Thomson, D.: Existence and properties of k-normal elements over finite fields. Finite Fields Appl. **24**, 170–183 (2013). https://doi.org/10.1016/j.ffa.2013.07.004
10. Hyde, T.: Normal elements in finite fields. arXiv preprint arXiv:1809.02155 (2018)
11. Kapetanakis, G.: Normal bases and primitive elements over finite fields. Finite Fields Appl. **26**, 123–143 (2014). https://doi.org/10.1016/j.ffa.2013.12.002
12. Kapetanakis, G., Reis, L.: Variations of the primitive normal basis theorem. Des. Codes Crypt. **87**(7), 1459–1480 (2018). https://doi.org/10.1007/s10623-018-0543-9
13. Lenstra Jr., H.W., Schoof, R.J.: Primitive normal bases for finite fields. Math. Comput. **48**(177), 217–231 (1987). https://doi.org/10.2307/2007886
14. Lidl, R., Niederreiter, H.: Finite Fields. Encyclopedia of Mathematics and its Applications, 2nd edn, vol. 20. Cambridge University Press, Cambridge (1997). With a foreword by P. M. Cohn
15. Lüneburg, H.: Translation Planes. Springer, Heidelberg (2012). https://books.google.ch/books?id=UuTrCAAAQBAJ
16. Menezes, A.J., Blake, I.F., Gao, X., Mullin, R.C., Vanstone, S.A., Yaghoobian, T.: Applications of Finite Fields. The Kluwer International Series in Engineering and Computer Science, vol. 199. Kluwer Academic Publishers, Boston (1993). https://doi.org/10.1007/978-1-4757-2226-0
17. Mullen, G.L.: Some open problems arising from my recent finite research. In: Contemporary Developments in Finite Fields and Applications, pp. 254–269. World Sci. Publ., Hackensack (2016)
18. Ore, O.: Contributions to the theory of finite fields. Trans. Am. Math. Soc. **36**(2), 243–274 (1934). https://doi.org/10.2307/1989836
19. Reis, L.: Existence results on k-normal elements over finite fields. Rev. Mat. Iberoam. **35**(3), 805–822 (2019). https://doi.org/10.4171/rmi/1070

20. Reis, L., Thomson, D.: Existence of primitive 1-normal elements in finite fields. Finite Fields Appl. **51**, 238–269 (2018). https://doi.org/10.1016/j.ffa.2018.02.002
21. Saygı, Z., Tilenbaev, E., Ürtiş, C.: On the number of k-normal elements over finite fields. Turkish J. Math. **43**(2), 795–812 (2019). https://doi.org/10.3906/mat-1805-113
22. Sozaya-Chan, J.A., Tapia-Recillas, H.: On k-normal elements over finite fields. Finite Fields Appl. **52**, 94–107 (2018). https://doi.org/10.1016/j.ffa.2018.03.006
23. The Sage Developers: SageMath, the Sage Mathematics Software System (Version 8.6) (2020). https://www.sagemath.org

Author Index

Batoul, Aicha 115
Bereg, Sergey 234
Bernal, José Joaquín 134
Borello, Martino 147
Boucher, Delphine 115
Boulanouar, Ranya Djihad 115

Caminata, Alessio 3
Cathébras, Joël 75
Chotin, Roselyne 75

Davidova, D. 195

Göloğlu, Faruk 207
Gomez, Ana I. 163
Gomez-Perez, Domingo 163
Gorla, Elisa 3
Gravel, Claude 174

Jamous, Abdelillah 147

Kaleyski, N. 195
Kölsch, Lukas 207
Kyureghyan, Gohar 207

Malouf, Brian 234
McGuire, Gary 37
Morales, Linda 234

Orsini, Emmanuela 42

Panario, Daniel 174
Perrin, Léo 207

Randriambololona, Hugues 92
Rigault, Bastien 174
Rosenthal, Joachim 255
Rousseau, Édouard 92

Sheekey, John 37
Simón, Juan Jacobo 134
Stanley, Thomas 234
Sudborough, I. Hal 234

Tinani, Simran 255
Tirkel, Andrew 163

Vega, Gerardo 222

Printed in the United States
By Bookmasters